T0323934

# Microbiota Brain Axis

# Microbiota Brain Axis

## A Neuroscience Primer

**JANE FOSTER**

Center for Depression Research and Clinical Care
Department of Psychiatry
UT Southwestern Medical Center
Dallas, TX

**GERARD CLARKE**

Department of Psychiatry and
Neurobehaioural Science & APC
Microbiome Ireland, University College
Cork, Cork, Ireland

ELSEVIER

**ACADEMIC PRESS**

An imprint of Elsevier

Academic Press is an imprint of Elsevier
125 London Wall, London EC2Y 5AS, United Kingdom
525 B Street, Suite 1650, San Diego, CA 92101, United States
50 Hampshire Street, 5th Floor, Cambridge, MA 02139, United States

**Notices**
Knowledge and best practice in this field are constantly changing. As new research and experience broaden our understanding, changes in research methods, professional practices, or medical treatment may become necessary.

Practitioners and researchers must always rely on their own experience and knowledge in evaluating and using any information, methods, compounds, or experiments described herein. In using such information or methods they should be mindful of their own safety and the safety of others, including parties for whom they have a professional responsibility.

To the fullest extent of the law, neither the Publisher nor the authors, contributors, or editors, assume any liability for any injury and/or damage to persons or property as a matter of products liability, negligence or otherwise, or from any use or operation of any methods, products, instructions, or ideas contained in the material herein.

ISBN: 978-0-12-814800-6

For information on all Academic Press publications visit our website at https://www.elsevier.com/books-and-journals

*Publisher:* Nikki Levy
*Acquisitions Editor:* Nikki Levy
*Editorial Project Manager:* Shivangi Mishra
*Production Project Manager:* Fahmida Sultana
*Cover Designer:* Mark Rogers

Cover art was conceptualized, designed, and created by Nicholas Kabitsis.

Working together
to grow libraries in
developing countries

www.elsevier.com • www.bookaid.org

Typeset by TNQ Technologies

# Contents

# About the authors

**Dr. Jane Foster**, Ph.D., is a Professor in the Department of Psychiatry, at the Center for Depression Research and Clinical Care at UT Southwestern Medical Center in Dallas, Texas. In the past 20 years, Dr. Foster has developed an internationally recognized translational research program that uses a 'bench to bedside' and back again approach to studying microbiota—brain and immune—brain systems. As one of the early proponents of the role of gut—brain axis in mental health, her lab produced important data demonstrating a role for the microbiome in brain development and behavior in animal studies and recently has extended this work to study the complex neurobiological underpinnings of microbiota—brain and neuro-immune systems in psychiatric illness in clinical populations. Her multi-disciplinary expertise includes behavioral neuroscience, molecular biology, immunology, neuroimaging, microbiome, and bioinformatics in both preclinical and clinical research domains. Dr. Foster's research program has developed high-quality analytical pipelines for biological data and novel analytical tools for integrating data across modalities.

**Prof. Gerard Clarke**, Ph.D., is a Professor of Neurobehavioral Science in the Department of Psychiatry and Neurobehavioral Science and a Principal Investigator in APC Microbiome Ireland at University College Cork. His research program includes a focus on translational biomarkers of stress-related neuropsychiatric disorders, the impact of the gut microbiome on brain and behavior across the life span, and microbial regulation of tryptophan metabolism. Key achievements of his lab in the generation of knowledge around the microbiota—gut—brain axis include the demonstration that the gut microbiome regulates the hippocampal serotonergic system in a sex-dependent manner, findings that paved the way for numerous lines of inquiry on the effects of the gut microbiome on neurodevelopment, brain function, and behavior. He is regularly included in Clarivate Analytics Highly Cited Researchers list, placing him among the world's top one percent of researchers by citation. His current approach is based on advancing frontier knowledge in microbiome research to yield potential new therapeutic targets for the effective treatment of the central nervous system and gastrointestinal disorders.

# Foreword

Over the last 2 decades, great strides have been made to document the scope of influence of the gut microbiota on brain function and behavior across the life span as well as understand the mechanisms underpinning this impact. This acceleration of research into the microbiota—gut—brain axis continues apace and the results demand the incorporation of a new framework into neuroscience curricula to equip the researchers of tomorrow with the knowledge and skills to evaluate and advance these emerging relationships.

This book provides critical and comprehensive discussions about the most significant areas of research and the key developments in the field, as well as providing an overview of the advances in microbiota analysis tools and techniques for future neuroscience-related research. Foster and Clarke are leading international authorities on these subjects and have crafted an unrivaled and translational resource, with each chapter offering informative and contemporary accounts of the significance of the microbiota in neuroscience. These chapters can be accessed either individually or as a whole to gain a broad overview of the current status of a rapidly evolving and exciting field. For novices to the field, it will serve as an in-depth introduction to this area of research, while simultaneously being of immense value to basic scientists and clinicians alike.

This book begins with an introduction to the gut—brain axis and the wide array of topics covered includes both preclinical and clinical observations that emphasize the importance of microbiota—brain communication in mental health, neurodevelopment, and neurodegeneration. I have no doubt that such a comprehensive guide will leave the reader with substantial insights into the conduct of effective microbiota—brain research, armed with an understanding of the appropriate methodologies that are currently used in the collection and analysis of microbiota data. In addition, this book discusses the potential for microbiota-related biomarkers to inform precision medicine approaches and evaluates the most promising microbial-related treatments.

This primer integrates the complexity of the microbiome with that of the brain and nervous system, an important first step for the interested reader in what promises to be a long, challenging and exciting journey. The progress so expertly documented by Foster and Clarke has not been achieved without difficulty and many open questions remain to be answered as we move toward therapeutic targeting of the gut to treat the disorders of the brain. As we continue to look to the microbiome for answers, it is important that the information obtained thus far is interpreted appropriately to fully harness the translational potential. Foster and Clarke chart the way forward in this regard while also discussing how to overcome the pitfalls and challenges accompanying such opportunities. I anticipate that the foundations and approaches described in this book will be further supplemented with a more granular appreciation of mechanisms and increased efforts to establish causality. Such developments and expansion will be key to drive the advances in microbiome science that will expedite the delivery of a new era of human translational neuroscience research.

Professor John F. Cryan MRIA,
Vice President for Research and Innovation, University College Cork
Professor, Department Anatomy and Neuroscience,
Principal Investigator, APC Microbiome Ireland
http://publish.ucc.ie/researchprofiles/C003/jcryan

Office of the Vice President for Research and Innovation, 4th Floor Food Science Building, University College Cork, College Rd., Cork, Ireland; j.cryan@ucc.ie

# CHAPTER 1

# The gut-brain axis

The gut-brain axis refers to the biological systems that connect the body's gastrointestinal (GI) tract and the brain. The components of the gut-brain axis now include the microbes that reside in our gastrointestinal (GI) tract, as well as the classical framework provided by the enteric nervous system that wraps the GI tract and regulates gut homeostasis and physiology, the endocrine and neural connections between our body and our brain, and the immune system (mucosal and systemic) (Fig. 1.1). Gut-brain communication is bidirectional, continual, and plays an important role in physical and mental health (see Chapter 4; Foster and Clarke, 2023). There are many historical examples of top-down and bottom-up control of the gut-brain axis. A well-cited example of early evidence of top-down control comes from the American army surgeon Beaumont who conducted a case study of an injured patient that had received a gunshot wound to the stomach that lead to a permanent fistula or opening into his stomach. Beaumont observed and recorded the relationship between "psychological stress" and an increase in gastric juices (Beaumont, 1833). An example of historical work that considered bottom-up control of the gut-brain axis was the work of Elie Metchnikoff that was published in 1908 where he suggested that long life was associated with the consumption of fermented foods (Metchnikoff, 1908). As microbiome research has expanded, this bottom-up perspective has been gaining momentum in biomedical research, as well as garnering attention from the media and the public (Robson, 2019; Willyard, 2021).

The concept that the gut influences the brain is not a new concept. Over 2000 years ago Hippocrates is credited to say that the "all diseases begin in the gut" (Hippocrates, n.d.) and while he was not referring to the microbes themselves the current research that is focused on the microbiota-brain axis developed from a rich body of historical research that considered the gut-brain connection. The high level of comorbidity of GI disorders with anxiety and depression has held the attention of gastroenterologists and neuroscientists for several decades (Goodoory et al., 2021; Wilmes

*Microbiota Brain Axis*
ISBN 978-0-12-814800-6
https://doi.org/10.1016/B978-0-12-814800-6.00008-X

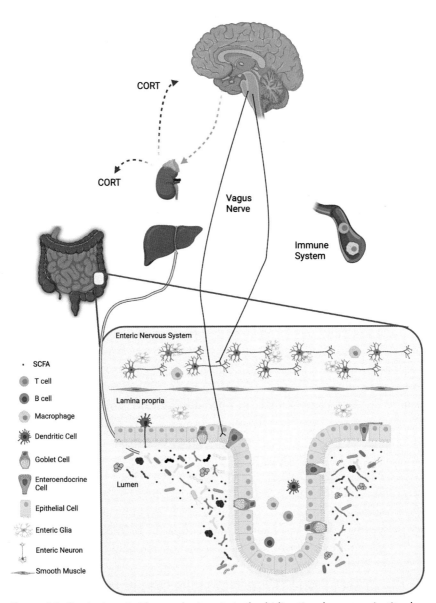

**Figure 1.1** Gut-brain axis. The gut-brain axis is the bidirectional communication between the gut and the brain, and its communication is facilitated by different pathways with the autonomic nervous system, the enteric nervous system, the neuroendocrine system, and the immune system. Trillions of gut microbes and their metabolites reside in the lumen of the gastrointestinal tract and influence the local mucosal system (visualized in the inset for the barrier in the colon). Signaling pathways between the gut microbiota and the brain include neural pathways such as the vagus

et al., 2021). Research aimed at understanding the contribution of functional GI issues compared to the contribution of activation of brain stress systems, such as the hypothalamic pituitary adrenal (HPA) axis, to the presence of both GI and psychiatric symptoms provided much of the early evidence that gut-brain communication is important to health and disease (Cryan et al., 2019; Foster et al., 2017). From the neuroscience perspective, researchers examining the neurobiology of stress were the first to consider the role of the microbiome (Bailey and Coe, 1999; Bailey et al., 2011; Bailey et al., 2010; Foster et al., 2017; Frankiensztajn et al., 2020; Lyte et al., 2011; O'Mahony et al., 2009; Sudo et al., 2004). In parallel, as the microbiome is essential for the development and function of the immune system (Cebra, 1999; Hooper and Macpherson, 2010; Macpherson and Harris, 2004), researchers studying neuroimmunology and immune-brain communication provided early evidence of microbiota-immune-brain signaling in behavior and brain function (Cryan et al., 2019; Foster, 2016; Foster and McVey Neufeld, 2013; Goehler et al., 2007; Hsiao et al., 2012; Lyte et al., 2011).

In the past decade, research related to the gut-brain axis has shifted to examine gut microbiota and their importance in health and disease. It has been established that microbe-host interactions at the gastrointestinal barrier play a role in host-pathogen defense, development of the immune system, energy metabolism, and gastrointestinal physiology and homeostasis. In the context of the gut-brain axis, neural, endocrine, and immune pathways are known to play a role in microbiota to brain signaling (Cryan et al., 2019). The human intestine harbors trillions of microbiota (Gill et al., 2006; Luckey, 1972; Savage, 1977). Research efforts to determine what constitutes a healthy microbiome in the past decade have advanced our understanding of microbiota composition, diversity, and function (See Chapter 2; Foster and Clarke, 2023). In parallel, many studies have worked to determine how genetic, environmental, social, and other factors influence the microbiome (See Chapter 3; Foster and Clarke, 2023). The gut microbiota and its human host interact in a mutualistic relationship. The host provides

nerve, immune pathways, and humoral pathways. Gut microbiota influence host metabolism through local interactions, peripheral systems, and bidirectional communication with the liver. *(Schematic created by Biorender.com and reproduced from Foster, J.A., Baker, G.B., Dursun, S.M., 2021. The relationship between the gut microbiom-immune system-brain axis and major depressive disorder. Front Neurol 12, 721126. https://doi.org/10.3389/fneur.2021.721126 (in press).)*

microbes with a rich environment to grow while microbiota contribute to healthy metabolism and play a critical role in the normal development of the immune, endocrine, and nervous system (Hooper et al., 2001; Macpherson and Harris, 2004; Macpherson et al., 2002; Macpherson and Uhr, 2004a; Tlaskalova-Hogenova et al., 2004). Excitement about the microbiome has spread across many domains of biomedical research, and captured the interest of the media and the public. Notably, popular magazines including *Scientific American, The Economist,* and *The New York Times Magazine* featured articles on the microbiome (Ackerman, 2012; Economist, 2012; Pollan, 2013). Cover images accompanied these articles, drawing attention to the emerging work on the microbiome (Fig. 1.2).

The increase in media and public interest in 2012/2013 was coincident with an increase in interest in the microbiome by neuroscientists, as evident by the cover of *Trends in Neuroscience* in May 2013 (Fig. 1.2) (Foster and McVey Neufeld, 2013). Prior to this time, a small number of key studies in germ-free (GF) mice attracted attention and stimulated interest from neuroscience, psychology, psychiatry, and neurology (Bercik et al., 2011; Diaz Heijtz et al., 2011; Neufeld et al., 2011b; Sudo et al., 2004). GF mice have no commensal microbiota and exhibit an altered immune system (Boman, 2000; Macpherson and Harris, 2004; Macpherson and Uhr, 2004b; Tlaskalova-Hogenova et al., 2005). The use of mice raised in a GF environment provides an opportunity to assess the impact of the absence of microbiota to the development of brain and body systems. In a landmark study, Sudo et al. (2004) demonstrated that male GF Balb/C mice exposed to acute restraint stress showed an enhanced activation of the hypothalamic pituitary adrenal (HPA) axis that included increased stress-related levels of plasma adrenocorticotrophic hormone (ACTH) and corticosterone (CORT) (Sudo et al., 2004). Moreover, monocolonization of parent GF mice with the commensal bacteria *Bifidobacteria infantis* prevented the exaggerated stress reactivity in their offspring (Sudo et al., 2004). In addition, conventionalization of GF pups with feces from conventionally housed specific pathogen-free (SPF) mice at 6 weeks of age attenuated the exaggerated stress reactivity in adulthood, whereas conventionalization of GF adult mice with SPF feces did not prevent the exaggerated stress response (Sudo et al., 2004). Importantly, these initial findings were reproduced in a study that showed exaggerated HPA activation in response to a novel environmental stressor in male and female Swiss Webster mice (Clarke et al., 2013). Several studies have examined stress and the microbiome advancing our understanding about the importance of microbiota to

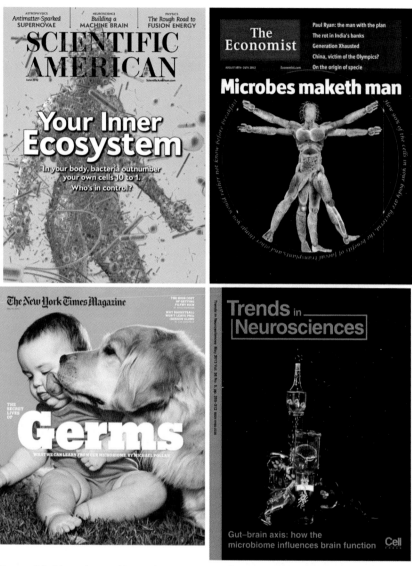

**Figure 1.2** Magazine and journal cover images featuring the microbiome. In 2012 and 2013, popular magazines, including *Scientific American, The Economist*, and *The New York Times Magazine*, featured articles and covers demonstrating the attention and interest in the microbiome. Attention to the microbiota-brain axis was increasing, and notably was featured on the cover of *Trends in Neuroscience* in May 2013. *(Covers reproduced with permission.)*

stress-related behavior, central nervous system (CNS) stress circuitry, and stress reactivity (See Chapters 5 and 6; Foster and Clarke, 2023).

With respect to neuroscience, advances in how the microbiome influences the brain have expanded and key papers representative of this work are visualized in the microbiota-brain axis timeline (Fig. 1.3). Following the report by Sudo et al. (2004), key behavioral neuroscience studies using GF mice broadened the scope of work by showing that microbiota influenced stress-related behavior (Bercik et al., 2011; Clarke et al., 2013; Diaz Heijtz et al., 2011; Neufeld et al., 2011b). In comparison to enhanced stress reactivity observed, behavioral studies showed reduced anxiety-like behavior in GF mice (Clarke et al., 2013; Diaz Heijtz et al., 2011; Neufeld et al., 2011b). Reduced anxiety-like behavior in GF Swiss Webster mice was observed in the elevated plus maze (Neufeld et al., 2011b). Colonization of GF mice with SPF feces at 8 weeks of age did not alter the behavioral phenotype when retested at 12 weeks of age, suggesting that microbiota influence stress circuitry early in life (Neufeld et al., 2011a). Similarly, male GF NMRI mice also showed reduced anxiety-like behavior when tested in the open field (OF), light dark box, and EPM (Diaz Heijtz et al., 2011). Conventionalization of parent GF mice in this study normalized offspring behavior in the OF and EPM, however, reduced anxiety-like behavior in the light dark box was observed for offspring of conventionalized GF mice, similar to GF mice (Diaz Heijtz et al., 2011).

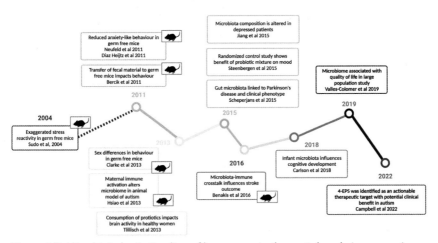

**Figure 1.3** Microbiota-brain timeline of key papers in the past decade in neuroscience and psychiatry. Foundational work in animal studies (denoted with rodent emoji) provided the impetus to translate findings to the clinic. *(Created with Biorender.com.)*

Additional evidence for a role for microbiota in anxiety-like behavior was provided by Clarke et al. (2013) who showed reduced anxiety-like behavior in male GF Swiss Webster mice in the light dark box. Moreover, normalization of this behavioral phenotype was observed in mice conventionalized with SPF feces at weaning (Clarke et al., 2013). In addition to behavioral changes, alterations in gene expression and other signaling systems were observed in these studies (Clarke et al., 2013; Diaz Heijtz et al., 2011; Neufeld et al., 2011b; Sudo et al., 2004).

GF mice have many biological differences from conventional mice including an altered immune system that could contribute to the behavioral differences observed in the above noted studies. Evidence that behavior changes were mediated by the alterations in microbiota was provided by a fecal matter transplantation study (Bercik et al., 2011). Using two different strains of GF mice that naturally show differences in anxiety-like behavior, the investigators provided evidence for a role for the microbiota in behavior (Bercik et al., 2011). Notably, inbred Balb/C mice normally show increased anxiety-like behavior in approach avoidance tests in comparison to outbred Swiss Webster mice. Additionally, microbiota composition differs between Balb/C and Swiss Webster mice (Bercik et al., 2011). When GF male Swiss Webster mice were conventionalized with fecal matter from SPF Balb/C mice, an increased anxiety-like phenotype was observed (Bercik et al., 2011). In the reverse experiment, GF Balb/C mice that received microbiota from SPF Swiss Webster mice showed a reduction in anxiety-like behavior (Bercik et al., 2011). The behavioral differences observed in these reconstitution experiments were associated with strain-specific microbiota profiles demonstrating an important role for the microbes themselves in anxiety-like behavior (Bercik et al., 2011). The association of microbiota and anxiety-like behavior has been demonstrated in several additional animal studies to date and more recently has been demonstrated in clinical populations (Chen et al., 2019; Jiang et al., 2018). The above-noted initial series of behavioral studies in GF mice ignited the interest and attention of the broader neuroscience community and the research quickly expanded to other behavioral domains (Chapter 5; Foster and Clarke, 2023) and to investigations into how microbiota influence brain systems directly (Chapter 6; Foster and Clarke, 2023).

A natural extension of the above-noted studies to consider the microbiome in animal models of psychiatric and neurological disorders is highlighted by key papers using a maternal immune activation (MIA) animal model of autism and an animal model of stroke (Benakis et al., 2016; Hsiao et al., 2013).

The MIA model of autism is based on epidemiological evidence linking maternal infection during pregnancy to increased incidence of autism spectrum disorders (ASD) in offspring (Chess et al., 1978; Yamashita et al., 2003). Maternal viral infection is modeled using maternal administration of polyriboinosinic-polyriboocytidilic acid (poly I:C), which mimics viral RNA (Cunningham et al., 2007; Gandhi et al., 2007; Gilmore et al., 2005). Exposure to poly I:C in utero on embryonic (E) day 9, 9.5, or 10 has been shown to result in the development of autistic-like behaviors in mice (Meyer et al., 2008; Shi et al., 2003). Comparison of MIA offspring to control offspring showed differential abundance of several gut bacterial taxa, as well as increased gastrointestinal permeability, increased intestinal interleukin-6 levels, and an altered profile of serum metabolites (Hsiao et al., 2013). Notably, this work highlighted that host-microbe interactions influence microbiota-brain communication. The potential for microbiota-related treatments to ameliorate the impact of MIA was explored, and it was demonstrated that oral treatment with *Bacteroides fragilis* at weaning attenuated the impact of MIA on behavioral and host physiological outcomes, as well as microbiota composition (Hsiao et al., 2013). The line of research extending from this early work includes many additional animal studies but also has been translated and explored in clinical ASD populations (Kang et al., 2017, 2018; Morais et al., 2018; Needham et al., 2021; Saunders et al., 2020; Sharon et al., 2019; Stewart Campbell et al., 2022). Building on previous work that linked the peripheral immune system to ischemic injury, investigators manipulated the microbiome using antibiotics and then employed a transient middle cerebral artery occlusion (MCAO) model of cerebral ischemia to demonstrate the importance of microbiota-immune communication in stroke outcome (Benakis et al., 2016). These foundational animal studies that showed the importance of microbiota to brain function, behavior, and disease have expanded in the past 5 years to include several more animal models of psychiatric and neurological disorders (Cryan et al., 2019; Sherwin et al., 2018).

Translating rodent studies to people is an essential step in advancing our understanding of the neurobiological basis of brain health and the systems that are altered in disease. Early research linking microbes to mental health examined the impact of probiotic consumption on mood in healthy individuals (Benson et al., 2010; Messaoudi et al., 2011). Probiotics are defined as "live microorganisms that, when administered in adequate amounts, confer a health benefit on the host" (Hill et al., 2014; Food and Agriculture Organization of the United Nations and World Health

Organization, 2006). Initial studies showed a benefit of probiotic consumption in healthy individuals including improved mood (Benton et al., 2007) and a beneficial effect on anxiety and depressive measures and reduced stress hormone levels (Messaoudi et al., 2011). A direct link between probiotics and brain function was demonstrated using functional magnetic resonance imaging (fMRI) (Tillisch et al., 2013). Administration of a fermented milk product containing *Bifidobacterium animalis* subsp *Lactis*, *Streptococcus thermophiles, Lactobacillus bulgaricus* and *Lactococcus lactis* subsp *Lactis* was associated with reduced fMRI response during an emotional face task in healthy women (Tillisch et al., 2013). Additional evidence of an impact of microbiota on brain function was generated in a randomized control trial that administered a probiotic containing *Bifidobacterium* W23, *Bifidobacterium lactic* W52, *Lactobacillus acidophilus* W37, *Lactobacillus brevis* W63, *Lactobacillus casei* W56, *Lactobacillus salivarius* W24, and *Lactococcus lactis* (W19, W58) to healthy participants, and demonstrated a postprobiotic reduction in cognitive reactivity to sad mood (Steenbergen et al., 2015). Microbiota-targeted therapies are attracting attention and have the potential to have clinical benefits (see Chapter 11; Foster and Clarke, 2023). The above-noted studies were the first to clearly demonstrate an influence of microbiota on brain function and were followed by clinical studies in depression, initially demonstrating compositional differences in microbiota in depressed individuals compared to nondepressed individuals (Jiang et al., 2015). Since that first report, several studies have demonstrated differences in both compositional and functional readouts of the microbiome in depression (Zhang et al., 2021). A recent large population study reported an association between the microbiome and quality of life and depression (Valles-Colomer et al., 2019). The relative abundances of *Faecalibacterium* and *Coprococcus* bacteria were associated with higher quality of life, and a reduction of *Coprococcus* and *Dialister* spp were linked to depression, an observation that was validated in a second cohort (Valles-Colomer et al., 2019). While there have been several studies in depressed individuals, research has now extended into several psychiatric disorders (see Chapter 8; Foster and Clarke, 2023). A key study that reported microbiota associations with Parkinson disease (Scheperjans et al., 2015) led the way as the field expanded to additional neurological disorders (See Chapter 9; Foster and Clarke, 2023).

Over the past decade, preclinical and clinical studies have demonstrated a role for the microbiome in neurodevelopment, particularly in autism (Vuong and Hsiao, 2017) (see Chapter 7; Foster and Clarke, 2023).

Evidence of a role for microbiota in normal brain development includes studies linking bacterial taxa to cognitive development (Carlson et al., 2018). Case-control studies demonstrate alterations in microbiota composition as well as altered plasma and fecal metabolic phenotypes in individuals with autism compared to typically developing individuals (Liu et al., 2019; Needham et al., 2021). Moreover, several bacterial taxa and signaling pathways are associated with clinical symptoms and severity (Needham et al., 2021). Reverse translation approaches that consider mechanistic findings in animal models are starting to translate to the clinic (Blacher et al., 2019; Needham et al., 2022; Stewart Campbell et al., 2022). Advances in understanding the mechanism of how microbiota influences the brain are at the forefront of neuroscience, psychiatry, and neurology. The revelation that several pharmacological drugs, including both antibiotic and nonantibiotic drugs, can impact microbiota growth and composition, reveals the importance of the microbiome to an individual's clinical response (see Chapter 10; Foster and Clarke, 2023). The clinical potential of microbiota-related biomarkers as precision medicine tools to improve treatment selection is an active area of research (see Chapter 12; Foster and Clarke, 2023). Overall, an enormous amount of preclinical and clinical research in the past decade has advanced our understanding of the importance of microbiota to brain health. To help the reader navigate this exciting area of research, this book provides a window into key research findings that have advanced the field. These discoveries make possible new opportunities to understand the gut-brain axis that has the potential to improve brain health across the lifespan.

## References

Ackerman, J., 2012. How Bacteria in Our Bodies Protect Our Health. Scientific American.

Bailey, M.T., Coe, C.L., 1999. Maternal separation disrupts the integrity of the intestinal microflora in infant rhesus monkeys. Dev Psychobiol 35, 146—155.

Bailey, M.T., Dowd, S.E., Galley, J.D., Hufnagle, A.R., Allen, R.G., Lyte, M., 2011. Exposure to a social stressor alters the structure of the intestinal microbiota: implications for stressor-induced immunomodulation. Brain Behav Immun 25, 397—407.

Bailey, M.T., Dowd, S.E., Parry, N.M., Galley, J.D., Schauer, D.B., Lyte, M., 2010. Stressor exposure disrupts commensal microbial populations in the intestines and leads to increased colonization by Citrobacter rodentium. Infect Immun 78, 1509—1519.

Beaumont, W., 1833. Experiments and Observations on the Gastric Juice and the Physiology of Digestion. F.P. Allen, Plattsburg.

Benakis, C., Brea, D., Caballero, S., Faraco, G., Moore, J., Murphy, M., Sita, G., Racchumi, G., Ling, L., Pamer, E.G., Iadecola, C., Anrather, J., 2016. Commensal microbiota affects ischemic stroke outcome by regulating intestinal gammadelta T cells. Nat Med 22, 516–523.

Benson, A.K., Kelly, S.A., Legge, R., Ma, F., Low, S.J., Kim, J., Zhang, M., Oh, P.L., Nehrenberg, D., Hua, K., Kachman, S.D., Moriyama, E.N., Walter, J., Peterson, D.A., Pomp, D., 2010. Individuality in gut microbiota composition is a complex polygenic trait shaped by multiple environmental and host genetic factors. Proc Natl Acad Sci U S A 107, 18933–18938.

Benton, D., Williams, C., Brown, A., 2007. Impact of consuming a milk drink containing a probiotic on mood and cognition. Eur J Clin Nutr 61, 355–361.

Bercik, P., Denou, E., Collins, J., Jackson, W., Lu, J., Jury, J., Deng, Y., Blennerhassett, P., Macri, J., McCoy, K.D., Verdu, E.F., Collins, S.M., 2011. The intestinal microbiota affect central levels of brain-derived neurotropic factor and behavior in mice. Gastroenterology 141, 599–609, 609 e591–593.

Blacher, E., Bashiardes, S., Shapiro, H., Rothschild, D., Mor, U., Dori-Bachash, M., Kleimeyer, C., Moresi, C., Harnik, Y., Zur, M., Zabari, M., Brik, R.B., Kviatcovsky, D., Zmora, N., Cohen, Y., Bar, N., Levi, I., Amar, N., Mehlman, T., Brandis, A., Biton, I., Kuperman, Y., Tsoory, M., Alfahel, L., Harmelin, A., Schwartz, M., Israelson, A., Arike, L., Johansson, M.E.V., Hansson, G.C., Gotkine, M., Segal, E., Elinav, E., 2019. Potential roles of gut microbiome and metabolites in modulating ALS in mice. Nature 572, 474–480.

Boman, H.G., 2000. Innate immunity and the normal microflora. Immunol Rev 173, 5–16.

Carlson, A.L., Xia, K., Azcarate-Peril, M.A., Goldman, B.D., Ahn, M., Styner, M.A., Thompson, A.L., Geng, X., Gilmore, J.H., Knickmeyer, R.C., 2018. Infant gut microbiome associated with cognitive development. Biol Psychiatry 83, 148–159.

Cebra, J.J., 1999. Influences of microbiota on intestinal immune system development. Am J Clin Nutr 69, 1046S–1051S.

Chen, Y.H., Bai, J., Wu, D., Yu, S.F., Qiang, X.L., Bai, H., Wang, H.N., Peng, Z.W., 2019. Association between fecal microbiota and generalized anxiety disorder: severity and early treatment response. J Affect Disord 259, 56–66.

Chess, S., Fernandez, P., Korn, S., 1978. Behavioral consequences of congenital rubella. J Pediatr 93, 699–703.

Clarke, G., Grenham, S., Scully, P., Fitzgerald, P., Moloney, R.D., Shanahan, F., Dinan, T.G., Cryan, J.F., 2013. The microbiome-gut-brain axis during early life regulates the hippocampal serotonergic system in a sex-dependent manner. Mol Psychiatry 18, 666–673.

Cryan, J.F., O'Riordan, K.J., Cowan, C.S.M., Sandhu, K.V., Bastiaanssen, T.F.S., Boehme, M., Codagnone, M.G., Cussotto, S., Fulling, C., Golubeva, A.V., Guzzetta, K.E., Jaggar, M., Long-Smith, C.M., Lyte, J.M., Martin, J.A., Molinero-Perez, A., Moloney, G., Morelli, E., Morillas, E., O'Connor, R., Cruz-Pereira, J.S., Peterson, V.L., Rea, K., Ritz, N.L., Sherwin, E., Spichak, S., Teichman, E.M., van de Wouw, M., Ventura-Silva, A.P., Wallace-Fitzsimons, S.E., Hyland, N., Clarke, G., Dinan, T.G., 2019. The microbiota-gut-brain axis. Physiol Rev 99, 1877–2013.

Cunningham, C., Campion, S., Teeling, J., Felton, L., Perry, V.H., 2007. The sickness behaviour and CNS inflammatory mediator profile induced by systemic challenge of mice with synthetic double-stranded RNA (poly I:C). Brain Behav Immun 21, 490–502.

Diaz Heijtz, R., Wang, S., Anuar, F., Qian, Y., Bjorkholm, B., Samuelsson, A., Hibberd, M.L., Forssberg, H., Pettersson, S., 2011. Normal gut microbiota modulates brain development and behavior. Proc Natl Acad Sci U S A 108, 3047–3052.

Economist, T., 2012. Microbes maketh man. In: The Economist. The Economist Group Limited.

Food and Agriculture Organization of the United Nations, World Health Organization, 2006. Health and nutrition properties of probiotics in food including powder milk with live lactic acid bacteria. Guidelines for the evaluation of probiotics in food. In: Probiotics in Food : Health and Nutritional Properties and Guidelines for Evaluation. Food and Agriculture Organization of the United Nations : World Health Organization, Italy: Rome.

Foster, J.A., Baker, G.B., Dursun, S.M., 2021. The relationship between the gut microbiom-immune system-brain axis and major depressive disorder. Front Neurol 12, 721126. https://doi.org/10.3389/fneur.2021.721126. In press.

Foster, J., Clarke, G., 2023. Microbiota Brain Axis. Elsevier.

Foster, J.A., 2016. Gut microbiome and behavior: focus on neuroimmune interactions. Int Rev Neurobiol 131, 49—65.

Foster, J.A., McVey Neufeld, K.A., 2013. Gut-brain axis: how the microbiome influences anxiety and depression. Trends Neurosci 36, 305—312.

Foster, J.A., Rinaman, L., Cryan, J.F., 2017. Stress & the gut-brain axis: regulation by the microbiome. Neurobiol Stress 7, 124—136.

Frankiensztajn, L.M., Elliott, E., Koren, O., 2020. The microbiota and the hypothalamus-pituitary-adrenocortical (HPA) axis, implications for anxiety and stress disorders. Curr Opin Neurobiol 62, 76—82.

Gandhi, R., Hayley, S., Gibb, J., Merali, Z., Anisman, H., 2007. Influence of poly I:C on sickness behaviors, plasma cytokines, corticosterone and central monoamine activity: moderation by social stressors. Brain Behav Immun 21, 477—489.

Gill, S.R., Pop, M., Deboy, R.T., Eckburg, P.B., Turnbaugh, P.J., Samuel, B.S., Gordon, J.I., Relman, D.A., Fraser-Liggett, C.M., Nelson, K.E., 2006. Metagenomic analysis of the human distal gut microbiome. Science 312, 1355—1359.

Gilmore, J.H., Jarskog, L.F., Vadlamudi, S., 2005. Maternal poly I:C exposure during pregnancy regulates TNFa, BDNF, and NGF expression in neonatal brain and the maternal-fetal unit of the rat. J Neuroimmunol 159.

Goehler, L.E., Park, S.M., Opitz, N., Lyte, M., Gaykema, R.P., 2007. Campylobacter jejuni infection increases anxiety-like behavior in the holeboard: possible anatomical substrates for viscerosensory modulation of exploratory behavior. Brain Behav Immun.

Goodoory, V.C., Mikocka-Walus, A., Yiannakou, Y., Houghton, L.A., Black, C.J., Ford, A.C., 2021. Impact of psychological comorbidity on the prognosis of irritable bowel syndrome. Am J Gastroenterol 116, 1485—1494.

Hill, C., Guarner, F., Reid, G., Gibson, G.R., Merenstein, D.J., Pot, B., Morelli, L., Canani, R.B., Flint, H.J., Salminen, S., Calder, P.C., Sanders, M.E., 2014. Expert consensus document. The international scientific association for probiotics and pre-biotics consensus statement on the scope and appropriate use of the term probiotic. Nat Rev Gastroenterol Hepatol 11, 506—514.

Hippocrates, n.d. AZQuotes.com. https://www.azquotes.com/quote/823835.

Hooper, L.V., Macpherson, A.J., 2010. Immune adaptations that maintain homeostasis with the intestinal microbiota. Nat Rev Immunol 10, 159—169.

Hooper, L.V., Wong, M.H., Thelin, A., Hansson, L., Falk, P.G., Gordon, J.I., 2001. Molecular analysis of commensal host-microbial relationships in the intestine. Science 291, 881—884.

Hsiao, E.Y., McBride, S.W., Chow, J., Mazmanian, S.K., Patterson, P.H., 2012. Modeling an autism risk factor in mice leads to permanent immune dysregulation. Proc Natl Acad Sci U S A 109, 12776—12781.

Hsiao, E.Y., McBride, S.W., Hsien, S., Sharon, G., Hyde, E.R., McCue, T., Codelli, J.A., Chow, J., Reisman, S.E., Petrosino, J.F., Patterson, P.H., Mazmanian, S.K., 2013. Microbiota modulate behavioral and physiological abnormalities associated with neu-rodevelopmental disorders. Cell 155, 1451—1463.

Jiang, H., Ling, Z., Zhang, Y., Mao, H., Ma, Z., Yin, Y., Wang, W., Tang, W., Tan, Z., Shi, J., Li, L., Ruan, B., 2015. Altered fecal microbiota composition in patients with major depressive disorder. Brain Behav Immun 48, 186—194.

Jiang, H.Y., Zhang, X., Yu, Z.H., Zhang, Z., Deng, M., Zhao, J.H., Ruan, B., 2018. Altered gut microbiota profile in patients with generalized anxiety disorder. J Psychiatr Res 104, 130—136.

Kang, D.W., Adams, J.B., Gregory, A.C., Borody, T., Chittick, L., Fasano, A., Khoruts, A., Geis, E., Maldonado, J., McDonough-Means, S., Pollard, E.L., Roux, S., Sadowsky, M.J., Lipson, K.S., Sullivan, M.B., Caporaso, J.G., Krajmalnik-Brown, R., 2017. Microbiota transfer therapy alters gut ecosystem and improves gastrointestinal and autism symptoms: an open-label study. Microbiome 5, 10.

Kang, D.W., Ilhan, Z.E., Isern, N.G., Hoyt, D.W., Howsmon, D.P., Shaffer, M., Lozupone, C.A., Hahn, J., Adams, J.B., Krajmalnik-Brown, R., 2018. Differences in fecal microbial metabolites and microbiota of children with autism spectrum disorders. Anaerobe 49, 121—131.

Liu, F., Li, J., Wu, F., Zheng, H., Peng, Q., Zhou, H., 2019. Altered composition and function of intestinal microbiota in autism spectrum disorders: a systematic review. Transl Psychiatry 9, 43.

Luckey, T.D., 1972. Introduction to intestinal microecology. Am J Clin Nutr 25, 1292—1294.

Lyte, M., Vulchanova, L., Brown, D.R., 2011. Stress at the intestinal surface: catecholamines and mucosa-bacteria interactions. Cell Tissue Res 343, 23—32.

Macpherson, A.J., Harris, N.L., 2004. Interactions between commensal intestinal bacteria and the immune system. Nat Rev Immunol 4, 478—485.

Macpherson, A.J., Martinic, M.M., Harris, N., 2002. The functions of mucosal T cells in containing the indigenous commensal flora of the intestine. Cell Mol Life Sci 59, 2088—2096.

Macpherson, A.J., Uhr, T., 2004a. Compartmentalization of the mucosal immune responses to commensal intestinal bacteria. Ann N Y Acad Sci 1029, 36—43.

Macpherson, A.J., Uhr, T., 2004b. Induction of protective IgA by intestinal dendritic cells carrying commensal bacteria. Science 303, 1662—1665.

Messaoudi, M., Violle, N., Bisson, J.F., Desor, D., Javelot, H., Rougeot, C., 2011. Beneficial psychological effects of a probiotic formulation (Lactobacillus helveticus R0052 and Bifidobacterium longum R0175) in healthy human volunteers. Gut Microb 2, 256—261.

Metchnikoff, E., 1908. The Prolongation of Life: Optimistic Studies. G.P. Putnam's Sons, New York.

Meyer, U., Murray, P.J., Urwyler, A., Yee, B.K., Schedlowski, M., Feldon, J., 2008. Adult behavioral and pharmacological dysfunctions following disruption of the fetal brain balance between pro-inflammatory and IL-10-mediated anti-inflammatory signaling. Mol Psychiatry 13, 208—221.

Morais, L.H., Felice, D., Golubeva, A.V., Moloney, G., Dinan, T.G., Cryan, J.F., 2018. Strain differences in the susceptibility to the gut-brain axis and neurobehavioural alterations induced by maternal immune activation in mice. Behav Pharmacol 29, 181—198.

Needham, B.D., Adame, M.D., Serena, G., Rose, D.R., Preston, G.M., Conrad, M.C., Campbell, A.S., Donabedian, D.H., Fasano, A., Ashwood, P., Mazmanian, S.K., 2021. Plasma and fecal metabolite profiles in autism spectrum disorder. Biol Psychiatry 89, 451—462.

Needham, B.D., Funabashi, M., Adame, M.D., Wang, Z., Boktor, J.C., Haney, J., Wu, W.L., Rabut, C., Ladinsky, M.S., Hwang, S.J., Guo, Y., Zhu, Q., Griffiths, J.A., Knight, R., Bjorkman, P.J., Shapiro, M.G., Geschwind, D.H., Holschneider, D.P.,

Fischbach, M.A., Mazmanian, S.K., 2022. A gut-derived metabolite alters brain activity and anxiety behaviour in mice. Nature 602, 647–653.

Neufeld, K.A., Kang, N., Bienenstock, J., Foster, J.A., 2011a. Effects of intestinal microbiota on anxiety-like behavior. Commun Integr Biol 4, 492–494.

Neufeld, K.M., Kang, N., Bienenstock, J., Foster, J.A., 2011b. Reduced anxiety-like behavior and central neurochemical change in germ-free mice. Neuro Gastroenterol Motil 23, 255–264 e119.

O'Mahony, S.M., Marchesi, J.R., Scully, P., Codling, C., Ceolho, A.M., Quigley, E.M., Cryan, J.F., Dinan, T.G., 2009. Early life stress alters behavior, immunity, and microbiota in rats: implications for irritable bowel syndrome and psychiatric illnesses. Biol Psychiatry 65, 263–267.

Pollan, M., 2013. Some of my best freinds are germs. In: New York Times Magazine. New York Times.

Robson, D., 2019. Microbes & Me. BBC Future: BBC.

Saunders, J.M., Moreno, J.L., Ibi, D., Sikaroodi, M., Kang, D.J., Munoz-Moreno, R., Dalmet, S.S., Garcia-Sastre, A., Gillevet, P.M., Dozmorov, M.G., Bajaj, J.S., Gonzalez-Maeso, J., 2020. Gut microbiota manipulation during the prepubertal period shapes behavioral abnormalities in a mouse neurodevelopmental disorder model. Sci Rep 10, 4697.

Savage, D.C., 1977. Microbial ecology of the gastrointestinal tract. Annu Rev Microbiol 31, 107–133.

Scheperjans, F., Aho, V., Pereira, P.A., Koskinen, K., Paulin, L., Pekkonen, E., Haapaniemi, E., Kaakkola, S., Eerola-Rautio, J., Pohja, M., Kinnunen, E., Murros, K., Auvinen, P., 2015. Gut microbiota are related to Parkinson's disease and clinical phenotype. Mov Disord 30, 350–358.

Sharon, G., Cruz, N.J., Kang, D.W., Gandal, M.J., Wang, B., Kim, Y.M., Zink, E.M., Casey, C.P., Taylor, B.C., Lane, C.J., Bramer, L.M., Isern, N.G., Hoyt, D.W., Noecker, C., Sweredoski, M.J., Moradian, A., Borenstein, E., Jansson, J.K., Knight, R., Metz, T.O., Lois, C., Geschwind, D.H., Krajmalnik-Brown, R., Mazmanian, S.K., 2019. Human gut microbiota from autism spectrum disorder promote behavioral symptoms in mice. Cell 177, 1600–1618 e1617.

Sherwin, E., Dinan, T.G., Cryan, J.F., 2018. Recent developments in understanding the role of the gut microbiota in brain health and disease. Ann N Y Acad Sci 1420, 5–25.

Shi, L., Fatemi, S.H., Sidwell, R.W., Patterson, P.H., 2003. Maternal influenza infection causes marked behavioral and pharmacological changes in the offspring. J Neurosci 23, 297–302.

Steenbergen, L., Sellaro, R., van Hemert, S., Bosch, J.A., Colzato, L.S., 2015. A randomized controlled trial to test the effect of multispecies probiotics on cognitive reactivity to sad mood. Brain Behav Immun 48, 258–264.

Stewart Campbell, A., Needham, B.D., Meyer, C.R., Tan, J., Conrad, M., Preston, G.M., Bolognani, F., Rao, S.G., Heussler, H., Griffith, R., Guastella, A.J., Janes, A.C., Frederick, B., Donabedian, D.H., Mazmanian, S.K., 2022. Safety and target engagement of an oral small-molecule sequestrant in adolescents with autism spectrum disorder: an open-label phase 1b/2a trial. Nat Med 28, 528–534.

Sudo, N., Chida, Y., Aiba, Y., Sonoda, J., Oyama, N., Yu, X.N., Kubo, C., Koga, Y., 2004. Postnatal microbial colonization programs the hypothalamic-pituitary-adrenal system for stress response in mice. J Physiol 558, 263–275.

Tillisch, K., Labus, J., Kilpatrick, L., Jiang, Z., Stains, J., Ebrat, B., Guyonnet, D., Legrain-Raspaud, S., Trotin, B., Naliboff, B., Mayer, E.A., 2013. Consumption of fermented milk product with probiotic modulates brain activity. Gastroenterology 144, 1394–1401, 1401 e1391–1394.

Tlaskalova-Hogenova, H., Stepankova, R., Hudcovic, T., Tuckova, L., Cukrowska, B., Lodinova-Zadnikova, R., Kozakova, H., Rossmann, P., Bartova, J., Sokol, D., Funda, D.P., Borovska, D., Rehakova, Z., Sinkora, J., Hofman, J., Drastich, P., Kokesova, A., 2004. Commensal bacteria (normal microflora), mucosal immunity and chronic inflammatory and autoimmune diseases. Immunol Lett 93, 97—108.

Tlaskalova-Hogenova, H., Tuckova, L., Stepankova, R., Hudcovic, T., Palova-Jelinkova, L., Kozakova, H., Rossmann, P., Sanchez, D., Cinova, J., Hrncir, T., Kverka, M., Frolova, L., Uhlig, H., Powrie, F., Bland, P., 2005. Involvement of innate immunity in the development of inflammatory and autoimmune diseases. Ann N Y Acad Sci 1051, 787—798.

Valles-Colomer, M., Falony, G., Darzi, Y., Tigchelaar, E.F., Wang, J., Tito, R.Y., Schiweck, C., Kurilshikov, A., Joossens, M., Wijmenga, C., Claes, S., Van Oudenhove, L., Zhernakova, A., Vieira-Silva, S., Raes, J., 2019. The neuroactive potential of the human gut microbiota in quality of life and depression. Nat Microbiol 4, 623—632.

Vuong, H.E., Hsiao, E.Y., 2017. Emerging roles for the gut microbiome in autism spectrum disorder. Biol Psychiatry 81, 411—423.

Willyard, C., 2021. How gut microbes could drive brain disorders. Nature 590, 22—25.

Wilmes, L., Collins, J.M., O'Riordan, K.J., O'Mahony, S.M., Cryan, J.F., Clarke, G., 2021. Of bowels, brain and behavior: a role for the gut microbiota in psychiatric comorbidities in irritable bowel syndrome. Neuro Gastroenterol Motil 33, e14095.

Yamashita, Y., Fujimoto, C., Nakajima, E., Isagai, T., Matsuishi, T., 2003. Possible association between congenital cytomegalovirus infection and autistic disorder. J Autism Dev Disord 33, 455—459.

Zhang, Q., Yun, Y., An, H., Zhao, W., Ma, T., Wang, Z., Yang, F., 2021. Gut microbiome composition associated with major depressive disorder and sleep quality. Front Psychiatr 12, 645045.

# CHAPTER 2

# What is a healthy microbiome?

## Introduction

The "microbiome" is a term used to refer to the microbes that cover or inhabit the different surfaces and niches/areas of the body and their related genetic material (Turnbaugh et al., 2007), and includes bacteria, viruses, parasites, fungi, and protozoa. Biomedical research to date has focused primarily on bacteria, although attention to other microbes is increasing. Advances in sequencing technologies and metagenomic tools provide researchers with the opportunity to examine the composition, function, active gene expression, and metabolic output of the bacteria in healthy individuals and clinical populations (Goodrich et al., 2014). With the advances in assessment and analytic tools, knowledge about the importance of the microbiome to health and disease is at the forefront of biomedical research. In parallel with understanding the importance of microbes and microbiota-brain communication to disease, researchers have endeavored to determine what constitutes a healthy microbiome. The resulting body of literature has provided the framework for understanding what constitutes a healthy microbiome and advances our understanding of the factors that influence microbiota composition and function in healthy individuals.

## Microbiota composition and function in healthy individuals

Two large consortiums, the MetaHIT consortium and the Human Microbiome Project (HMP), conducted large-scale analysis of microbial diversity and composition in healthy individuals (Human Microbiome Project, 2012a, b; Qin et al., 2010). The Human Microbiome Project enrolled 300 individuals, including 149 men and 151 women between the ages 18 and 40 years old, from two clinical sites to examine the healthy microbiome (Aagaard et al., 2013; Gevers et al., 2012). Importantly, the

*Microbiota Brain Axis*
ISBN 978-0-12-814800-6
https://doi.org/10.1016/B978-0-12-814800-6.00011-X

establishment of standardized procedures for collection and analysis of samples from multiple body sites has provided a comprehensive series of reference articles for studying the human microbiome (Human Microbiome Project, 2012a, b). To determine the community structure, specimens were collected from different body areas, including the oral cavity (9 subsites sampled), skin (4 subsites sampled), nasal cavity, and vagina (3 subsites sampled in women) and a stool specimen was collected to represent the gastrointestinal tract (Aagaard et al., 2013). A detailed overview of the clinical sample demographics is provided in Aagaard et al. (2013) and published results from the Human Microbiome Project are listed at https://hmpdacc.org/hmp/publications.php. The MetaHit consortium focused on identifying microbial genes of human gut microbiota in healthy individuals but also in inflammatory bowel disease and obesity (https://cordis.europa.eu/project/id/201052). Results from the MetaHit consortium, the HMP consortium, and other research groups are integrated here to provide an understanding of the healthy microbiome.

## Microbiota diversity across body sites in healthy individuals

Measuring microbial diversity is the first step to identifying the similarities and differences between the microbial communities that occupy different body sites (for primer see (Finotello et al., 2018)). Prior to the work of the Human Microbiome Project, investigators examined bacteria present across many body sites over time in a small number of adults (Costello et al., 2009). This work showed that much of the bacterial composition was determined by the body site of the sample, that there was significant interpersonal variability in taxa composition, and that the diversity of the communities varied across body sites (Costello et al., 2009). The main findings of this early work were confirmed and expanded in the studies of the Human Microbiome Project. Differences in both alpha diversity and beta diversity are evident across body sites (Costello et al., 2009; Human Microbiome Project, 2012b; Li et al., 2012). Alpha diversity is a within-sample diversity measure that can be calculated using several approaches. Shannon and Simpson diversity examine richness (number) and evenness of the community, whereas Chao1 estimates species richness. The total observed species count is also a metric for alpha diversity. In samples from healthy individuals, alpha diversity was measured using 16S rRNA sequencing and the Simpson Index showed the highest diversity for oral and stool samples, while the lowest alpha diversity was observed in vaginal samples (Human Microbiome Project, 2012b; Li et al., 2012). The low alpha diversity in vaginal samples reflects the presence of predominantly *Lactobacillus* spp in this body site (Human Microbiome Project, 2012b). Nasal and skin samples showed an intermediate alpha diversity. In addition to examining bacterial composition using 16S rRNA sequencing, researchers showed similar differences in alpha diversity were observed using shotgun metagenomics (Human Microbiome Project, 2012b). Using the total number of operational taxonomic units (OTUs) to estimate total richness of bacterial samples, Huse et al. (2012) showed that stool had the greatest richness composition, followed by oral sites, then nasal, skin, and vagina, although there was more variability in oral communities than other body sites (Fig. 2.1). In this analysis, few OTUs or taxa were identified to be present across all individuals (Huse et al., 2012). The interpretation of alpha diversity readouts in the context of defining a healthy microbiome thus depends on both the sampling site and age. In adulthood at least, higher gut

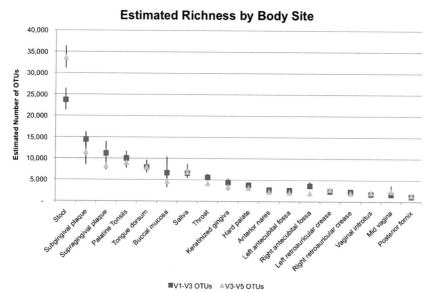

**Figure 2.1** Bacterial richness estimates for each body site. Observed OTU counts show estimated richness calculated using both the V1—V3 and the V3—V5 tag data for each body site. Bars represent the upper and lower confidence bounds. Stool samples showed the most richness followed by oral samples then by the skin and vaginal samples. *(Reproduced with permission Huse, S.M., Ye, Y., Zhou, Y., Fodor, A.A., 2012. A core human microbiome as viewed through 16S rRNA sequence clusters. PLoS One 7, e34242. https://doi.org/10.1371/journal.pone.0034242.g006. CC BY.)*

microbiota diversity is one of the metrics most consistently associated with better host health.

Variations in community structure, measured by beta diversity, were also distinct for each body site (Human Microbiome Project, 2012b). Beta diversity is a measure of between-sample diversity in composition that takes into account the specific taxa in the sample and is often visualized using a principle coordinate plot. As displayed in Fig. 2.2, clustering of samples was distinct for oral, gastrointestinal, and vaginal microbiota; overlap between the skin and nasal bacterial communities was observed revealing similarity in community structure between these sites (Human Microbiome Project, 2012b). A comparison of beta diversity across the 18 body sites sampled showed some subsite differences. Of note, the low beta diversity measured using genus-level phylotype in the vagina was accompanied by much higher beta diversity measured using OTUs. *Lactobacillus* sp. are the predominant bacteria found in the vagina, however, as noted by the OTU beta

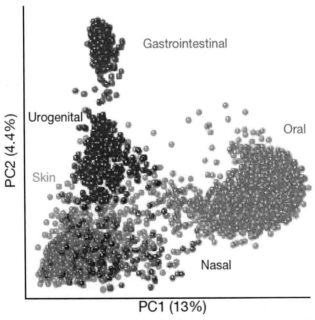

**Figure 2.2** Body sites differ in microbiome diversity and composition. Principal coordinates plot showing variation among samples collected from different body sites. Clustering of individuals by body area demonstrates intraindividual differences between the oral, gastrointestinal, skin and urogenital sites. *(Reproduced with permission from Huttenhower et al. Nature 486, 207–214 (2012) https://doi.org/10.1038/nature11234, CC BY-NC-SA 3.0C.)*

diversity, the number of distinct species present is high and this level of species diversity is not captured in the analysis of beta diversity using 16S phylotypes (Human Microbiome Project, 2012b). The findings from 16S rRNA sequencing for composition were confirmed in a subset of samples that were analyzed using shotgun-sequencing for metagenomics data and metabolically profiled. The patterns of diversity observed using these alternative approaches were similar to those obtained by 16S rRNA sequencing (Human Microbiome Project, 2012b).

## Interindividual differences in bacterial taxa in healthy individuals

In addition to diversity, several researchers have examined the relative abundance of bacterial taxa at the phyla, order, class, family, and genus level using 16S rRNA sequencing approaches (Bolyen et al., 2019; Callahan et al., 2016; Navas-Molina et al., 2013; Whelan and Surette, 2017). High throughput sequencing of the 16S rRNA gene in bacteria allows researchers to profile microbiota; and through bioinformatic tools and pipelines, sequence data can be processed to determine the composition of a sample. Mapping operational taxonomical units (OTUs) or amplicon sequence variants (ASVs) to bacteria taxa provides a readout of the relative abundance of bacterial groups at each taxonomical level (Bolyen et al., 2019; Callahan et al., 2016; Navas-Molina et al., 2013; Whelan and Surette, 2017). At the phyla level, most bacteria species identified are members of the Firmicutes, Bacteroides, and Actinobacteria (Eckburg et al., 2005; Zhernakova et al., 2016). Less abundant phyla include Proteobacteria, Fusobacteria, and Verrucomicrobia (Eckburg et al., 2005; Zhernakova et al., 2016). As noted above for diversity, the greatest amount of variability in the composition of microbiota is related to interindividual differences, particularly at the phyla level (Eckburg et al., 2005). As part of the MetaHIT consortium, researchers mapped microbial genes in the healthy human gut and identified 18 bacterial taxa that were detected in all individuals in their cohort, suggesting these taxa as a "core" microbiome; however, the abundance of these abundant taxa was highly variable between individuals (Qin et al., 2010). These taxa are classified to eight different genera (Table 2.1). Several studies have endeavored to identified core taxa using a variety of approaches, analytical tools, and study populations, an expanded list of core taxa are summarized in Table 2.1 (Falony et al., 2016; Huse et al., 2012; Jalanka-Tuovinen et al., 2011; Li et al., 2012; Mahnic and Rupnik, 2018; Qin et al., 2010; Zhang et al., 2014).

**Table 2.1** Core bacterial genera in the human gut.

| | Qin et al. (2010) | Jalanka-Tuovinen et al. (2011) | Huse et al. (2012) | Li et al. (2012) | Zhang et al. (2014) | Falony et al. (2016) | Mahnic and Rupnik (2018) |
|---|---|---|---|---|---|---|---|
| *Akkermansia* | | ★ | | | | ★ | |
| *Alistipes* | | ★ | | | | ★ | |
| *Anaerostipes* | ★ | | | | | | ★ |
| *Anaerotruncus* | | ★ | | | | ★ | |
| *Bacteroides* | | ★ | ★ | ★ | ★ | ★ | ★ |
| *Bifidobacterium* | | ★ | | | | ★ | |
| *Blautia* | | | | | | ★ | ★ |
| *Bryantella* | | ★ | | | | | |
| *Butyricicoccus* | | ★ | | | | | |
| *Butyrivibrio* | | ★ | | | | | |
| *Clostridium* | ★ | ★ | | ★ | | | ★ |
| *Collinsella* | | ★ | | | ★ | | |
| *Coprococcus* | ★ | ★ | ★ | | ★ | ★ | |
| *Dorea* | ★ | ★ | | | ★ | ★ | |
| *Eubacterium* | ★ | ★ | | | | | |
| *Faecalibacterium* | ★ | ★ | ★ | | | ★ | ★ |
| *Lachnobacillus* | | ★ | | | | | |
| *Lachnospira* | | ★ | | | | | |
| *Lachnospiraceae* | | | | | | ★ | |
| *Oscillospira* | | | | | | | ★ |
| *Oscillibacter* | | ★ | ★ | | ★ | ★ | |
| *Oxalobacter* | | ★ | | | | | |
| *Papillibacter* | | ★ | | | | | |

| | MetaHit | Finish | HMP | HMP | Mongolian | Finish, Dutch | Slovenian |
|---|---|---|---|---|---|---|---|
| *Parabacteroides* | | | | | | | |
| *Prevotella* | ★ | ★ | | | ★ | | |
| *Pseudobutyrivibrio* | ★ | | | | | | |
| *Roseburia* | | ★ | ★ | | | ★ | ★ |
| *Ruminococcus* | | ★ | ★ | | ★ | ★ | |
| *Sporobacter* | | ★ | | | | | |
| *Streptococcus* | | ★ | ★ | | ★ | ★ | |
| *Subdoligranulum* | | ★ | | | | | |
| *Criteria* | 100% | 100% | 95% | 97.5% | 90% | 95% | 95% |
| *Population* | MetaHit | Finish | HMP | HMP | Mongolian | Finish, Dutch | Slovenian |
| *Sample Size* | 124 | 9 | 238 | 208 | 64 | 1106, 1135 | 186 |

For example, four common taxa that included three members of the Class Clostridia and one classified to genus level *Bacteroides* were identified in the Human Microbiome Project samples. Although the three Clostridia taxa were present in greater than 97.5% of the stool samples, the abundance in most samples was quite low (Li et al., 2012). In contrast, *Bacteroides* was detected in greater than 97.5% of the most samples and highly abundant in most samples (Li et al., 2012).

Another approach the MetaHIT consortium used to assess individual differences was to cluster samples into subgroups based on their taxonomic similarity in metagenomes, the cohort included samples from Japanese, American, Danish, French, Italian, and Spanish individuals (Arumugam et al., 2011). Interindividual differences based on enrichment of bacterial taxa at the genus level clustered individuals into subgroups which were referred to as enterotypes. Three enterotypes for gut microbiota were identified that were enriched in *Bacteriodes, Prevotella,* or *Ruminococcus* (Arumugam et al., 2011). These enterotypes were not explained by body mass index (BMI), age, or gender (Arumugam et al., 2011), however, it must be noted that the sample size was relatively small and there has been some debate on whether they should be reframed as gradients (Jeffery et al., 2012). In a similar analysis using data from the Human Microbiome Project, researchers identified two enterotypes, one enriched in *Bacteriodes* and one enriched in *Prevotella* (Wu et al., 2011). In this analysis, the *Ruminococcus* cluster was a component of the *Bacteriodes* enterotype (Wu et al., 2011). Of interest, these investigators associated enterotypes with long-term diet habits (Wu et al., 2011). Related to this, diet and housing in a group of elderly Irish individuals was associated with the same two enterotypes, such that the *Prevotella* enterotype was more frequent in community-dwelling individuals in parallel with a more diverse (healthy) diet (Claesson et al., 2012). Distinct microbiota compositions were observed in community-dwelling individuals compared to individuals in long-term care housing, with greater abundance of *Coprococcus* and *Roseburia* associated with community-dwelling and greater abundance of *Parabacteroides, Eubacterium, Anaerotruncus, Lactonifactor,* and *Coprobacillus* associated with long-term care subjects (Claesson et al., 2012). A study of 41 healthy adult Chinese individuals identified a *Bacteroides* enterotype and a *Prevotella* enterotype (Yin et al., 2017). The bacterial taxa profile in Chinese individuals was similar to African individuals, whereas evident differences in abundant taxa were observed between Chinese and European individuals. In comparison to Chinese individuals, European individuals showed higher levels of *Faecalibacterium* and lower levels of *Bacteroides* (Yin et al., 2017).

While there continues to be a debate in the field about the term "enterotype" (Gorvitovskaia et al., 2016; Koren et al., 2013), studies using this approach have repeatedly identified "enterotypes" or clusters in their populations; some reports have identified the original three enterotypes (de Moraes et al., 2017) while others have identified two clusters that are represented by high relative abundance of *Bacteriodes* and *Prevotella* (Tillisch et al., 2017; Yin et al., 2017). The potential to use a few key taxa or cluster types to stratify populations into more homogeneous groups for research purposes is an attractive option and may provide better clinical or biological associations. For example, a study examining the structure and composition of gut microbiota and brain structure in healthy women identified two enterotypes/clusters, a *Bacteroides* cluster and a *Prevotella* cluster (Tillisch et al., 2017). Using functional MRI, structural MRI, and diffusion tensor imaging, the investigators identified association between these subgroups and emotional response, white matter connectivity, and brain volume (Tillisch et al., 2017). Further, in a cohort of Brazilian adults, researchers identified the same enterotypes (*Bacteroides, Prevotella,* and *Ruminococcus*) previously identified by Arumugam et al. (2011) and identified an association between microbiota and cardiometabolic risk factors that was linked to the identified enterotypes (de Moraes et al., 2017). In the above noted elderly Irish cohort, analysis of microbiome-health connections, in community compared to long-term care housing individuals, identified associations between microbiome and inflammation, nutritional quality, frailty, and other health outcomes including geriatric depression measures (Claesson et al., 2012).

The above examples demonstrate how key findings related to the healthy microbiome have informed and influenced work targeted at understanding functional connections between microbiota and health, and provided a framework for understanding interindividual differences in clinical populations. In addition, a more detailed understanding of human microbiome communities and the factors that influence intra- and interindividual variation in composition and function is at the forefront of this research field. In an attempt to associate bacterial taxa with demographic and life history characteristics, investigators used Dirichlet multinomial mixture models to identify community clusters of taxa in 300 individuals at multiple body sites (Ding and Schloss, 2014). This analysis recognized the previously identified enterotypes but suggested four community types and suggested the relative abundance of five key taxa including *Bacteroides, Prevotella, Ruminococcaceae,*

*Alistipes,* and *Faecalibacterium* contributed to more than half of the difference between the four community types identified (Ding and Schloss, 2014). Further, they associated breastfeeding with a community type that had high abundance of *Bacteroides* but low abundance of the other four key taxa, whereas individuals not breastfed were more likely to have a community type with lower *Bacteroides* but higher levels of *Prevotella* (Ding and Schloss, 2014). Both gender and level of education were also associated with bacterial composition. Similar to previous studies noted above, these researchers demonstrated that different body sites had distinct bacterial compositions; however, they observed that the community type at one body site was predictive of the community type at other sites, particularly comparisons between the oral cavity and stool samples (Ding and Schloss, 2014). Groups of bacterial taxa are sometimes referred to as "clades" and consideration of how specific phyla or taxa cooccur in the same body habitat and more specifically in niche microenvironments within a body site represents another feature that may contribute to interindividual differences and the functioning of the healthy microbiome (Faust et al., 2012).

A key question to consider is "how stable is the composition of an individual's microbiome over time?" This is an essential question as research studies are most often cross-sectional in nature, with a single sample collected for microbiome analysis. It is important to determine the reliability of a single sample in accurately reflecting the microbiota signature of an individual. Moreover, as precision medicine approaches are advanced in both treatment and drug discovery domains, biomarkers related to the microbiota-brain axis have great potential. Consensus across studies and research areas indicates that an individual's microbiome is their own and interpersonal variability in microbiota is estimated to be in the range of 90% at the Phyla level (Dorrestein et al., 2014; Ursell et al., 2012). In the Human Microbiome Project report discussed above, repeated sampling from the same individuals over time showed that the diversity of the community structure of an individual's microbiome was relatively stable over time compared to the differences in diversity found in the population as a whole (Human Microbiome Project, 2012b). This was also observed when community types were considered; however, the stability of the oral cavity was the least stable while the samples from gut and vagina were the most stable over time (Ding and Schloss, 2014; Jalanka-Tuovinen et al., 2011).

# Sex influences microbiota composition and function

Sex differences between women and men influence health and recently, more attention has been directed at deciphering sex differences in preclinical and clinical research (Clayton and Collins, 2014; Tannenbaum et al., 2016). Sex differences are important to risk of disease as sexual dichotomy exists in the development, presentation, and course of neuropsychiatric and neurological disorders. With respect to the healthy microbiome, only a few studies to date have reported sex differences. In the Human Microbiome Project samples, sex differences were observed in bacterial diversity and in specific bacterial taxa (Aagaard et al., 2013; Human Microbiome Project, 2012b). Reduced alpha diversity in females was observed in skin (retroauricular crease) and oral cavity (throat, hard palate, tongue dorsum) samples (Aagaard et al., 2013; Human Microbiome Project, 2012b). In addition, several sex differences in specific taxa were observed at the phyla, class, order, family, and genus level (Human Microbiome Project, 2012b). Notably, most of these differences were found in skin samples or in samples of the oral cavity. *Clostridium clostridium* was the only taxa in stool samples reported to have a significant association with sex in that study (Human Microbiome Project, 2012b). Ding and Schloss (2014) showed an association between sex and community type or cluster of bacterial taxa in stool, tongue, retroauricular crease, and antecubital fossa. Specifically, there was a greater likelihood that females would have a stool community with fewer *Bacteriodes* and higher levels of *Prevotella* (Ding and Schloss, 2014). In contrast, in older men with BMI greater than 33, *Bacteroides* was lower in men than women, emphasizing the known fact that several factors can influence the microbiome (Haro et al., 2016). In another study, sex was significantly associated with gut microbiota composition in 94 healthy individuals (Dominianni et al., 2015). Both sex and age were observed to be significantly associated with interindividual differences in the microbiome in 186 healthy Slovenian individuals (Mahnic and Rupnik, 2018). Certainly, these studies demonstrate that sex differences may contribute to differences in the microbiome, however, since sex has an impact itself on so many biological factors, it is possible that individual differences in sex-related biological factors mask or confound our ability to detect sex differences in microbiome-related outcomes.

## Age influences microbiota composition and function

Microbiota composition and function changes over the life span. Studies have examined changes in diversity, composition, and gene function at different ages (Odamaki et al., 2016; Yatsunenko et al., 2012). Interindividual differences in diversity in gut microbiota are greatest in the first few years of life as the child's microbiome matures (Yatsunenko et al., 2012) and the trajectory of microbiome acquisition likely plays a key role in the subsequent establishment of a healthy gut microbiome. Bacterial transmission from mother to infant is particularly important in seeding the initial assembly (Ferretti et al., 2018). Although the infant acquires bacteria from a number of maternal sites and other sources, it is the maternal gut strains that are ecologically better adapted for stable colonization (Yassour et al., 2018). In a comparison of fecal samples from 531 individuals from US and non-US (Amazonas of Venezuela and rural Malawi) sites, geographical differences were observed between the US and non-US sites in individuals older than 3 years of age, and yet it was evident that bacterial diversity increased with age across all populations (Yatsunenko et al., 2012). Interestingly, in adults *Prevotella* was overrepresented in non-US samples in the study by Yatsunenko et al. (2012), which mirrors the observation of higher abundance of *Prevotella* in West African children when compared to Italian children (De Filippo et al., 2010). *Prevotella* is of interest as it was identified as one of the taxa that distinguishes enterotypes or clusters in Western European populations, however, discrete clusters were not detected in this cohort (Yatsunenko et al., 2012). Distinct changes in diversity were also observed with age in a cohort of 371 Japanese individuals from 0 to 104 years of age (Odamaki et al., 2016). At the phyla level, age-related changes in relative abundance of the four main phyla were observed. Actinobacteria was dominant in the first 2 years of life but reduced to a lower stable level at 3 years of age, and then reduced again in individual over 60 years of age, and very low in centenarians. Firmicutes was the most abundant phyla from 3 years old throughout adult life, although a reduction in relative abundance in elderly populations was evident that was accompanied by a parallel increase in both Bacteroidetes and Proteobacteria (Odamaki et al., 2016). At the taxa level, *Bifidobacteria* dominated fecal communities in the first year of life in this study (Yatsunenko et al., 2012), similar to other studies (Cheng et al., 2016). Further *Bifidobacterium* abundance declined with increasing age (Mahnic and Rupnik, 2018; Yatsunenko et al., 2012). To examine changes in bacteria genera over the life span, Odamaki et al. (2016) clustered genera into coabundance clusters (nine CAGs identified) and examined changes in abundance of these cluster groups at different ages. This approach identified adult-enriched taxa as well as clusters of taxa that were similarly enriched in infants and elderly populations (Odamaki et al., 2016).

## Diet influences microbiota composition and function

Our relationship with our microbiome begins for the most part at birth, and the initial microbiota composition is influenced by mode of delivery (Chu et al., 2017; Dominguez-Bello et al., 2010; Fallani et al., 2011; Mueller et al., 2016) and feeding choices (breastmilk or formula) (Bergstrom et al., 2014; Bokulich et al., 2016; Fallani et al., 2011; Stearns et al., 2017). Temporal development (9, 18, and 36 months) of the composition and function of the microbiota was examined in a longitudinal cohort of 300 healthy Danish infants (Bergstrom et al., 2014). Interestingly, this study examined enterotypes and observed the presence of *Bacteroides* and *Prevotella* enterotypes at 36 months of age but not at earlier time points, an observation that was related to a delayed colonization of infant gut by *Prevotella* (Bergstrom et al., 2014). As in previous studies, *Bifidobacterium longum*, as well as *B. breve,* were abundant early in life and decreased in relative abundance between 9 and 36 months, however, other *Bifidobacterium* spp. including *B. adolescentis and B. catenulatum* increased during this period. The most notable changes in bacterial taxa were observed between 9 and 18 months, with increased abundance of *Bacteroides thetaiotaomicron, Alistipes* spp., *Akkermansia muciniphila,* and *Desulfovibrio* spp. (Bergstrom et al., 2014). *Lactobacillus* spp. and *Enterobacteriaceae* decreased in abundance between 9 and 18 months. Key butyrate-producing bacteria increased between 9 and 36 months, including *Clostridium leptum* and *Roseburia* spp. Short chain fatty acids (SCFAs) including acetate, proprionate, and butyrate are metabolites that are products of commensal fermentation (Rios-Covian et al., 2016). SCFAs produced by gut bacteria influence other commensals, are important to gut physiology but are also part of microbiota-host signaling systems that extend beyond the gut, including important roles in microbiota-brain communication (Dalile et al., 2019; Rios-Covian et al., 2016). Several of the temporal changes in bacterial taxa were associated with breastfeeding. The transition from higher levels of *Lactobacillus* spp. and *Bifidobacterium* spp. to the butyrate-producing taxa noted above was directly associated with continued breastfeeding at 9 months. Similarly, lower levels of *Desulfovibrio* spp. and *A. muciniphila* were also observed in infants that continued breastfeeding at 9 months of age. The TEDDY study, the Environmental Determinants of Diabetes in the Young, is a prospective study that examined microbiome development in early life in American and European children (Stewart et al., 2018; Vatanen et al., 2018). This study identified 121 bacterial species that were associated with breastfeeding and their data demonstrated that breastfeeding was the primary factor that influenced microbiota composition in the first year of life, and that birth mode and geography were identified as key covariates (Stewart et al., 2018).

Diet continues to be a key factor to influence microbiota composition and function beyond the early postnatal period. Understanding the role of diet in the healthy microbiome is essential prior to considering how diet-microbiome interactions may influence disease. One naturalistic approach to understand the impact of diet on the microbiome has been to compare individuals from geographically distinct areas that are known to have different diets. A comparison of children aged 1−6 years from a rural village of Boulpon in Africa with a more grain-based, high fiber diet compared to European children in Florence, Italy showed significant differences in bacterial composition at the phyla level with Firmicutes more abundant in European children (De Filippo et al., 2010). Key taxa detected only in African children including *Xylanibacter, Prevotella, Butyrivibrio,* and *Treponema* and their presence is suggested to be directly related to the high fiber diet and the ability of these bacteria to ferment dietary fiber, starch, and carbohydrates to produce SCFAs (De Filippo et al., 2010). Although other SCFA-producing bacteria were present in both populations, SCFA levels were increased in African samples. Further, the increased SCFAs were associated with decreased abundance of potentially pathogenic bacteria in the rural samples (De Filippo et al., 2010). The association of dietary fiber and microbiota composition has also been demonstrated in a cohort of 82 healthy adults. Of note, different fiber types were differentially associated with microbiome composition in men (bean fiber) and women (fruit and vegetable fiber and grain fiber) (Dominianni et al., 2015).

Several studies show that habitual long-term diet influences microbiota composition and function. In a study of 98 healthy volunteers, researchers showed an association of gut microbiota and diet, identifying 97 microbiome-associated nutrients (Wu et al., 2011). Enterotype analysis revealed both *Bacteroides* and *Prevotella* enterotypes in this cohort; however, the *Bacteroides* enterotype was further distinguished by the presence of *Alistipes* and *Parabacteroides,* while the *Prevotella* enterotype was associated with the presence of *Paraprevotella* and Catenibacterium. Diet analysis demonstrated that individuals in the *Bacteroides* enterotype consumed a diet that included more animal protein, a mix of amino acids and saturated fats, and in contrast a more carbohydrate-based diet was associated with the *Prevotella* enterotype (Wu et al., 2011). A similar relationship between diet and microbiota composition was observed in children in the Philippines (Nakayama et al., 2017). Here investigators showed an association of preferred carbohydrate diet with the *Prevotella* enterotype and a preferred fats diet with the *Bacteroides* enterotype (Nakayama et al., 2017).

Based on the diet-related differences reported by Wu et al. (2011), targeted analysis of individual with vegan, vegetarian, and omnivore diets has confirmed that habitual long-term diet patterns influence microbiota composition (Franco-de-Moraes et al., 2017; Losasso et al., 2018; Wu et al., 2016). In the comparison of vegans to omnivores, the differences in microbiota composition are subtle with only a small number of taxa showing significant differences in abundance (Wu et al., 2016). In contrast, the profile of plasma metabolites was distinct in vegans and omnivores, mirroring the more distinct dietary macronutrients and micronutrients associated with these diets (Wu et al., 2016). The impact of diet-related differences in the microbiome on health and health risks is of great interest. Recent work has demonstrated that different bacterial taxa can produce different amounts and types of SCFAs from the same dietary source; the fermentation of dietary fiber into SCFAs differs in healthy individuals with the *Prevotella* enterotype compared to the *Bacteroides* enterotype (Chen et al., 2017). Further, higher abundance of *Prevotella* in individuals with vegan or plant-based diets has been associated with reduced inflammatory tone and reduced cardiovascular risk factors (de Moraes et al., 2017; Franco-de-Moraes et al., 2017).

## Key taxa in the healthy microbiome

With more than 2000 bacterial species present in the healthy human microbiome (Hugon et al., 2015) and interindividual differences in the range of 90% (Dorrestein et al., 2014; Ursell et al., 2012), consideration of geography, ethnicity, sex, age, and diet on microbiome composition and function is essential. Research in the past decade has advanced our understanding of the importance of the microbiome to health with differences in diversity, composition, and function contributing to interindividual differences in the healthy microbiome. Efforts aimed at understanding if there are specific taxa that contribute in a general way to maintaining a healthy microbiome are also important. To this end, it is useful to highlight specific genera that have emerged to have a key role in the healthy microbiome.

## Faecalibacterium prausnitzii

*F. prausnitzii* is a member of the phylum Firmicutes and is one of the most abundant commensals in the healthy human gut and plays an important role in gut physiology (Lopez-Siles et al., 2017; Miquel et al., 2014). *F. prausnitzii* is an oxygen-sensitive bacteria that is primarily localized to the duodenum and the terminal ileum (Duncan et al., 2002). *F. prausnitzii* is a key butyrate producer and can utilize carbohydrate sources that are both host-derived and diet-derived (Barcenilla et al., 2000; Duncan et al., 2002; Heinken et al., 2014). In addition to the production of butyrate, its presence promotes healthy gut physiology through antiinflammatory actions (Quevrain et al., 2016; Sokol et al., 2008) and influence on gut barrier function through regulation of tight junction proteins (Carlsson et al., 2013) and modulation of mucus production (Wrzosek et al., 2013). *F. prausnitzii* cooccurs with other butyrate-producing commensals, such as *C. coccoides,* and *Bacteroides thetaiotaomicron* (Qin et al., 2010; Wrzosek et al., 2013). *F. prausnitzii* abundance is sensitive to oxygen levels, pH, and nutrient availability. It produces butyrate but utilizes acetate, and it is thought that cooccurrence with other commensals provides necessary nutrients by cross-feeding (Lopez-Siles et al., 2012; Wrzosek et al., 2013). Notably, reduced relative abundance of *F. prausnitzii* has been reported in several clinical populations and may be an attractive biomarker of gut health.

## Akkermansia muciniphila

*A. muciniphila* is often referred to as a health-promoting bacterial species (Bressa et al., 2017). *A. muciniphila* is an abundant (1%–4%), recently identified mucin-degrading bacteria that is the only commensal representative of the phyla Verrucomicrobia (Derrien et al., 2004, 2017). *A. muciniphila* is found in the mucosal layer of the colon and utilizes mucus as an energy source to produce both proprionate and acetate (Belzer et al., 2017). *A. muciniphila* cooccurs with butyrate-producing commensals such as *Anaerostipes caccae, Eubacterium* spp., *Faecalibacterium prausnitzii,* and *Roseburia intestinalis,* and by producing acetate could stimulate butyrate production independent of dietary carbohydrates (Belzer et al., 2017). Increased abundance of *A. muciniphila* in parallel with overall increased microbiome gene richness is related to a healthier metabolic status (Dao et al., 2016) and higher relative abundance of *A. muciniphila* has been observed in active women compared to sedentary women (Bressa et al., 2017). *A. muciniphila* is detected in fecal samples from infants as early as 1 month of age and abundance continues to increase through adulthood and stabilizes in elder individuals (Collado et al., 2007). To note, centenarians greater than 105 years of age show increased relative abundance of *Akkermansia* as well as *Bifidobacterium* in parallel with a reduction of other health-promoting taxa such as *F. prausnitzii* (Belzer et al., 2017). Interest in the potential use of *Akkermansia* as a therapeutic agent is gaining momentum in the area of metabolic disorders (Plovier et al., 2017) but its use may also be beneficial in overall physical and mental health, as well as healthy aging.

## Bacteroides

*Bacteroides* sp. are abundant, gram-negative, anaerobic bacteria that belong to the phyla Bacteroidetes. *Bacteroides* are cultivable from human feces and their ability to colonize the human intestinal tract as well as their influence on health and disease has been studied for over a century (Eggerth and Gagnon, 1933). *Bacteroides* are abundant in the distal large intestine and there is a remarkable diversity of *Bacteroides* sp. (Li et al., 2009). As glycan degraders, *Bacteroides* are able to recognize and metabolize many plant-derived and host-derived complex polysaccharides (Salyers et al., 1977; Wexler and Goodman, 2017). Their unique molecular adaptations include the presence of polysaccharide utilization loci (PULs) that allow them to process complex polysaccharides more readily than other commensals. The presence of these PUL gene clusters are distinct in each species of *Bacteroides* and influence the metabolic niches that they can occupy (Wexler and Goodman, 2017). *Bacteroides* also assemble polysaccharides to their cell surface which contains glycoproteins and capsular polysaccharides. These contribute to microbe-microbe and microbe-host interactions in the gut (Martens et al., 2009). *Bacteroides* sp. produce short chain fatty acids to promote gut health and recently, the production of sphingolipids by *Bacteroides* sp. has been shown to contribute to healthy gut homeostasis (Brown et al., 2019). Several bacterial commensals are important to immune development and function (Macpherson and Harris, 2004). Work in germ-free mice show that *Bacteroides* colonization can correct many of the immune defects (Mazmanian et al., 2005). Notably, *Bacteroides* sp. are known to be antibiotic resistant and a role for some *Bacteroides* sp. in human inflammatory disease has been consistently reported (Wexler and Goodman, 2017). The pathogenic role of *Bacteroides* is thought to in part be related to their ability to cleave sugars including fucose and sialic acid from mucosal glycoproteins; these products can be utilized by pathogenic bacteria such as *E. coli* and contribute to inflammatory diseases (Ng et al., 2013). Importantly, these diverse genera include many different species that can play a beneficial role and at times a pathogenic role. As such, it is necessary to consider the species of interest in research that aims to determine the role of this abundant taxa in health and disease.

## Prevotella

*Prevotella* are gram-negative, anaerobic bacteria that belong to the Bacteroidetes phylum and are found at mucosal sites including the oral cavity, gastrointestinal tract, urogenital tract, and skin (Gupta et al., 2015). Health benefits associated with increased abundance of *Prevotella* include increased production of short-chain fatty acids and improved glucose metabolism (De Filippis et al., 2016; De Vadder et al., 2016; Kovatcheva-Datchary et al., 2015). *Prevotella* and *Bacteroides* represent the dominant taxa in reported enterotypes across numerous studies (Arumugam et al., 2011; Chen et al., 2017; Costea et al., 2018). A higher abundance of *Prevotella* has been associated with a plant-rich diet and also higher abundance of *Prevotella* is more common in non-Western populations (De Filippo et al., 2010; de Moraes et al., 2017; Tett et al., 2019; Wu et al., 2011). Considering the species diversity of *Prevotella*, more recent studies have considered the importance of genomic diversity in this genera and the influence of *Prevotella,* diet, and lifestyle on variations in human health (De Filippis et al., 2019). While much of the literature has identified the beneficial effects of *Prevotella*, several studies have demonstrated an association between higher abundance of some *Prevotella* spp. with inflammatory disease (Dillon et al., 2016; Scher et al., 2013).

## Future directions

Across all areas of biomedical research, investigators are interested in determining the features of a healthy microbiome. Extensive work in the past decade has contributed significantly to advances in our knowledge about the bacterial microbiome across different body sites. Further, research has informed us about how many host, environmental, and lifestyle factors influence the composition and function of the microbiome. Much of the analysis to date has examined the composition of the microbiota at the genus level and has indicated that there are likely multiple stable states with much variation at the individual level. The extensive diversity of bacterial species within some key genera, and the genetic diversity and functional diversity of microbes depends on the microbial environment. Future studies that examine species level analysis as well as functional read outs using metagenomics, metatranscriptomics, and metabolomics are necessary to complete the picture of the healthy microbiome and to move beyond the current overreliance on compositional data described in terms of relative abundance alterations (Shanahan and Hill, 2019). There also remain many unknowns relating to undetected microbiota members, sequences which are detected but without a matching microbial genome, and the detection of microbial genes without a match in functional databases (Thomas and Segata, 2019). The awareness of potential confounding factors and host variables, such as alcohol consumption, influencing microbial composition across health and disease requires constant monitoring for research updates in a fast evolving field (Vujkovic-Cvijin et al., 2020). Cross-talk between microbes and host systems is bidirectional and a more comprehensive understanding of the mechanisms involved is needed. The potential for the microbiome and proxies of microbe-host cross-talk as biomarkers of individual differences in health and disease is significant. Further, a more comprehensive understanding of the dynamic nature of the healthy microbiome will allow for the expansion of microbiota-targeted treatment to improve gut health, and improve the overall health of the microbiota-brain axis.

## References

Aagaard, K., Petrosino, J., Keitel, W., Watson, M., Katancik, J., Garcia, N., Patel, S., Cutting, M., Madden, T., Hamilton, H., Harris, E., Gevers, D., Simone, G., McInnes, P., Versalovic, J., 2013. The human microbiome project strategy for comprehensive sampling of the human microbiome and why it matters. FASEB J 27, 1012–1022.

Arumugam, M., Raes, J., Pelletier, E., Le Paslier, D., Yamada, T., Mende, D.R., Fernandes, G.R., Tap, J., Bruls, T., Batto, J.M., Bertalan, M., Borruel, N., Casellas, F., Fernandez, L., Gautier, L., Hansen, T., Hattori, M., Hayashi, T., Kleerebezem, M., Kurokawa, K., Leclerc, M., Levenez, F., Manichanh, C., Nielsen, H.B., Nielsen, T., Pons, N., Poulain, J., Qin, J., Sicheritz-Ponten, T., Tims, S., Torrents, D., Ugarte, E., Zoetendal, E.G., Wang, J., Guarner, F., Pedersen, O., de Vos, W.M., Brunak, S., Dore, J., Antolin, M., Artiguenave, F., Blottiere, H.M., Almeida, M., Brechot, C., Cara, C., Chervaux, C., Cultrone, A., Delorme, C., Denariaz, G., Dervyn, R., Foerstner, K.U., Friss, C., van de Guchte, M., Guedon, E., Haimet, F., Huber, W., van Hylckama-Vlieg, J., Jamet, A., Juste, C., Kaci, G., Knol, J., Lakhdari, O., Layec, S., Le Roux, K., Maguin, E., Merieux, A., Melo Minardi, R., M'Rini, C., Muller, J., Oozeer, R., Parkhill, J., Renault, P., Rescigno, M., Sanchez, N., Sunagawa, S.,

Torrejon, A., Turner, K., Vandemeulebrouck, G., Varela, E., Winogradsky, Y., Zeller, G., Weissenbach, J., Ehrlich, S.D., Bork, P., 2011. Enterotypes of the human gut microbiome. Nature 473, 174–180.

Barcenilla, A., Pryde, S.E., Martin, J.C., Duncan, S.H., Stewart, C.S., Henderson, C., Flint, H.J., 2000. Phylogenetic relationships of butyrate-producing bacteria from the human gut. Appl Environ Microbiol 66, 1654–1661.

Belzer, C., Chia, L.W., Aalvink, S., Chamlagain, B., Piironen, V., Knol, J., de Vos, W.M., 2017. Microbial metabolic networks at the mucus layer lead to diet-independent butyrate and vitamin B12 production by intestinal symbionts. mBio 8.

Bergstrom, A., Skov, T.H., Bahl, M.I., Roager, H.M., Christensen, L.B., Ejlerskov, K.T., Molgaard, C., Michaelsen, K.F., Licht, T.R., 2014. Establishment of intestinal microbiota during early life: a longitudinal, explorative study of a large cohort of Danish infants. Appl Environ Microbiol 80, 2889–2900.

Bokulich, N.A., Chung, J., Battaglia, T., Henderson, N., Jay, M., Li, H., Lieber, A.D., Wu, F., Perez-Perez, G.I., Chen, Y., Schweizer, W., Zheng, X., Contreras, M., Dominguez-Bello, M.G., Blaser, M.J., 2016. Antibiotics, birth mode, and diet shape microbiome maturation during early life. Sci Transl Med 8, 343ra382.

Bolyen, E., Rideout, J.R., Dillon, M.R., Bokulich, N.A., Abnet, C.C., Al-Ghalith, G.A., Alexander, H., Alm, E.J., Arumugam, M., Asnicar, F., Bai, Y., Bisanz, J.E., Bittinger, K., Brejnrod, A., Brislawn, C.J., Brown, C.T., Callahan, B.J., Caraballo-Rodriguez, A.M., Chase, J., Cope, E.K., Da Silva, R., Diener, C., Dorrestein, P.C., Douglas, G.M., Durall, D.M., Duvallet, C., Edwardson, C.F., Ernst, M., Estaki, M., Fouquier, J., Gauglitz, J.M., Gibbons, S.M., Gibson, D.L., Gonzalez, A., Gorlick, K., Guo, J., Hillmann, B., Holmes, S., Holste, H., Huttenhower, C., Huttley, G.A., Janssen, S., Jarmusch, A.K., Jiang, L., Kaehler, B.D., Kang, K.B., Keefe, C.R., Keim, P., Kelley, S.T., Knights, D., Koester, I., Kosciolek, T., Kreps, J., Langille, M.G.I., Lee, J., Ley, R., Liu, Y.X., Loftfield, E., Lozupone, C., Maher, M., Marotz, C., Martin, B.D., McDonald, D., McIver, L.J., Melnik, A.V., Metcalf, J.L., Morgan, S.C., Morton, J.T., Naimey, A.T., Navas-Molina, J.A., Nothias, L.F., Orchanian, S.B., Pearson, T., Peoples, S.L., Petras, D., Preuss, M.L., Pruesse, E., Rasmussen, L.B., Rivers, A., Robeson 2nd, M.S., Rosenthal, P., Segata, N., Shaffer, M., Shiffer, A., Sinha, R., Song, S.J., Spear, J.R., Swafford, A.D., Thompson, L.R., Torres, P.J., Trinh, P., Tripathi, A., Turnbaugh, P.J., Ul-Hasan, S., van der Hooft, J.J.J., Vargas, F., Vazquez-Baeza, Y., Vogtmann, E., von Hippel, M., Walters, W., Wan, Y., Wang, M., Warren, J., Weber, K.C., Williamson, C.H.D., Willis, A.D., Xu, Z.Z., Zaneveld, J.R., Zhang, Y., Zhu, Q., Knight, R., Caporaso, J.G., 2019. Reproducible, interactive, scalable and extensible microbiome data science using QIIME 2. Nat Biotechnol 37, 852–857.

Bressa, C., Bailen-Andrino, M., Perez-Santiago, J., Gonzalez-Soltero, R., Perez, M., Montalvo-Lominchar, M.G., Mate-Munoz, J.L., Dominguez, R., Moreno, D., Larrosa, M., 2017. Differences in gut microbiota profile between women with active lifestyle and sedentary women. PLoS One 12, e0171352.

Brown, E.M., Ke, X., Hitchcock, D., Jeanfavre, S., Avila-Pacheco, J., Nakata, T., Arthur, T.D., Fornelos, N., Heim, C., Franzosa, E.A., Watson, N., Huttenhower, C., Haiser, H.J., Dillow, G., Graham, D.B., Finlay, B.B., Kostic, A.D., Porter, J.A., Vlamakis, H., Clish, C.B., Xavier, R.J., 2019. Bacteroides-derived sphingolipids are critical for maintaining intestinal homeostasis and symbiosis. Cell Host Microbe 25, 668–680.

Callahan, B.J., McMurdie, P.J., Rosen, M.J., Han, A.W., Johnson, A.J., Holmes, S.P., 2016. DADA2: high-resolution sample inference from Illumina amplicon data. Nat Methods 13, 581–583.

Carlsson, A.H., Yakymenko, O., Olivier, I., Hakansson, F., Postma, E., Keita, A.V., Soderholm, J.D., 2013. Faecalibacterium prausnitzii supernatant improves intestinal barrier function in mice DSS colitis. Scand J Gastroenterol 48, 1136–1144.

Chen, T., Long, W., Zhang, C., Liu, S., Zhao, L., Hamaker, B.R., 2017. Fiber-utilizing capacity varies in Prevotella- versus Bacteroides-dominated gut microbiota. Sci Rep 7, 2594.

Cheng, J., Ringel-Kulka, T., Heikamp-de Jong, I., Ringel, Y., Carroll, I., de Vos, W.M., Salojarvi, J., Satokari, R., 2016. Discordant temporal development of bacterial phyla and the emergence of core in the fecal microbiota of young children. ISME J 10, 1002–1014.

Chu, D.M., Ma, J., Prince, A.L., Antony, K.M., Seferovic, M.D., Aagaard, K.M., 2017. Maturation of the infant microbiome community structure and function across multiple body sites and in relation to mode of delivery. Nat Med 23, 314–326.

Claesson, M.J., Jeffery, I.B., Conde, S., Power, S.E., O'Connor, E.M., Cusack, S., Harris, H.M., Coakley, M., Lakshminarayanan, B., O'Sullivan, O., Fitzgerald, G.F., Deane, J., O'Connor, M., Harnedy, N., O'Connor, K., O'Mahony, D., van Sinderen, D., Wallace, M., Brennan, L., Stanton, C., Marchesi, J.R., Fitzgerald, A.P., Shanahan, F., Hill, C., Ross, R.P., O'Toole, P.W., 2012. Gut microbiota composition correlates with diet and health in the elderly. Nature 488, 178–184.

Clayton, J.A., Collins, F.S., 2014. Policy: NIH to balance sex in cell and animal studies. Nature 509, 282–283.

Collado, M.C., Derrien, M., Isolauri, E., de Vos, W.M., Salminen, S., 2007. Intestinal integrity and Akkermansia muciniphila, a mucin-degrading member of the intestinal microbiota present in infants, adults, and the elderly. Appl Environ Microbiol 73, 7767–7770.

Costea, P.I., Hildebrand, F., Arumugam, M., Backhed, F., Blaser, M.J., Bushman, F.D., de Vos, W.M., Ehrlich, S.D., Fraser, C.M., Hattori, M., Huttenhower, C., Jeffery, I.B., Knights, D., Lewis, J.D., Ley, R.E., Ochman, H., O'Toole, P.W., Quince, C., Relman, D.A., Shanahan, F., Sunagawa, S., Wang, J., Weinstock, G.M., Wu, G.D., Zeller, G., Zhao, L., Raes, J., Knight, R., Bork, P., 2018. Enterotypes in the landscape of gut microbial community composition. Nat Microbiol 3, 8–16.

Costello, E.K., Lauber, C.L., Hamady, M., Fierer, N., Gordon, J.I., Knight, R., 2009. Bacterial community variation in human body habitats across space and time. Science 326, 1694–1697.

Dalile, B., Van Oudenhove, L., Vervliet, B., Verbeke, K., 2019. The role of short-chain fatty acids in microbiota-gut-brain communication. Nat Rev Gastroenterol Hepatol 16, 461–478.

Dao, M.C., Everard, A., Aron-Wisnewsky, J., Sokolovska, N., Prifti, E., Verger, E.O., Kayser, B.D., Levenez, F., Chilloux, J., Hoyles, L., Consortium, M.I.-O., Dumas, M.E., Rizkalla, S.W., Dore, J., Cani, P.D., Clement, K., 2016. Akkermansia muciniphila and improved metabolic health during a dietary intervention in obesity: relationship with gut microbiome richness and ecology. Gut 65, 426–436.

De Filippis, F., Pasolli, E., Tett, A., Tarallo, S., Naccarati, A., De Angelis, M., Neviani, E., Cocolin, L., Gobbetti, M., Segata, N., Ercolini, D., 2019. Distinct genetic and functional traits of human intestinal Prevotella copri strains are associated with different habitual diets. Cell Host Microbe 25, 444–453 e443.

De Filippis, F., Pellegrini, N., Vannini, L., Jeffery, I.B., La Storia, A., Laghi, L., Serrazanetti, D.I., Di Cagno, R., Ferrocino, I., Lazzi, C., Turroni, S., Cocolin, L., Brigidi, P., Neviani, E., Gobbetti, M., O'Toole, P.W., Ercolini, D., 2016. High-level adherence to a Mediterranean diet beneficially impacts the gut microbiota and associated metabolome. Gut 65, 1812–1821.

De Filippo, C., Cavalieri, D., Di Paola, M., Ramazzotti, M., Poullet, J.B., Massart, S., Collini, S., Pieraccini, G., Lionetti, P., 2010. Impact of diet in shaping gut microbiota revealed by a comparative study in children from Europe and rural Africa. Proc Natl Acad Sci U S A 107, 14691–14696.

de Moraes, A.C., Fernandes, G.R., da Silva, I.T., Almeida-Pititto, B., Gomes, E.P., Pereira, A.D., Ferreira, S.R., 2017. Enterotype may drive the dietary-associated cardiometabolic risk factors. Front Cell Infect Microbiol 7, 47.

De Vadder, F., Kovatcheva-Datchary, P., Zitoun, C., Duchampt, A., Backhed, F., Mithieux, G., 2016. Microbiota-produced succinate improves glucose homeostasis via intestinal gluconeogenesis. Cell Metab 24, 151–157.

Derrien, M., Belzer, C., de Vos, W.M., 2017. Akkermansia muciniphila and its role in regulating host functions. Microb Pathog 106, 171–181.

Derrien, M., Vaughan, E.E., Plugge, C.M., de Vos, W.M., 2004. Akkermansia muciniphila gen. nov., sp. nov., a human intestinal mucin-degrading bacterium. Int J Syst Evol Microbiol 54, 1469–1476.

Dillon, S.M., Lee, E.J., Kotter, C.V., Austin, G.L., Gianella, S., Siewe, B., Smith, D.M., Landay, A.L., McManus, M.C., Robertson, C.E., Frank, D.N., McCarter, M.D., Wilson, C.C., 2016. Gut dendritic cell activation links an altered colonic microbiome to mucosal and systemic T-cell activation in untreated HIV-1 infection. Mucosal Immunol 9, 24–37.

Ding, T., Schloss, P.D., 2014. Dynamics and associations of microbial community types across the human body. Nature 509, 357–360.

Dominguez-Bello, M.G., Costello, E.K., Contreras, M., Magris, M., Hidalgo, G., Fierer, N., Knight, R., 2010. Delivery mode shapes the acquisition and structure of the initial microbiota across multiple body habitats in newborns. Proc Natl Acad Sci U S A 107, 11971–11975.

Dominianni, C., Sinha, R., Goedert, J.J., Pei, Z., Yang, L., Hayes, R.B., Ahn, J., 2015. Sex, body mass index, and dietary fiber intake influence the human gut microbiome. PLoS One 10, e0124599.

Dorrestein, P.C., Mazmanian, S.K., Knight, R., 2014. Finding the missing links among metabolites, microbes, and the host. Immunity 40, 824–832.

Duncan, S.H., Hold, G.L., Harmsen, H.J., Stewart, C.S., Flint, H.J., 2002. Growth requirements and fermentation products of Fusobacterium prausnitzii, and a proposal to reclassify it as Faecalibacterium prausnitzii gen. nov., comb. nov. Int J Syst Evol Microbiol 52, 2141–2146.

Eckburg, P.B., Bik, E.M., Bernstein, C.N., Purdom, E., Dethlefsen, L., Sargent, M., Gill, S.R., Nelson, K.E., Relman, D.A., 2005. Diversity of the human intestinal microbial flora. Science 308, 1635–1638.

Eggerth, A.H., Gagnon, B.H., 1933. The Bacteroides of human feces. J Bacteriol 25, 389–413.

Fallani, M., Amarri, S., Uusijarvi, A., Adam, R., Khanna, S., Aguilera, M., Gil, A., Vieites, J.M., Norin, E., Young, D., Scott, J.A., Dore, J., Edwards, C.A., 2011. Determinants of the human infant intestinal microbiota after the introduction of first complementary foods in infant samples from five European centres. Microbiology 157, 1385–1392.

Falony, G., Joossens, M., Vieira-Silva, S., Wang, J., Darzi, Y., Faust, K., Kurilshikov, A., Bonder, M.J., Valles-Colomer, M., Vandeputte, D., Tito, R.Y., Chaffron, S., Rymenans, L., Verspecht, C., De Sutter, L., Lima-Mendez, G., D'Hoe, K., Jonckheere, K., Homola, D., Garcia, R., Tigchelaar, E.F., Eeckhaudt, L., Fu, J., Henckaerts, L., Zhernakova, A., Wijmenga, C., Raes, J., 2016. Population-level analysis of gut microbiome variation. Science 352, 560–564.

Faust, K., Sathirapongsasuti, J.F., Izard, J., Segata, N., Gevers, D., Raes, J., Huttenhower, C., 2012. Microbial co-occurrence relationships in the human microbiome. PLoS Comput Biol 8, e1002606.

Ferretti, P., Pasolli, E., Tett, A., Asnicar, F., Gorfer, V., Fedi, S., Armanini, F., Truong, D.T., Manara, S., Zolfo, M., Beghini, F., Bertorelli, R., De Sanctis, V., Bariletti, I., Canto, R., Clementi, R., Cologna, M., Crifo, T., Cusumano, G., Gottardi, S., Innamorati, C., Mase, C., Postai, D., Savoi, D., Duranti, S., Lugli, G.A., Mancabelli, L., Turroni, F., Ferrario, C., Milani, C., Mangifesta, M., Anzalone, R., Viappiani, A., Yassour, M., Vlamakis, H., Xavier, R., Collado, C.M., Koren, O., Tateo, S., Soffiati, M., Pedrotti, A., Ventura, M., Huttenhower, C., Bork, P., Segata, N., 2018. Mother-to-Infant microbial transmission from different body sites shapes the developing infant gut microbiome. Cell Host Microbe 24, 133–145.

Finotello, F., Mastrorilli, E., Di Camillo, B., 2018. Measuring the diversity of the human microbiota with targeted next-generation sequencing. Brief Bioinform 19, 679–692.

Franco-de-Moraes, A.C., de Almeida-Pititto, B., da Rocha Fernandes, G., Gomes, E.P., da Costa Pereira, A., Ferreira, S.R.G., 2017. Worse inflammatory profile in omnivores than in vegetarians associates with the gut microbiota composition. Diabetol Metab Syndr 9, 62.

Gevers, D., Knight, R., Petrosino, J.F., Huang, K., McGuire, A.L., Birren, B.W., Nelson, K.E., White, O., Methe, B.A., Huttenhower, C., 2012. The human microbiome project: a community resource for the healthy human microbiome. PLoS Biol 10, e1001377.

Goodrich, J.K., Di Rienzi, S.C., Poole, A.C., Koren, O., Walters, W.A., Caporaso, J.G., Knight, R., Ley, R.E., 2014. Conducting a microbiome study. Cell 158, 250–262.

Gorvitovskaia, A., Holmes, S.P., Huse, S.M., 2016. Interpreting Prevotella and Bacteroides as biomarkers of diet and lifestyle. Microbiome 4, 15.

Gupta, V.K., Chaudhari, N.M., Iskepalli, S., Dutta, C., 2015. Divergences in gene repertoire among the reference Prevotella genomes derived from distinct body sites of human. BMC Genom 16, 153.

Haro, C., Rangel-Zuniga, O.A., Alcala-Diaz, J.F., Gomez-Delgado, F., Perez-Martinez, P., Delgado-Lista, J., Quintana-Navarro, G.M., Landa, B.B., Navas-Cortes, J.A., Tena-Sempere, M., Clemente, J.C., Lopez-Miranda, J., Perez-Jimenez, F., Camargo, A., 2016. Intestinal microbiota is influenced by gender and body mass index. PLoS One 11, e0154090.

Heinken, A., Khan, M.T., Paglia, G., Rodionov, D.A., Harmsen, H.J., Thiele, I., 2014. Functional metabolic map of Faecalibacterium prausnitzii, a beneficial human gut microbe. J Bacteriol 196, 3289–3302.

Hugon, P., Dufour, J.C., Colson, P., Fournier, P.E., Sallah, K., Raoult, D., 2015. A comprehensive repertoire of prokaryotic species identified in human beings. Lancet Infect Dis 15, 1211–1219.

Human Microbiome Project, C., 2012a. A framework for human microbiome research. Nature 486, 215–221.

Human Microbiome Project, C., 2012b. Structure, function and diversity of the healthy human microbiome. Nature 486, 207–214.

Huse, S.M., Ye, Y., Zhou, Y., Fodor, A.A., 2012. A core human microbiome as viewed through 16S rRNA sequence clusters. PLoS One 7, e34242.

Jalanka-Tuovinen, J., Salonen, A., Nikkila, J., Immonen, O., Kekkonen, R., Lahti, L., Palva, A., de Vos, W.M., 2011. Intestinal microbiota in healthy adults: temporal analysis reveals individual and common core and relation to intestinal symptoms. PLoS One 6, e23035.

Jeffery, I.B., Claesson, M.J., O'Toole, P.W., Shanahan, F., 2012. Categorization of the gut microbiota: enterotypes or gradients? Nat Rev Microbiol 10, 591–592.

Koren, O., Knights, D., Gonzalez, A., Waldron, L., Segata, N., Knight, R., Huttenhower, C., Ley, R.E., 2013. A guide to enterotypes across the human body: meta-analysis of microbial community structures in human microbiome datasets. PLoS Comput Biol 9, e1002863.

Kovatcheva-Datchary, P., Nilsson, A., Akrami, R., Lee, Y.S., De Vadder, F., Arora, T., Hallen, A., Martens, E., Bjorck, I., Backhed, F., 2015. Dietary fiber-induced improvement in glucose metabolism is associated with increased abundance of Prevotella. Cell Metab 22, 971–982.

Li, K., Bihan, M., Yooseph, S., Methe, B.A., 2012. Analyses of the microbial diversity across the human microbiome. PLoS One 7, e32118.

Li, M., Zhou, H., Hua, W., Wang, B., Wang, S., Zhao, G., Li, L., Zhao, L., Pang, X., 2009. Molecular diversity of Bacteroides spp. in human fecal microbiota as determined by group-specific 16S rRNA gene clone library analysis. Syst Appl Microbiol 32, 193–200.

Lopez-Siles, M., Duncan, S.H., Garcia-Gil, L.J., Martinez-Medina, M., 2017. Faecalibacterium prausnitzii: from microbiology to diagnostics and prognostics. ISME J 11, 841–852.

Lopez-Siles, M., Khan, T.M., Duncan, S.H., Harmsen, H.J., Garcia-Gil, L.J., Flint, H.J., 2012. Cultured representatives of two major phylogroups of human colonic Faecalibacterium prausnitzii can utilize pectin, uronic acids, and host-derived substrates for growth. Appl Environ Microbiol 78, 420–428.

Losasso, C., Eckert, E.M., Mastrorilli, E., Villiger, J., Mancin, M., Patuzzi, I., Di Cesare, A., Cibin, V., Barrucci, F., Pernthaler, J., Corno, G., Ricci, A., 2018. Assessing the influence of vegan, vegetarian and omnivore oriented westernized dietary styles on human gut microbiota: a cross sectional study. Front Microbiol 9, 317.

Macpherson, A.J., Harris, N.L., 2004. Interactions between commensal intestinal bacteria and the immune system. Nat Rev Immunol 4, 478–485.

Mahnic, A., Rupnik, M., 2018. Different host factors are associated with patterns in bacterial and fungal gut microbiota in Slovenian healthy cohort. PLoS One 13, e0209209.

Martens, E.C., Roth, R., Heuser, J.E., Gordon, J.I., 2009. Coordinate regulation of glycan degradation and polysaccharide capsule biosynthesis by a prominent human gut symbiont. J Biol Chem 284, 18445–18457.

Mazmanian, S.K., Liu, C.H., Tzianabos, A.O., Kasper, D.L., 2005. An immunomodulatory molecule of symbiotic bacteria directs maturation of the host immune system. Cell 122, 107–118.

Miquel, S., Martin, R., Bridonneau, C., Robert, V., Sokol, H., Bermudez Humaran, L.G., Thomas, M., Langella, P., 2014. Ecology and metabolism of the beneficial intestinal commensal bacterium Faecalibacterium prausnitzii. Gut Microb 5, 146–151.

Mueller, N.T., Shin, H., Pizoni, A., Werlang, I.C., Matte, U., Goldani, M.Z., Goldani, H.A., Dominguez-Bello, M.G., 2016. Birth mode-dependent association between pre-pregnancy maternal weight status and the neonatal intestinal microbiome. Sci Rep 6, 23133.

Nakayama, J., Yamamoto, A., Palermo-Conde, L.A., Higashi, K., Sonomoto, K., Tan, J., Lee, Y.K., 2017. Impact of westernized diet on gut microbiota in children on leyte island. Front Microbiol 8, 197.

Navas-Molina, J.A., Peralta-Sanchez, J.M., Gonzalez, A., McMurdie, P.J., Vazquez-Baeza, Y., Xu, Z., Ursell, L.K., Lauber, C., Zhou, H., Song, S.J., Huntley, J., Ackermann, G.L., Berg-Lyons, D., Holmes, S., Caporaso, J.G., Knight, R., 2013. Advancing our understanding of the human microbiome using QIIME. Methods Enzymol 531, 371–444.

Ng, K.M., Ferreyra, J.A., Higginbottom, S.K., Lynch, J.B., Kashyap, P.C., Gopinath, S., Naidu, N., Choudhury, B., Weimer, B.C., Monack, D.M., Sonnenburg, J.L., 2013.

Microbiota-liberated host sugars facilitate post-antibiotic expansion of enteric pathogens. Nature 502, 96—99.

Odamaki, T., Kato, K., Sugahara, H., Hashikura, N., Takahashi, S., Xiao, J.Z., Abe, F., Osawa, R., 2016. Age-related changes in gut microbiota composition from newborn to centenarian: a cross-sectional study. BMC Microbiol 16, 90.

Plovier, H., Everard, A., Druart, C., Depommier, C., Van Hul, M., Geurts, L., Chilloux, J., Ottman, N., Duparc, T., Lichtenstein, L., Myridakis, A., Delzenne, N.M., Klievink, J., Bhattacharjee, A., van der Ark, K.C., Aalvink, S., Martinez, L.O., Dumas, M.E., Maiter, D., Loumaye, A., Hermans, M.P., Thissen, J.P., Belzer, C., de Vos, W.M., Cani, P.D., 2017. A purified membrane protein from Akkermansia muciniphila or the pasteurized bacterium improves metabolism in obese and diabetic mice. Nat Med 23, 107—113.

Qin, J., Li, R., Raes, J., Arumugam, M., Burgdorf, K.S., Manichanh, C., Nielsen, T., Pons, N., Levenez, F., Yamada, T., Mende, D.R., Li, J., Xu, J., Li, S., Li, D., Cao, J., Wang, B., Liang, H., Zheng, H., Xie, Y., Tap, J., Lepage, P., Bertalan, M., Batto, J.M., Hansen, T., Le Paslier, D., Linneberg, A., Nielsen, H.B., Pelletier, E., Renault, P., Sicheritz-Ponten, T., Turner, K., Zhu, H., Yu, C., Jian, M., Zhou, Y., Li, Y., Zhang, X., Qin, N., Yang, H., Wang, J., Brunak, S., Dore, J., Guarner, F., Kristiansen, K., Pedersen, O., Parkhill, J., Weissenbach, J., Bork, P., Ehrlich, S.D., 2010. A human gut microbial gene catalogue established by metagenomic sequencing. Nature 464, 59—65.

Quevrain, E., Maubert, M.A., Michon, C., Chain, F., Marquant, R., Tailhades, J., Miquel, S., Carlier, L., Bermudez-Humaran, L.G., Pigneur, B., Lequin, O., Kharrat, P., Thomas, G., Rainteau, D., Aubry, C., Breyner, N., Afonso, C., Lavielle, S., Grill, J.P., Chassaing, G., Chatel, J.M., Trugnan, G., Xavier, R., Langella, P., Sokol, H., Seksik, P., 2016. Identification of an anti-inflammatory protein from Faecalibacterium prausnitzii, a commensal bacterium deficient in Crohn's disease. Gut 65, 415—425.

Rios-Covian, D., Ruas-Madiedo, P., Margolles, A., Gueimonde, M., de Los Reyes-Gavilan, C.G., Salazar, N., 2016. Intestinal short chain fatty acids and their link with diet and human health. Front Microbiol 7, 185.

Salyers, A.A., Vercellotti, J.R., West, S.E., Wilkins, T.D., 1977. Fermentation of mucin and plant polysaccharides by strains of Bacteroides from the human colon. Appl Environ Microbiol 33, 319—322.

Scher, J.U., Sczesnak, A., Longman, R.S., Segata, N., Ubeda, C., Bielski, C., Rostron, T., Cerundolo, V., Pamer, E.G., Abramson, S.B., Huttenhower, C., Littman, D.R., 2013. Expansion of intestinal Prevotella copri correlates with enhanced susceptibility to arthritis. Elife 2, e01202.

Shanahan, F., Hill, C., 2019. Language, numeracy and logic in microbiome science. Nat Rev Gastroenterol Hepatol 16, 387—388.

Sokol, H., Pigneur, B., Watterlot, L., Lakhdari, O., Bermudez-Humaran, L.G., Gratadoux, J.J., Blugeon, S., Bridonneau, C., Furet, J.P., Corthier, G., Grangette, C., Vasquez, N., Pochart, P., Trugnan, G., Thomas, G., Blottiere, H.M., Dore, J., Marteau, P., Seksik, P., Langella, P., 2008. Faecalibacterium prausnitzii is an anti-inflammatory commensal bacterium identified by gut microbiota analysis of Crohn disease patients. Proc Natl Acad Sci U S A 105, 16731—16736.

Stearns, J.C., Zulyniak, M.A., de Souza, R.J., Campbell, N.C., Fontes, M., Shaikh, M., Sears, M.R., Becker, A.B., Mandhane, P.J., Subbarao, P., Turvey, S.E., Gupta, M., Beyene, J., Surette, M.G., Anand, S.S., NutriGen, A., 2017. Ethnic and diet-related differences in the healthy infant microbiome. Genome Med 9, 32.

Stewart, C.J., Ajami, N.J., O'Brien, J.L., Hutchinson, D.S., Smith, D.P., Wong, M.C., Ross, M.C., Lloyd, R.E., Doddapaneni, H., Metcalf, G.A., Muzny, D., Gibbs, R.A., Vatanen, T., Huttenhower, C., Xavier, R.J., Rewers, M., Hagopian, W., Toppari, J.,

Ziegler, A.G., She, J.X., Akolkar, B., Lernmark, A., Hyoty, H., Vehik, K., Krischer, J.P., Petrosino, J.F., 2018. Temporal development of the gut microbiome in early childhood from the TEDDY study. Nature 562, 583−588.

Tannenbaum, C., Schwarz, J.M., Clayton, J.A., de Vries, G.J., Sullivan, C., 2016. Evaluating sex as a biological variable in preclinical research: the devil in the details. Biol Sex Differ 7, 13.

Tett, A., Huang, K.D., Asnicar, F., Fehlner-Peach, H., Pasolli, E., Karcher, N., Armanini, F., Manghi, P., Bonham, K., Zolfo, M., De Filippis, F., Magnabosco, C., Bonneau, R., Lusingu, J., Amuasi, J., Reinhard, K., Rattei, T., Boulund, F., Engstrand, L., Zink, A., Collado, M.C., Littman, D.R., Eibach, D., Ercolini, D., Rota-Stabelli, O., Huttenhower, C., Maixner, F., Segata, N., 2019. The Prevotella copri complex comprises four distinct clades underrepresented in westernized populations. Cell Host Microbe 26, 666−679.

Thomas, A.M., Segata, N., 2019. Multiple levels of the unknown in microbiome research. BMC Biol 17, 48.

Tillisch, K., Mayer, E.A., Gupta, A., Gill, Z., Brazeilles, R., Le Neve, B., van Hylckama Vlieg, J.E.T., Guyonnet, D., Derrien, M., Labus, J.S., 2017. Brain structure and response to emotional stimuli as related to gut microbial profiles in healthy women. Psychosom Med 79, 905−913.

Turnbaugh, P.J., Ley, R.E., Hamady, M., Fraser-Liggett, C.M., Knight, R., Gordon, J.I., 2007. The human microbiome project. Nature 449, 804−810.

Ursell, L.K., Clemente, J.C., Rideout, J.R., Gevers, D., Caporaso, J.G., Knight, R., 2012. The interpersonal and intrapersonal diversity of human-associated microbiota in key body sites. J Allergy Clin Immunol 129, 1204−1208.

Vatanen, T., Franzosa, E.A., Schwager, R., Tripathi, S., Arthur, T.D., Vehik, K., Lernmark, A., Hagopian, W.A., Rewers, M.J., She, J.X., Toppari, J., Ziegler, A.G., Akolkar, B., Krischer, J.P., Stewart, C.J., Ajami, N.J., Petrosino, J.F., Gevers, D., Lahdesmaki, H., Vlamakis, H., Huttenhower, C., Xavier, R.J., 2018. The human gut microbiome in early-onset type 1 diabetes from the TEDDY study. Nature 562, 589−594.

Vujkovic-Cvijin, I., Sklar, J., Jiang, L., Natarajan, L., Knight, R., Belkaid, Y., 2020. Host variables confound gut microbiota studies of human disease. Nature 587, 448−454.

Wexler, A.G., Goodman, A.L., 2017. An insider's perspective: Bacteroides as a window into the microbiome. Nat Microbiol 2, 17026.

Whelan, F.J., Surette, M.G., 2017. A comprehensive evaluation of the sl1p pipeline for 16S rRNA gene sequencing analysis. Microbiome 5, 100.

Wrzosek, L., Miquel, S., Noordine, M.L., Bouet, S., Joncquel Chevalier-Curt, M., Robert, V., Philippe, C., Bridonneau, C., Cherbuy, C., Robbe-Masselot, C., Langella, P., Thomas, M., 2013. Bacteroides thetaiotaomicron and Faecalibacterium prausnitzii influence the production of mucus glycans and the development of goblet cells in the colonic epithelium of a gnotobiotic model rodent. BMC Biol 11, 61.

Wu, G.D., Chen, J., Hoffmann, C., Bittinger, K., Chen, Y.Y., Keilbaugh, S.A., Bewtra, M., Knights, D., Walters, W.A., Knight, R., Sinha, R., Gilroy, E., Gupta, K., Baldassano, R., Nessel, L., Li, H., Bushman, F.D., Lewis, J.D., 2011. Linking long-term dietary patterns with gut microbial enterotypes. Science 334, 105−108.

Wu, G.D., Compher, C., Chen, E.Z., Smith, S.A., Shah, R.D., Bittinger, K., Chehoud, C., Albenberg, L.G., Nessel, L., Gilroy, E., Star, J., Weljie, A.M., Flint, H.J., Metz, D.C., Bennett, M.J., Li, H., Bushman, F.D., Lewis, J.D., 2016. Comparative metabolomics in vegans and omnivores reveal constraints on diet-dependent gut microbiota metabolite production. Gut 65, 63−72.

Yassour, M., Jason, E., Hogstrom, L.J., Arthur, T.D., Tripathi, S., Siljander, H., Selvenius, J., Oikarinen, S., Hyoty, H., Virtanen, S.M., Ilonen, J., Ferretti, P.,

Pasolli, E., Tett, A., Asnicar, F., Segata, N., Vlamakis, H., Lander, E.S., Huttenhower, C., Knip, M., Xavier, R.J., 2018. Strain-level analysis of mother-to-child bacterial transmission during the first few months of life. Cell Host Microbe 24, 146–154.

Yatsunenko, T., Rey, F.E., Manary, M.J., Trehan, I., Dominguez-Bello, M.G., Contreras, M., Magris, M., Hidalgo, G., Baldassano, R.N., Anokhin, A.P., Heath, A.C., Warner, B., Reeder, J., Kuczynski, J., Caporaso, J.G., Lozupone, C.A., Lauber, C., Clemente, J.C., Knights, D., Knight, R., Gordon, J.I., 2012. Human gut microbiome viewed across age and geography. Nature 486, 222–227.

Yin, Y., Fan, B., Liu, W., Ren, R., Chen, H., Bai, S., Zhu, L., Sun, G., Yang, Y., Wang, X., 2017. Investigation into the stability and culturability of Chinese enterotypes. Sci Rep 7, 7947.

Zhang, J., Guo, Z., Lim, A.A., Zheng, Y., Koh, E.Y., Ho, D., Qiao, J., Huo, D., Hou, Q., Huang, W., Wang, L., Javzandulam, C., Narangerel, C., Lee, Y.K., Zhang, H., 2014. Mongolians core gut microbiota and its correlation with seasonal dietary changes. Sci Rep 4, 5001.

Zhernakova, A., Kurilshikov, A., Bonder, M.J., Tigchelaar, E.F., Schirmer, M., Vatanen, T., Mujagic, Z., Vila, A.V., Falony, G., Vieira-Silva, S., Wang, J., Imhann, F., Brandsma, E., Jankipersadsing, S.A., Joossens, M., Cenit, M.C., Deelen, P., Swertz, M.A., LifeLines cohort, s., Weersma, R.K., Feskens, E.J., Netea, M.G., Gevers, D., Jonkers, D., Franke, L., Aulchenko, Y.S., Huttenhower, C., Raes, J., Hofker, M.H., Xavier, R.J., Wijmenga, C., Fu, J., 2016. Population-based metagenomics analysis reveals markers for gut microbiome composition and diversity. Science 352, 565–569.

## CHAPTER 3

# Gene-environment factors influence microbiota composition, diversity, and function

## Introduction

The debate on the importance of nature versus nurture on health and disease has roots back to ancient Greek theories of personality. Nature refers to what is influenced by genetic and biological factors, whereas nurture refers to external or environmental factors. As researchers attempt to understand how the microbiome influences brain function, understanding how nature (gene) and nurture (environment) contribute to differences in the microbiome in healthy individuals and in disease is an important consideration. Genetic and environmental factors known to influence microbiota composition and diversity also influence microbe-host interactions. While it is common for researchers to suggest that environmental factors, such as diet, are the most important factors influencing microbiome composition, an individual's genetics and host systems also influence microbiome features and associated host-microbe interactions in response to environmental, societal, and other factors. It is the dynamic interplay between these factors that contribute to individual differences in microbiota composition and function.

This chapter examines several factors that are known to influence the microbiome, some in the "nature" category, some in the "nurture" category (Fig. 3.1).

Host genetics, that is the DNA we are born with, influence many physiological systems including immune and gut function which can in turn impact microbiome composition, diversity, and function at different body sites (Blekhman et al., 2015; Bonder et al., 2016; Kolde et al., 2018).

*Microbiota Brain Axis*
ISBN 978-0-12-814800-6
https://doi.org/10.1016/B978-0-12-814800-6.00005-4

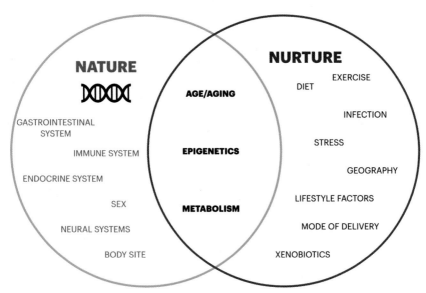

**Figure 3.1** Factors that influence the microbiome—nature versus nurture. Nature refers to what is influenced by genetic and biological factors, whereas nurture refers to external or environmental factors. Both environmental and host factors influence microbiota composition and function—together these impact gut-brain axis communication and brain health.

Studies in twins highlight the contribution of genetic variation to microbiome composition, demonstrating that microbiota composition was more similar in monzygotic compared to dizygotic twins (Goodrich et al., 2014, 2016; Turnbaugh et al., 2009). Environmental factors also influence microbiome composition and function and headlines related to new results in this domain continue to fuel the debate related to the influence of nature versus nurture on the microbiome (Cell_Press, 2016; Goodrich et al., 2017; Rothschild et al., 2018; Saez, 2022; Weizmann_Institute_of_Science, 2018). Environmental exposures and lifestyle choices can impact the microbiome. Geographic location, diet, exercise, body mass index, medications, hygiene, and stress have all been shown to influence microbiome composition and function (Cheng and Ning, 2019; Ding and Schloss, 2014; Gupta et al., 2017; Mobeen et al., 2018; Tang et al., 2019; Yatsunenko et al., 2012). This chapter will provide an overview of research examining how genetic and environmental factors influence the microbiome, and will highlight the importance of these influences for microbial regulation of brain function and brain health.

# Animal models

The use of different inbred and outbred strains of mice in research studies, as well as investigations in other mammals (Weinstein et al., 2021), provides an avenue to dissect the genetic basis of complex traits. Different strains of mice have distinct behavioral phenotypes and biological differences, including many linked to alterations in microbiome-gut-brain axis signaling such as depression, anxiety, and stress. Several studies have demonstrated that microbiota composition and diversity is dependent on genetic background (Benson et al., 2010; Campbell et al., 2012; Horne et al., 2019; O'Connor et al., 2014). An advantage in these studies, in comparison to human studies, is that several environmental factors can be controlled for that might otherwise complicate interpretations. Research over a decade ago showed that the composition of the gut microbiota behaves as a polygenic trait in the murine model and identified 18 host quantitative trait loci (QTL) that correlated to relative abundances of microbial taxa (Benson et al., 2010). Many of those QTL identified in that first study have been replicated (Leamy et al., 2014; Org et al., 2015). Additional evidence of the importance of host genetics to the microbiome, and its influence on the brain, arises from mapping natural genetic variation in different strains of mice as well as studies using genetically modified mice.

The Collaborative Cross (CC) mouse population is a collection of re-combinant, inbred mouse strains that have been widely used to examine the genetic and environmental basis of complex traits (Chesler et al., 2008; Churchill et al., 2004; Philip et al., 2011). An examination of 10 inbred mouse lines from the CC population showed strain-related differences in microbiota diversity in cecal samples using 16S rRNA sequencing, with much greater interstrain differences in beta diversity compared to intrastrain differences (Campbell et al., 2012). Differential abundance analysis showed strain-specific profiles of taxa; however, the data did not reveal strain-specific taxa that could predict strain based on bacterial abundance (Campbell et al., 2012). In a separate study, pairwise comparison between inbred strains, Balb/C and C57Bl/6 mice, revealed bacterial taxa that were present in one strain and not the other, suggesting that strain-specific taxa may be present, however, the role of other factors such as housing and supplier could be involved (Campbell et al., 2012; Horne et al., 2019). In most studies, mice are co-housed with littermates after weaning, and since mice are coprophagic, the impact of maternal effects (that would impact each litter) and cohousing is a confounding factor of many microbiome mouse studies (Campbell et al., 2012). Certainly, the source of experimental animals from in-house breeding or an outside supplier, as well as time in a particular facility or housing room, and co-caging impacts microbiota composition (Brinkman et al., 2013; Ericsson et al., 2015; Hoy et al., 2015). Identifying which bacterial taxa and signaling cascades are regulated by the interplay between genotype, housing, maternal effects as well as stress, sex and other experimental factors is needed. Further, consideration of these effects through balanced experimental design will improve reliability and reproducibility of the results and interpretations.

Using 30 CC mouse strains, a recent study identified 169 joint QTL intervals that were significantly associated with gut microbiota abundance (Snijders et al., 2016). While diet did not show a significant impact on microbiota composition in this study, it did have a significant impact on the gut metabolome highlighting the importance of assessing functional read outs in microbiome studies (Snijders et al., 2016). Interestingly, the immune gene, major histocompatibility complex (MHC), was significantly associated with the abundance of *Lactobacillaceae*, which was further predictive of T cell counts in peripheral blood, particularly T helper cells (Snijders et al., 2016). Microbiota-immune signaling influences the CNS, and a recent study demonstrated that cross-talk between gut microbiota and peripheral T cells influence the postnatal development of both the gut and the brain metabolome (Caspani et al., 2022).

Linking host-microbial interactions to disease phenotypes using systems genetics and data-driven analytical approaches revealed an association between *Odoribacter* abundance and sleep phenotypes in a diverse CC breeding population (Bubier et al., 2020). This study integrated data sets, including host transcriptome, behavioral phenotypes, and microbiota data, from more than 100 CC mouse strains (Bubier et al., 2020). Correlation of microbiota abundance and host phenotypes revealed 46 significant microbiota-host correlations, 26 of which were significant following FDR correction for multiple testing (Bubier et al., 2020). Sleep phenotypes were predominant in the significant microbe-host associations. Further, with a specific association found between the sleep-related taxa, *Odoribacter*, and the QTL *Micab7*, a region on chromosome 7 that contains 42 genes that include the sleep related gene *Igf1r* (Bubier et al., 2020). This relationship was further validated using the db/db *Lepr* mutant mouse model that has been reported to have altered sleep phenotypes (Bubier et al., 2020). Similar approaches can be utilized to better understand the association between microbes and host phenotypes, and to determine the mechanisms involved.

Host genetics has an impact on many biological features that in turn may influence the microbiome, for example, production of secretory IgA and its relationship to microbiota composition. Mucosal IgA plays an important role in maintaining homeostasis within the intestine as well as in the barrier integrity (Corthesy, 2013; Fransen et al., 2015). Germ-free (GF) mice have been shown to have smaller and fewer Peyer's Patches (lymphoid nodules which play a role in bacterial surveillance within the gut) as well as exhibit a reduction in the number of mucosal IgA-producing cells and that through colonization with commensal bacteria was reversed (Macpherson and Harris, 2004). Importantly, to date relatively few strains of mice have been used in GF studies and therefore there are many gaps in the knowledge related to links between host genetics and the microbiome. That said, host genetics is suggested to influence IgA as abundance of IgA is strain dependent in mice (Fransen et al., 2015). Moreover, distinct difference in IgA production between inbred strains of mice BALB/c and C57Bl/6 was shown to relate to differences observed in gut microbiota composition (Fransen et al., 2015).

microRNAs (miRNAs) are small noncoding RNA molecules with an important role in the posttranscriptional regulation of gene expression. Specifically, miRNA are 18–22 nucleotide long RNA molecules that regulate genes through selective binding to mRNA targets in the 3'-untranslated region (3'UTR), resulting in repression of translation or degradation of the transcript (Ha and Kim, 2014). A bidirectional relationship between host miRNA expression and gut microbiota composition has been reported (Dalmasso et al., 2011; Liu et al., 2016; Moloney et al., 2018; Singh et al., 2012), demonstrating a role for host miRNA expression in intestinal epithelial cells (IEC) on bacterial diversity (Liu et al., 2016). Evaluating the potential for eukaryotic host miRNA to target bacterial genes has also been considered and a few studies have provided evidence of potential regulation of bacterial composition by miRNA (Liu et al., 2016; Moloney et al., 2018; Teng et al., 2018). Previous work looking at bacterial small regulatory RNA (sRNA) gene regulation has found multiple possible binding sites for bacterial sRNA to bacterial genes (De Lay et al., 2013), and several studies have now demonstrated the presence of host-derived miRNAs in murine (Horne et al., 2019; Liu et al., 2016; Moloney et al., 2018) and human feces (Liu et al., 2016) that may bind to bacterial genes. Reduction of host fecal miRNAs has been shown to alter gut bacteria composition and influence bacterial growth (Liu et al., 2016; Moloney et al., 2018). The potential for miRNAs to enter bacterial cells was shown by Liu et al. (2016) and more recently, evidence demonstrated that plant-derived exosomal miRNAs are taken up by gut microbiota, target bacterial genes to influence cross-talk between gut microbiota and the host immune system (Teng et al., 2018). Interestingly, specific targeting of food-derived exosomal miRNAs to mucosal-associated bacteria, *Lactobacillus*, was highlighted (Teng et al., 2018). A recent study demonstrated a strong negative correlation with fecal miRNA let-7b-5p and the relative abundance of *Parabacteroides* (Horne et al., 2019). Let-7 family miRNA expression has previously been shown to be involved in the host antibacterial defense through modulation of immune response (Schulte et al., 2011). Intriguingly, *Parabacteroides* has also been associated with inflammatory markers (Conley et al., 2016). In addition, a positive association between fecal mir-21a-5p and the relative abundance of *Akkermansia* was reported (Horne et al., 2019). *A. muciniphila* has been shown to play a role in energy metabolism, showing a role in regulating adiposity, by regulating mRNA expression of markers of adipocyte differentiation and lipid oxidation (Everard et al., 2013). Mir-21a-5p has also been shown to play a critical role in regulating the proliferation of human adipose tissue (Kim et al., 2012), thus the correlation observed may be reflective of a mechanistic link between *Akkermansia* abundance and miR-21 expression in energy metabolism of the host (Horne et al., 2019).

The gut microbiota is well established as essential for development of the adaptive immune system, and an increasing body of evidence suggests that cross-talk between microbes, environment, and host immunity shape the maturation and function of the central nervous system (CNS). In particular, T-cells are heavily influenced by gut microbes, wherein specific bacterial taxa and their metabolites are known to promote proliferation of distinct T-cell subsets (Cheng et al., 2019; Lee and Kim, 2017). In adulthood, studies have demonstrated a variety of behavioral abnormalities in immunodeficient models lacking T-cells (McGowan et al., 2011; Quinnies et al., 2015), as well as the influence of T-cells on regulation of social behavior and synaptic plasticity (Derecki et al., 2010; Tanabe and Yamashita, 2018). Using a targeted genetic approach with T-cell-deficient mice, due to the double knockout of the β and δ chains of the T-cell receptor (*TCR β-/-δ-/-*), studies have revealed altered behaviors in the postnatal period including developmental delays in righting reflex, increased vocalizations in response to maternal separation, and reduced exploration and activity in the open field following exposure to early life stress (Francella et al., 2022). *TCR β-/-δ-/-* mice have altered maturation, diversity, and composition of the microbiome (Caspani et al., 2022; Francella et al., 2022). By examining both the microbiota (cecal and fecal) and the metabolome (cecal, fecal, gut, and brain) at several time points over postnatal development, key T-cell related taxa and their metabolites that are critical to microbe-immune-brain signaling in the postnatal period were identified. A novel finding of this work was that T–cell related changes in the gut microbiome affected neuroactive metabolites in the brain. In particular, the neurotransmitters, GABA and glycine, were elevated in the brains of T–cell-deficient mice (Caspani et al., 2022). At the level of the CNS, peripheral T cells contribute to synaptic pruning via their action on microglial maturation in the postnatal mouse brain (Pasciuto et al., 2020). Moreover, several studies have demonstrated that microbiota-T cell-microglial signaling also impacts acute brain injury in stroke models linking these developmental mechanisms to outcomes later in life (Benakis et al., 2016; Brea et al., 2021; Sadler et al., 2020; Singh et al., 2016, 2018). Together this work revealed the importance of the microbiome to neuroimmune interactions and emphasizes the need to consider both the peripheral immune system and the microbiome to gain a comprehensive understanding of the neuroimmune mechanisms that are important to brain health.

Host genetics influences disease risk and as noted above for the immune system, interactions between host genotype and gut microbiota also impact autoimmune disease (Montgomery et al., 2020). Experimental autoimmune encephalomyelitis (EAE) is a long-standing and well-established animal model of multiple sclerosis (Gold et al., 2006). Susceptibility of inbred strains to active induction of EAE by encephalitogenic peptides varies based on strain and the active peptide used (Racke, 2001). A recent study compared EAE disease susceptibility in C57Bl/6 mice and in the wild-derived PWD/Phj strain (Montgomery et al., 2020). These genetically divergent strains of mice have distinct microbiota and immune phenotypes (Bearoff et al., 2016; Horne et al., 2019). Using 27 B6Chr$^{PWD}$ consomic strains, susceptibility to EAE was linked to host genetic-related alterations in gut microbiota composition and function (Montgomery et al., 2020). Interestingly, *Lactobacillus* abundance emerged as a key taxa that cosegregated with EAE severity, that was shown to exacerbate EAE pathogenesis in parallel with modulation of T cell responses (Montgomery et al., 2020). A potential role for key microbial metabolites including short chain fatty acids were noted, highlighting the need to consider functional readouts as well as the compositional profile of microbiota (Montgomery et al., 2020).

A substantial body of preclinical and clinical evidence demonstrates a role for the gut microbiota in neurodevelopmental disorders (Vuong and Hsiao, 2017). Mouse models provide an opportunity to systematically investigate the contribution of host genetics and other factors to behavioral and disease phenotypes. Using a targeted genetic (*Cntnap2* knock out) mouse model for neurodevelopmental disorders, a recent study was able to dissociate the role of host genetics and the microbiome. In this study, hyperactive alterations in behavior were associated with host genetics, whereas social behavior deficits were mediated by gut microbiota (Buffington et al., 2021). This is a key finding as clinical approaches that target the microbiome may be more relevant for social deficits and alternative approaches may be more appropriate to treat hyperactivity (Buffington et al., 2021). Furthermore, these mechanistic animal studies provide the framework and insight to explore how host genetics and the microbiome interact in people.

## Human studies

Host genetics is a key factor in shaping the human microbiome composition. Studies in monozygotic and dizygotic twins first demonstrated a contribution of genetic variation to microbiome composition showing that microbiota composition was more similar in monozygotic compared to dizygotic twins (Goodrich et al., 2014, 2016; Turnbaugh et al., 2009). More recently, genome wide analyses have identified genes and pathways that are associated with microbiome composition. Genome-wide association studies (GWAS) have been extensively used in population genetics, genetic epidemiology, and to identify genetic variation associated with complex traits. In the past decade, the association of genetic variation with microbial features has been examined across healthy and to a lesser extent in disease (Sanna et al., 2022). For example, genome-wide analysis was conducted to identify human genes and pathways associated with the microbiome composition in a subset of individuals included in the Human Microbiome Project (Blekhman et al., 2015). Host genomic single nucleotide polymorphism (SNP) data was examined in 93 healthy individuals and associations between host genetic variation and alpha diversity and beta diversity of microbiota collected from several body sites including nares and stool samples were observed (Blekhman et al., 2015). Using a pathway analysis, an enrichment of leptin signaling was identified in association with microbiome composition in the nose, oral cavity and skin. Specific host gene to taxa associations found included *HLA-DRA* and *Selenomonas* in the throat, *Tlr1* and *Lautropia* in tongue dorsum, and *LCT* and *Bifidobacterium* in the gut (Blekhman et al., 2015). In addition, this analysis identified several immune-related pathways that were associated with microbiome composition (Blekhman et al., 2015). Several additional genome-wide association studies (GWAS) have been conducted to identify the association of host genetic single nucleotide polymorphisms (SNP) with microbial traits (Davenport et al., 2015; Hughes et al., 2020; Kurilshikov et al., 2021; Liu et al., 2021; Lopera-Maya et al., 2022; Qin et al., 2022; Rothschild et al., 2018; Ruhlemann et al., 2021; Scepanovic et al., 2019; Turpin et al., 2016; Wang et al., 2016; Xu et al., 2020). The analytical approaches and microbiome tools (16S rRNA sequencing and metagenomics) vary across these studies and yet, these studies clearly show host genetic variations are associated with microbial features in both the healthy and disease cohorts examined. Notably, within the microbiome research domain, metagenome-wide association studies, modeled after GWAS approaches, have associated microbiome metagenomics combined with metabolomics with several diseases (Wang and Jia, 2016).

A novel approach to consider uses bidirectional mendelian randomization to examine links between host genetics, microbiota, and clinical traits. In a recent study that examined relationships between the gut microbiota and metabolic disease, the bidirectional analysis included microbiome-genetics links to metabolic traits as well as metabolic-genetics to microbiome features (Sanna et al., 2019). Through a systematic analytical workflow this study identified a microbial pathway that had a causal influence on insulin secretion. Specifically, the analysis demonstrated that host genetic variation influenced gut microbiome composition to modulate butyrate, which increased the ability of islet cells to secret insulin in response to a physiological glucose challenge (Sanna et al., 2019). Butyrate producers *Eubacterium rectale* and *Roseburia intestinalis* were suggested as key taxa in this host genetic-microbe-clinical trait association. In addition, fecal propionate levels were linked to three genetic predictors and these were causally associated with body mass index (BMI) and increased risk of type 2 diabetes (T2D) in the primary cohort. This observed risk for T2D was replicated in a UK Biobank sample, however the BMI finding was not confirmed (Sanna et al., 2019).

Analytical tools to integrate across data domains continue to move to the forefront of biomedical research and there are many ways to view the association between host genetics and the microbiome. A key question that arises is how do these gene-microbe relationships associate with clinical traits? Moreover, are disease-related genetic variants associated with microbiota influencing disease in a causal manner?

While several genome-wide studies have considered host-microbe associations; to date none have linked these genetic-microbiota connections to brain function or brain-related disease. For now, the nurture versus nature debate related to the microbiome continues and there is a need for integrated and balanced studies to examine host genetics and environmental factors in the same cohort or potentially applying these questions to large consortium/multisite studies. In all cases, there is a need for state-of-the-art design features that integrate both discovery and validation.

## Future directions

The potential of microbial genetic association studies that utilize global analytical approaches have demonstrated that they may help us understand the complexity of the host-microbiome-health continuum compared to studies that treat the microbiome and host genome separately and which are aimed at identifying specific associations (Weissbrod et al., 2018). Global approaches include the use of polygenic models that consider how multiple genetic variants may contribute to a heritable trait and the use of genetic correlation, which examines genetic similarity between pairs of traits that may share a common biological mechanism (Weissbrod et al., 2018). Another global approach to consider in microbial-genetic association studies is construction of polygenic risk scores (PRS) linked to heritable microbial taxa and clinical or host phenotypes (Weissbrod et al., 2018). PRS have mainly been used to consider an individual's genetic predisposition for a particular disease; however, the potential to generate PRS linked to heritable microbiota taxa provides a new avenue to investigate the importance of host genetics in microbiome-related health outcomes. Further, the consideration of the microbial genome and the biochemical flexibility that it gives the host has yet to be considered in the context of genome-wide studies. As discussed in the section on animal studies, host genetics contributes to host physiological systems including gastrointestinal and immune systems that can in turn directly and indirectly influence the gut-brain axis. While it is evident that a strong argument exists that environmental factors are the main driving force of interindividual differences in gut microbial composition, we cannot ignore the contribution of host genetic-microbe interactions in shaping the impact of those environmental factors to influence individual differences in health and disease.

# References

Bearoff, F., Del Rio, R., Case, L.K., Dragon, J.A., Nguyen-Vu, T., Lin, C.Y., Blankenhorn, E.P., Teuscher, C., Krementsov, D.N., 2016. Natural genetic variation profoundly regulates gene expression in immune cells and dictates susceptibility to CNS autoimmunity. Gene Immun 17, 386–395.

Benakis, C., Brea, D., Caballero, S., Faraco, G., Moore, J., Murphy, M., Sita, G., Racchumi, G., Ling, L., Pamer, E.G., Iadecola, C., Anrather, J., 2016. Commensal microbiota affects ischemic stroke outcome by regulating intestinal gammadelta T cells. Nat Med 22, 516–523.

Benson, A.K., Kelly, S.A., Legge, R., Ma, F., Low, S.J., Kim, J., Zhang, M., Oh, P.L., Nehrenberg, D., Hua, K., Kachman, S.D., Moriyama, E.N., Walter, J., Peterson, D.A., Pomp, D., 2010. Individuality in gut microbiota composition is a complex polygenic trait shaped by multiple environmental and host genetic factors. Proc Natl Acad Sci U S A 107, 18933–18938.

Blekhman, R., Goodrich, J.K., Huang, K., Sun, Q., Bukowski, R., Bell, J.T., Spector, T.D., Keinan, A., Ley, R.E., Gevers, D., Clark, A.G., 2015. Host genetic variation impacts microbiome composition across human body sites. Genome Biol 16, 191.

Bonder, M.J., Kurilshikov, A., Tigchelaar, E.F., Mujagic, Z., Imhann, F., Vila, A.V., Deelen, P., Vatanen, T., Schirmer, M., Smeekens, S.P., Zhernakova, D.V., Jankipersadsing, S.A., Jaeger, M., Oosting, M., Cenit, M.C., Masclee, A.A., Swertz, M.A., Li, Y., Kumar, V., Joosten, L., Harmsen, H., Weersma, R.K., Franke, L., Hofker, M.H., Xavier, R.J., Jonkers, D., Netea, M.G., Wijmenga, C., Fu, J., Zhernakova, A., 2016. The effect of host genetics on the gut microbiome. Nat Genet 48, 1407–1412.

Brea, D., Poon, C., Benakis, C., Lubitz, G., Murphy, M., Iadecola, C., Anrather, J., 2021. Stroke affects intestinal immune cell trafficking to the central nervous system. Brain Behav Immun 96, 295–302.

Brinkman, B.M., Becker, A., Ayiseh, R.B., Hildebrand, F., Raes, J., Huys, G., Vandenabeele, P., 2013. Gut microbiota affects sensitivity to acute DSS-induced colitis independently of host genotype. Inflamm Bowel Dis 19, 2560–2567.

Bubier, J.A., Philip, V.M., Quince, C., Campbell, J., Zhou, Y., Vishnivetskaya, T., Duvvuru, S., Blair, R.H., Ndukum, J., Donohue, K.D., Foster, C.M., Mellert, D.J., Weinstock, G., Culiat, C.T., O'Hara, B.F., Palumbo, A.V., Podar, M., Chesler, E.J., 2020. A microbe associated with sleep revealed by a novel systems genetic analysis of the microbiome in collaborative cross mice. Genetics 214, 719–733.

Buffington, S.A., Dooling, S.W., Sgritta, M., Noecker, C., Murillo, O.D., Felice, D.F., Turnbaugh, P.J., Costa-Mattioli, M., 2021. Dissecting the contribution of host genetics and the microbiome in complex behaviors. Cell 184, 1740–1756 e1716.

Campbell, J.H., Foster, C.M., Vishnivetskaya, T., Campbell, A.G., Yang, Z.K., Wymore, A., Palumbo, A.V., Chesler, E.J., Podar, M., 2012. Host genetic and environmental effects on mouse intestinal microbiota. ISME J 6, 2033–2044.

Caspani, G., Green, M., Swann, J.R., Foster, J.A., 2022. Microbe-immune crosstalk: evidence that T cells influence the development of the brain metabolome. Int J Mol Sci 23, 3259.

Cell_Press, May 11, 2016. Twin Study Finds that Gut Microbiomes Run in Families. Science Daily.

Cheng, H., Guan, X., Chen, D., Ma, W., 2019. The Th17/Treg cell balance: a gut microbiota-modulated story. Microorganisms 7, 583.

Cheng, M., Ning, K., 2019. Stereotypes about enterotype: the old and new ideas. Dev Reprod Biol 17, 4–12.

Chesler, E.J., Miller, D.R., Branstetter, L.R., Galloway, L.D., Jackson, B.L., Philip, V.M., Voy, B.H., Culiat, C.T., Threadgill, D.W., Williams, R.W., Churchill, G.A., Johnson, D.K., Manly, K.F., 2008. The collaborative cross at Oak Ridge National Laboratory: developing a powerful resource for systems genetics. Mamm Genome 19, 382−389.

Churchill, G.A., Airey, D.C., Allayee, H., Angel, J.M., Attie, A.D., Beatty, J., Beavis, W.D., Belknap, J.K., Bennett, B., Berrettini, W., Bleich, A., Bogue, M., Broman, K.W., Buck, K.J., Buckler, E., Burmeister, M., Chesler, E.J., Cheverud, J.M., Clapcote, S., Cook, M.N., Cox, R.D., Crabbe, J.C., Crusio, W.E., Darvasi, A., Deschepper, C.F., Doerge, R.W., Farber, C.R., Forejt, J., Gaile, D., Garlow, S.J., Geiger, H., Gershenfeld, H., Gordon, T., Gu, J., Gu, W., de Haan, G., Hayes, N.L., Heller, C., Himmelbauer, H., Hitzemann, R., Hunter, K., Hsu, H.C., Iraqi, F.A., Ivandic, B., Jacob, H.J., Jansen, R.C., Jepsen, K.J., Johnson, D.K., Johnson, T.E., Kempermann, G., Kendziorski, C., Kotb, M., Kooy, R.F., Llamas, B., Lammert, F., Lassalle, J.M., Lowenstein, P.R., Lu, L., Lusis, A., Manly, K.F., Marcucio, R., Matthews, D., Medrano, J.F., Miller, D.R., Mittleman, G., Mock, B.A., Mogil, J.S., Montagutelli, X., Morahan, G., Morris, D.G., Mott, R., Nadeau, J.H., Nagase, H., Nowakowski, R.S., O'Hara, B.F., Osadchuk, A.V., Page, G.P., Paigen, B., Paigen, K., Palmer, A.A., Pan, H.J., Peltonen-Palotie, L., Peirce, J., Pomp, D., Pravenec, M., Prows, D.R., Qi, Z., Reeves, R.H., Roder, J., Rosen, G.D., Schadt, E.E., Schalkwyk, L.C., Seltzer, Z., Shimomura, K., Shou, S., Sillanpaa, M.J., Siracusa, L.D., Snoeck, H.W., Spearow, J.L., Svenson, K., Tarantino, L.M., Threadgill, D., Toth, L.A., Valdar, W., de Villena, F.P., Warden, C., Whatley, S., Williams, R.W., Wiltshire, T., Yi, N., Zhang, D., Zhang, M., Zou, F., Complex Trait, C, 2004. The collaborative cross, a community resource for the genetic analysis of complex traits. Nat Genet 36, 1133−1137.

Conley, M.N., Wong, C.P., Duyck, K.M., Hord, N., Ho, E., Sharpton, T.J., 2016. Aging and serum MCP-1 are associated with gut microbiome composition in a murine model. PeerJ 4, e1854.

Corthesy, B., 2013. Role of secretory IgA in infection and maintenance of homeostasis. Autoimmun Rev 12, 661−665.

Dalmasso, G., Nguyen, H.T., Yan, Y., Laroui, H., Charania, M.A., Ayyadurai, S., Sitaraman, S.V., Merlin, D., 2011. Microbiota modulate host gene expression via microRNAs. PLoS One 6, e19293.

Davenport, E.R., Cusanovich, D.A., Michelini, K., Barreiro, L.B., Ober, C., Gilad, Y., 2015. Genome-wide association studies of the human gut microbiota. PLoS One 10, e0140301.

De Lay, N., Schu, D.J., Gottesman, S., 2013. Bacterial small RNA-based negative regulation: Hfq and its accomplices. J Biol Chem 288, 7996−8003.

Derecki, N.C., Cardani, A.N., Yang, C.H., Quinnies, K.M., Crihfield, A., Lynch, K.R., Kipnis, J., 2010. Regulation of learning and memory by meningeal immunity: a key role for IL-4. J Exp Med 207, 1067−1080.

Ding, T., Schloss, P.D., 2014. Dynamics and associations of microbial community types across the human body. Nature 509, 357−360.

Ericsson, A.C., Davis, J.W., Spollen, W., Bivens, N., Givan, S., Hagan, C.E., McIntosh, M., Franklin, C.L., 2015. Effects of vendor and genetic background on the composition of the fecal microbiota of inbred mice. PLoS One 10, e0116704.

Everard, A., Belzer, C., Geurts, L., Ouwerkerk, J.P., Druart, C., Bindels, L.B., Guiot, Y., Derrien, M., Muccioli, G.G., Delzenne, N.M., de Vos, W.M., Cani, P.D., 2013. Cross-talk between Akkermansia muciniphila and intestinal epithelium controls diet-induced obesity. Proc Natl Acad Sci U S A 110, 9066−9071.

Francella, C., Green, M., Caspani, G., Lai, J., Rilett, K., Foster, J.A., 2022. Microbe-immune-stress interactions impact behaviour during postnatal development. Int. J. Mol. Sci. 23, 15064. https://doi.org/10.3390/ijms2323150604. In press.

Fransen, F., Zagato, E., Mazzini, E., Fosso, B., Manzari, C., El Aidy, S., Chiavelli, A., D'Erchia, A.M., Sethi, M.K., Pabst, O., Marzano, M., Moretti, S., Romani, L., Penna, G., Pesole, G., Rescigno, M., 2015. BALB/c and C57BL/6 mice differ in polyreactive IgA abundance, which impacts the generation of antigen-specific IgA and microbiota diversity. Immunity 43, 527–540.

Gold, R., Linington, C., Lassmann, H., 2006. Understanding pathogenesis and therapy of multiple sclerosis via animal models: 70 years of merits and culprits in experimental autoimmune encephalomyelitis research. Brain 129, 1953–1971.

Goodrich, J.K., Davenport, E.R., Beaumont, M., Jackson, M.A., Knight, R., Ober, C., Spector, T.D., Bell, J.T., Clark, A.G., Ley, R.E., 2016. Genetic determinants of the gut microbiome in UK twins. Cell Host Microbe 19, 731–743.

Goodrich, J.K., Davenport, E.R., Clark, A.G., Ley, R.E., 2017. The relationship between the human genome and microbiome comes into view. Annu Rev Genet 51, 413–433.

Goodrich, J.K., Waters, J.L., Poole, A.C., Sutter, J.L., Koren, O., Blekhman, R., Beaumont, M., Van Treuren, W., Knight, R., Bell, J.T., Spector, T.D., Clark, A.G., Ley, R.E., 2014. Human genetics shape the gut microbiome. Cell 159, 789–799.

Gupta, V.K., Paul, S., Dutta, C., 2017. Geography, ethnicity or subsistence-specific variations in human microbiome composition and diversity. Front Microbiol 8, 1162.

Ha, M., Kim, V.N., 2014. Regulation of microRNA biogenesis. Nat Rev Mol Cell Biol 15, 509–524.

Horne, R., St Pierre, J., Odeh, S., Surette, M., Foster, J.A., 2019. Microbe and host interaction in gastrointestinal homeostasis. Psychopharmacology (Berl) 236, 1623–1640.

Hoy, Y.E., Bik, E.M., Lawley, T.D., Holmes, S.P., Monack, D.M., Theriot, J.A., Relman, D.A., 2015. Variation in taxonomic composition of the fecal microbiota in an inbred mouse strain across individuals and time. PLoS One 10, e0142825.

Hughes, D.A., Bacigalupe, R., Wang, J., Ruhlemann, M.C., Tito, R.Y., Falony, G., Joossens, M., Vieira-Silva, S., Henckaerts, L., Rymenans, L., Verspecht, C., Ring, S., Franke, A., Wade, K.H., Timpson, N.J., Raes, J., 2020. Genome-wide associations of human gut microbiome variation and implications for causal inference analyses. Nat Microbiol 5, 1079–1087.

Kim, Y.J., Hwang, S.H., Cho, H.H., Shin, K.K., Bae, Y.C., Jung, J.S., 2012. MicroRNA 21 regulates the proliferation of human adipose tissue-derived mesenchymal stem cells and high-fat diet-induced obesity alters microRNA 21 expression in white adipose tissues. J Cell Physiol 227, 183–193.

Kolde, R., Franzosa, E.A., Rahnavard, G., Hall, A.B., Vlamakis, H., Stevens, C., Daly, M.J., Xavier, R.J., Huttenhower, C., 2018. Host genetic variation and its microbiome interactions within the Human Microbiome Project. Genome Med 10, 6.

Kurilshikov, A., Medina-Gomez, C., Bacigalupe, R., Radjabzadeh, D., Wang, J., Demirkan, A., Le Roy, C.I., Raygoza Garay, J.A., Finnicum, C.T., Liu, X., Zhernakova, D.V., Bonder, M.J., Hansen, T.H., Frost, F., Ruhlemann, M.C., Turpin, W., Moon, J.Y., Kim, H.N., Lull, K., Barkan, E., Shah, S.A., Fornage, M., Szopinska-Tokov, J., Wallen, Z.D., Borisevich, D., Agreus, L., Andreasson, A., Bang, C., Bedrani, L., Bell, J.T., Bisgaard, H., Boehnke, M., Boomsma, D.I., Burk, R.D., Claringbould, A., Croitoru, K., Davies, G.E., van Duijn, C.M., Duijts, L., Falony, G., Fu, J., van der Graaf, A., Hansen, T., Homuth, G., Hughes, D.A., Ijzerman, R.G., Jackson, M.A., Jaddoe, V.W.V., Joossens, M., Jorgensen, T., Keszthelyi, D., Knight, R., Laakso, M., Laudes, M., Launer, L.J., Lieb, W., Lusis, A.J., Masclee, A.A.M., Moll, H.A., Mujagic, Z., Qibin, Q., Rothschild, D., Shin, H.,

Sorensen, S.J., Steves, C.J., Thorsen, J., Timpson, N.J., Tito, R.Y., Vieira-Silva, S., Volker, U., Volzke, H., Vosa, U., Wade, K.H., Walter, S., Watanabe, K., Weiss, S., Weiss, F.U., Weissbrod, O., Westra, H.J., Willemsen, G., Payami, H., Jonkers, D., Arias Vasquez, A., de Geus, E.J.C., Meyer, K.A., Stokholm, J., Segal, E., Org, E., Wijmenga, C., Kim, H.L., Kaplan, R.C., Spector, T.D., Uitterlinden, A.G., Rivadeneira, F., Franke, A., Lerch, M.M., Franke, L., Sanna, S., D'Amato, M., Pedersen, O., Paterson, A.D., Kraaij, R., Raes, J., Zhernakova, A., 2021. Large-scale association analyses identify host factors influencing human gut microbiome composition. Nat Genet 53, 156−165.

Leamy, L.J., Kelly, S.A., Nietfeldt, J., Legge, R.M., Ma, F., Hua, K., Sinha, R., Peterson, D.A., Walter, J., Benson, A.K., Pomp, D., 2014. Host genetics and diet, but not immunoglobulin A expression, converge to shape compositional features of the gut microbiome in an advanced intercross population of mice. Genome Biol 15, 552.

Lee, N., Kim, W.-U., 2017. Microbiota in T-cell homeostasis and inflammatory diseases. Exp Mol Med 49, e340.

Liu, S., da Cunha, A.P., Rezende, R.M., Cialic, R., Wei, Z., Bry, L., Comstock, L.E., Gandhi, R., Weiner, H.L., 2016. The host shapes the gut microbiota via fecal MicroRNA. Cell Host Microbe 19, 32−43.

Liu, X., Tang, S., Zhong, H., Tong, X., Jie, Z., Ding, Q., Wang, D., Guo, R., Xiao, L., Xu, X., Yang, H., Wang, J., Zong, Y., Liu, W., Liu, X., Zhang, Y., Brix, S., Kristiansen, K., Hou, Y., Jia, H., Zhang, T., 2021. A genome-wide association study for gut metagenome in Chinese adults illuminates complex diseases. Cell Discov 7, 9.

Lopera-Maya, E.A., Kurilshikov, A., van der Graaf, A., Hu, S., Andreu-Sanchez, S., Chen, L., Vila, A.V., Gacesa, R., Sinha, T., Collij, V., Klaassen, M.A.Y., Bolte, L.A., Gois, M.F.B., Neerincx, P.B.T., Swertz, M.A., LifeLines Cohort, S., Harmsen, H.J.M., Wijmenga, C., Fu, J., Weersma, R.K., Zhernakova, A., Sanna, S., 2022. Effect of host genetics on the gut microbiome in 7,738 participants of the Dutch Microbiome Project. Nat Genet 54, 143−151.

Macpherson, A.J., Harris, N.L., 2004. Interactions between commensal intestinal bacteria and the immune system. Nat Rev Immunol 4, 478−485.

McGowan, P.O., Hope, T.A., Meck, W.H., Kelsoe, G., Williams, C.L., 2011. Impaired social recognition memory in recombination activating gene 1-deficient mice. Brain Res 1383, 187−195.

Mobeen, F., Sharma, V., Tulika, P., 2018. Enterotype variations of the healthy human gut microbiome in different geographical regions. Bioinformation 14, 560−573.

Moloney, G.M., Viola, M.F., Hoban, A.E., Dinan, T.G., Cryan, J.F., 2018. Faecal microRNAs: indicators of imbalance at the host-microbe interface? Benef Microbes 9, 175−183.

Montgomery, T.L., Kunstner, A., Kennedy, J.J., Fang, Q., Asarian, L., Culp-Hill, R., D'Alessandro, A., Teuscher, C., Busch, H., Krementsov, D.N., 2020. Interactions between host genetics and gut microbiota determine susceptibility to CNS autoimmunity. Proc Natl Acad Sci U S A 117, 27516−27527.

O'Connor, A., Quizon, P.M., Albright, J.E., Lin, F.T., Bennett, B.J., 2014. Responsiveness of cardiometabolic-related microbiota to diet is influenced by host genetics. Mamm Genome 25, 583−599.

Org, E., Parks, B.W., Joo, J.W., Emert, B., Schwartzman, W., Kang, E.Y., Mehrabian, M., Pan, C., Knight, R., Gunsalus, R., Drake, T.A., Eskin, E., Lusis, A.J., 2015. Genetic and environmental control of host-gut microbiota interactions. Genome Res 25, 1558−1569.

Pasciuto, E., Burton, O.T., Roca, C.P., Lagou, V., Rajan, W.D., Theys, T., Mancuso, T., Tito, R.Y., Kouser, L.G., Callaerts-Vegh, Z., de la Fuente, A.G., Prezzemolo, T., Mascali, L.G., Brajic, A., Whyte, C.E., Yshii, L., Martinez-Muriana, A., Naughton, M.,

Young, A., Moudra, A., Lemaitre, P., Poovathingal, S., Raes, J., De Strooper, B., Fitzgerald, D., Dooley, J., Liston, A., 2020. Microglia require CD4 T cells to complete the fetal-to-adult transition. Cell 182, 625−640. https://doi.org/10.1016/j.cell.2020.06.026.

Philip, V.M., Sokoloff, G., Ackert-Bicknell, C.L., Striz, M., Branstetter, L., Beckmann, M.A., Spence, J.S., Jackson, B.L., Galloway, L.D., Barker, P., Wymore, A.M., Hunsicker, P.R., Durtschi, D.C., Shaw, G.S., Shinpock, S., Manly, K.F., Miller, D.R., Donohue, K.D., Culiat, C.T., Churchill, G.A., Lariviere, W.R., Palmer, A.A., O'Hara, B.F., Voy, B.H., Chesler, E.J., 2011. Genetic analysis in the Collaborative Cross breeding population. Genome Res 21, 1223−1238.

Qin, Y., Havulinna, A.S., Liu, Y., Jousilahti, P., Ritchie, S.C., Tokolyi, A., Sanders, J.G., Valsta, L., Brozynska, M., Zhu, Q., Tripathi, A., Vazquez-Baeza, Y., Loomba, R., Cheng, S., Jain, M., Niiranen, T., Lahti, L., Knight, R., Salomaa, V., Inouye, M., Meric, G., 2022. Combined effects of host genetics and diet on human gut microbiota and incident disease in a single population cohort. Nat Genet 54, 134−142.

Quinnies, K.M., Cox, K.H., Rissman, E.F., 2015. Immune deficiency influences juvenile social behavior and maternal behavior. Behav Neurosci 129, 331−338.

Racke, M.K., 2001. Experimental autoimmune encephalomyelitis (EAE). Current protocols in neuroscience. Editorial board, Jacqueline N Crawley [et al Chapter 9, Unit9 7].

Rothschild, D., Weissbrod, O., Barkan, E., Kurilshikov, A., Korem, T., Zeevi, D., Costea, P.I., Godneva, A., Kalka, I.N., Bar, N., Shilo, S., Lador, D., Vila, A.V., Zmora, N., Pevsner-Fischer, M., Israeli, D., Kosower, N., Malka, G., Wolf, B.C., Avnit-Sagi, T., Lotan-Pompan, M., Weinberger, A., Halpern, Z., Carmi, S., Fu, J., Wijmenga, C., Zhernakova, A., Elinav, E., Segal, E., 2018. Environment dominates over host genetics in shaping human gut microbiota. Nature 555, 210−215.

Ruhlemann, M.C., Hermes, B.M., Bang, C., Doms, S., Moitinho-Silva, L., Thingholm, L.B., Frost, F., Degenhardt, F., Wittig, M., Kassens, J., Weiss, F.U., Peters, A., Neuhaus, K., Volker, U., Volzke, H., Homuth, G., Weiss, S., Grallert, H., Laudes, M., Lieb, W., Haller, D., Lerch, M.M., Baines, J.F., Franke, A., 2021. Genome-wide association study in 8,956 German individuals identifies influence of ABO histo-blood groups on gut microbiome. Nat Genet 53, 147−155.

Sadler, R., Cramer, J.V., Heindl, S., Kostidis, S., Betz, D., Zuurbier, K.R., Northoff, B.H., Heijink, M., Goldberg, M.P., Plautz, E.J., Roth, S., Malik, R., Dichgans, M., Holdt, L.M., Benakis, C., Giera, M., Stowe, A.M., Liesz, A., 2020. Short-chain fatty acids improve poststroke recovery via immunological mechanisms. J Neurosci 40, 1162−1173.

Saez, C., September 22, 2022. Are we closer to knowing what a healthy gut microbiota is? Gut Microbiota Health.

Sanna, S., Kurilshikov, A., van der Graaf, A., Fu, J., Zhernakova, A., 2022. Challenges and future directions for studying effects of host genetics on the gut microbiome. Nat Genet 54, 100−106.

Sanna, S., van Zuydam, N.R., Mahajan, A., Kurilshikov, A., Vich Vila, A., Vosa, U., Mujagic, Z., Masclee, A.A.M., Jonkers, D., Oosting, M., Joosten, L.A.B., Netea, M.G., Franke, L., Zhernakova, A., Fu, J., Wijmenga, C., McCarthy, M.I., 2019. Causal relationships among the gut microbiome, short-chain fatty acids and metabolic diseases. Nat Genet 51, 600−605.

Scepanovic, P., Hodel, F., Mondot, S., Partula, V., Byrd, A., Hammer, C., Alanio, C., Bergstedt, J., Patin, E., Touvier, M., Lantz, O., Albert, M.L., Duffy, D., Quintana-Murci, L., Fellay, J., Milieu Interieur, C, 2019. A comprehensive assessment of demographic, environmental, and host genetic associations with gut microbiome diversity in healthy individuals. Microbiome 7, 130.

Schulte, L.N., Eulalio, A., Mollenkopf, H.J., Reinhardt, R., Vogel, J., 2011. Analysis of the host microRNA response to Salmonella uncovers the control of major cytokines by the let-7 family. EMBO J 30, 1977−1989.

Singh, N., Shirdel, E.A., Waldron, L., Zhang, R.H., Jurisica, I., Comelli, E.M., 2012. The murine caecal microRNA signature depends on the presence of the endogenous microbiota. Int J Biol Sci 8, 171−186.

Singh, V., Roth, S., Llovera, G., Sadler, R., Garzetti, D., Stecher, B., Dichgans, M., Liesz, A., 2016. Microbiota dysbiosis controls the neuroinflammatory response after stroke. J Neurosci 36, 7428−7440.

Singh, V., Sadler, R., Heindl, S., Llovera, G., Roth, S., Benakis, C., Liesz, A., 2018. The gut microbiome primes a cerebroprotective immune response after stroke. J Cerebr Blood Flow Metabol 38, 1293−1298.

Snijders, A.M., Langley, S.A., Kim, Y.M., Brislawn, C.J., Noecker, C., Zink, E.M., Fansler, S.J., Casey, C.P., Miller, D.R., Huang, Y., Karpen, G.H., Celniker, S.E., Brown, J.B., Borenstein, E., Jansson, J.K., Metz, T.O., Mao, J.H., 2016. Influence of early life exposure, host genetics and diet on the mouse gut microbiome and metabolome. Nat Microbiol 2, 16221.

Tanabe, S., Yamashita, T., 2018. The role of immune cells in brain development and neurodevelopmental diseases. Int Immunol 30, 437−444.

Tang, Z.Z., Chen, G., Hong, Q., Huang, S., Smith, H.M., Shah, R.D., Scholz, M., Ferguson, J.F., 2019. Multi-omic analysis of the microbiome and metabolome in healthy subjects reveals microbiome-dependent relationships between diet and metabolites. Front Genet 10, 454.

Teng, Y., Ren, Y., Sayed, M., Hu, X., Lei, C., Kumar, A., Hutchins, E., Mu, J., Deng, Z., Luo, C., Sundaram, K., Sriwastva, M.K., Zhang, L., Hsieh, M., Reiman, R., Haribabu, B., Yan, J., Jala, V.R., Miller, D.M., Van Keuren-Jensen, K., Merchant, M.L., McClain, C.J., Park, J.W., Egilmez, N.K., Zhang, H.G., 2018. Plant-derived exosomal MicroRNAs shape the gut microbiota. Cell Host Microbe 24, 637−652 e638.

Turnbaugh, P.J., Hamady, M., Yatsunenko, T., Cantarel, B.L., Duncan, A., Ley, R.E., Sogin, M.L., Jones, W.J., Roe, B.A., Affourtit, J.P., Egholm, M., Henrissat, B., Heath, A.C., Knight, R., Gordon, J.I., 2009. A core gut microbiome in obese and lean twins. Nature 457, 480−484.

Turpin, W., Espin-Garcia, O., Xu, W., Silverberg, M.S., Kevans, D., Smith, M.I., Guttman, D.S., Griffiths, A., Panaccione, R., Otley, A., Xu, L., Shestopaloff, K., Moreno-Hagelsieb, G., , Consortium, G.E.M.P.R, Paterson, A.D., Croitoru, K., 2016. Association of host genome with intestinal microbial composition in a large healthy cohort. Nat Genet 48, 1413−1417.

Vuong, H.E., Hsiao, E.Y., 2017. Emerging roles for the gut microbiome in autism spectrum disorder. Biol Psychiatr 81, 411−423.

Wang, J., Jia, H., 2016. Metagenome-wide association studies: fine-mining the microbiome. Nat Rev Microbiol 14, 508−522.

Wang, J., Thingholm, L.B., Skieceviciene, J., Rausch, P., Kummen, M., Hov, J.R., Degenhardt, F., Heinsen, F.A., Ruhlemann, M.C., Szymczak, S., Holm, K., Esko, T., Sun, J., Pricop-Jeckstadt, M., Al-Dury, S., Bohov, P., Bethune, J., Sommer, F., Ellinghaus, D., Berge, R.K., Hubenthal, M., Koch, M., Schwarz, K., Rimbach, G., Hubbe, P., Pan, W.H., Sheibani-Tezerji, R., Hasler, R., Rosenstiel, P., D'Amato, M., Cloppenborg-Schmidt, K., Kunzel, S., Laudes, M., Marschall, H.U., Lieb, W., Nothlings, U., Karlsen, T.H., Baines, J.F., Franke, A., 2016. Genome-wide association analysis identifies variation in vitamin D receptor and other host factors influencing the gut microbiota. Nat Genet 48, 1396−1406.

Weinstein, S.B., Martinez-Mota, R., Stapleton, T.E., Klure, D.M., Greenhalgh, R., Orr, T.J., Dale, C., Kohl, K.D., Dearing, M.D., 2021. Microbiome stability and structure is governed by host phylogeny over diet and geography in woodrats (Neotoma spp.). Proc Natl Acad Sci U S A 118, e2108787118.

Weissbrod, O., Rothschild, D., Barkan, E., Segal, E., 2018. Host genetics and microbiome associations through the lens of genome wide association studies. Curr Opin Microbiol 44, 9–19.

Weizmann_Institute_of_Science, February 28, 2018. Genetics or Lifestyle: What Is it that Shapes Our Microbiome? A Study Brings New Hope for Improving Our Health. Science Daily.

Xu, F., Fu, Y., Sun, T.Y., Jiang, Z., Miao, Z., Shuai, M., Gou, W., Ling, C.W., Yang, J., Wang, J., Chen, Y.M., Zheng, J.S., 2020. The interplay between host genetics and the gut microbiome reveals common and distinct microbiome features for complex human diseases. Microbiome 8, 145.

Yatsunenko, T., Rey, F.E., Manary, M.J., Trehan, I., Dominguez-Bello, M.G., Contreras, M., Magris, M., Hidalgo, G., Baldassano, R.N., Anokhin, A.P., Heath, A.C., Warner, B., Reeder, J., Kuczynski, J., Caporaso, J.G., Lozupone, C.A., Lauber, C., Clemente, J.C., Knights, D., Knight, R., Gordon, J.I., 2012. Human gut microbiome viewed across age and geography. Nature 486, 222–227.

# CHAPTER 4

# Microbiota to brain communication

## Introduction

Preclinical and clinical research has provided evidence that the gut microbiome, via the gut–brain axis, is capable of influencing multiple facets of host physiology, brain function, and behavior (Cryan et al., 2019). Understanding the granular detail of the mechanisms underpinning microbiota to brain communication remains an important research objective to expedite the incorporation of microbiome-based innovations into clinical practice (Wilkinson et al., 2021). Currently, it is appreciated that host–microbe dialog is facilitated along the key pillars of the gut-brain axis (see Fig. 4.1) and that microbial metabolites (see Fig. 4.2) and other cellular components are key mediators in the complex signaling pathways recruited. The immune system and the hypothalamic-pituitary-adrenal (HPA) axis are important components of this network as is signaling facilitated by the vagus nerve and the enteric nervous system. In this chapter, we outline these key mechanisms for microbiota to brain communication and emphasize the important role for microbial metabolites such as short chain fatty acids and bacterial surface molecules such as peptidoglycans. Optimal microbiome-gut-brain axis signaling is also supported by microbial regulation of the bioprocessing of precursors such as tryptophan. We also pinpoint the mechanistic advances in our understanding of the actions of the gastrointestinal microbiota that will be essential to expedite delivery of the therapeutic opportunities arising from targeting of the gut microbiome.

## The vagus nerve in microbiota to brain communication

The vagus nerve is the longest nerve in the body and connects the visceral organs and the brain, running from the medulla oblongata to the digestive

*Microbiota Brain Axis*
ISBN 978-0-12-814800-6
https://doi.org/10.1016/B978-0-12-814800-6.00007-8

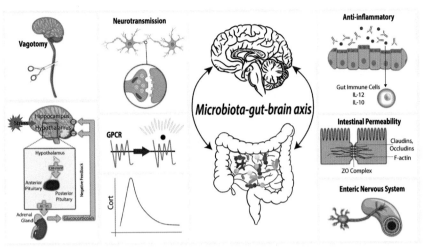

**Figure 4.1** *Signaling pathways for microbiota to brain communication.* Gut microbiota to brain communication is facilitated by the architectural pillars of the gut-brain axis. This allows the recruitment of a number of signaling pathways such as the vagus nerve, the HPA axis, the ENS, the immune system, and neurotransmission in the CNS. Much of this communication is reciprocal, with gut microbiota composition, structure, and function receptive to host stress or immune signals. Some microbial metabolites can also activate host GPCRs to impact functions such as gastrointestinal secretion and motility. Experimental interventions, such as vagotomy, allow for mechanistic insights into the specific pathways underpinning particular observations.

tract and containing a mixture of afferent sensory and efferent motor nerve fibers (Bonaz et al., 2021). It is regarded as an important communication pathway for the gut microbiota to influence brain function and behavior (Fulling et al., 2019), a view supported by a number of important experimental observations. Vagal sensory neurons in particular have been implicated in transmitting a variety of gut-derived signals to distant CNS sites with functional implications for mood and emotion, cognitive function, and reward in addition to energy homeostasis, digestive function, and the immune response (Yu et al., 2020).

The use of surgical vagotomy (usually bilateral subdiaphragmatic transection of ventral and dorsal truncal branches) has been a key preclinical experimental strategy to reveal the role of the vagus nerve in microbiota to brain communication (Fig. 4.1). This approach was used to demonstrate that the anxiolytic effect of *Bifidobacterium longum NCC3001* (Bercik et al., 2011b) and the anxiolytic and antidepressant effects of *Lactobacillus rhamnosus (JB-1)* (Bravo et al., 2011) involved vagal pathways of communication. More recently, it was demonstrated using this approach that *L. reuteri* rescues social behavior deficits in a number of different animal models via vagal pathways (Sgritta et al., 2019). It has also been reported that chronic stress-induced changes in the gut microbiota promote the expression of depression-like behavior via the vagus nerve (Siopi et al., 2019). The vagus nerve has further been implicated in the depression-like phenotype that emerges in mice after fecal microbiota transplantation from Chrna7 (codes for the $\alpha$7 subtype of the nicotinic acetylcholine receptor) knock-out mice (Pu et al., 2021). Of course not all signals are relayed via neural routes and other reports have used vagotomy to rule out a role for the vagus nerve in the impact of oral antimicrobials on exploratory behavior (Bercik et al., 2011a).

Although we do not always have a complete mechanistic picture of all the players in local gastrointestinal microbe-vagus nerve interactions, the pathways activated in the CNS include alterations in central GABA receptor expression (Bravo et al., 2011) and hippocampal neurogenesis (Siopi et al., 2019). These observations are consistent with studies which have indicated both a role for the microbiota and the vagus nerve in hippocampal neurogenesis (O'Leary et al., 2018; Ogbonnaya et al., 2015), and the increased GABA levels following *L. rhamnosus* JB-1 ingestion as indicated by magnetic resonance spectroscopy in mice (Janik et al., 2016). This strain was also able to rapidly increase the constitutive firing frequency of vagal afferents (Perez-Burgos et al., 2013). The anxiety-related behavior following administration of a subclinical dose of *Campylobacter jejuni*, a diarrhea-causing pathogen, was associated with increased levels of c-Fos immunoreactivity (a marker of neuronal activation) in cell bodies of vagal afferents and in the nucleus tractus solitarius (NTS), a key projection side of gut-related vagal afferents in the brain (Goehler et al., 2005).

The picture of how microbial signals influence vagus nerve signaling locally in the gastrointestinal tract lacks some definition. Gastrointestinal vagal afferents in the gut mucosa can detect chemicals absorbed across the epithelial layer or released following stimulation of epithelial cells (Fulling et al., 2019). This gives the gut microbiota the opportunity to alter vagus nerve signaling by directly and indirectly influencing the concentration or release of these chemicals from enteroendocrine or gastrointestinal immune cells (Bonaz et al., 2021). Furthermore, some bacterial strains can produce neurotransmitters such as GABA and serotonin (Patterson et al., 2014), and it is possible that once these agents are absorbed, they can act to stimulate vagus nerve signaling (Yu et al., 2020). It is thought that signaling via the secretions of enterochromaffin cells, for example, acting essentially as epithelial transducer cells, represents a key mechanism (Rhee et al., 2009).

New experimental approaches have recently been applied to gather additional information about novel vagal pathways. Viral tracing approaches and optical activation of gut-innervating vagal sensory neurons identified a neural gut-brain circuit for reward, a pathway only mediated by the right vagus nerve (Han et al., 2018). Virus-based tracing methods have also been deployed to identify a circuit linking gastrointestinal-derived vagal sensory signaling to hippocampal-dependent memory function via a brainstem-septal relay (Suarez et al., 2018). A neuroepithelial circuit that connects the intestinal lumen to the brainstem, in which "neuropod" entero-oendocrine cells synapse with vagal neurons using glutamate as a neuro-transmitter, has also been proposed (Kaelberer et al., 2018).

While vagotomy studies have provided important insights, a number of caveats exist and should feed into the interpretation of these results. There have been reports of retrograde degeneration (Szereda-Przestaszewska, 1985) and neurotransmitter receptor alterations (Manaker and Zucchi, 1993) within the brain following vagotomy, and the vagus nerve has an important antiinflammatory role (Matteoli and Boeckxstaens, 2013; Murray and Reardon, 2018). This has been noted as a potential confounder that may retune important features of gastrointestinal homeostasis (Breen et al., 2019). Bilateral transection may also obscure some of the nuances of vagal nerve signaling, such as that revealed by isolation of specific signaling pathways to the right vagus nerve by viral tracing methods (Han et al., 2018). The latter study is an important illustration of how state-of-the art technologies can be applied to follow up the hints from vagotomy studies and provide the necessary level of detail required to fully exploit the therapeutic opportunities provided by vagal nerve signaling in microbiota to brain communication (Yu et al., 2020). Consideration also needs to be given to the role of the sympathetic nervous system. IP injections of the selective adrenergic neurotoxin 6-hydroxydopamine has been used to ablate peripheral sympathetic nerve fibers and rule out a role for sympa-thetic nervous system innervation of the gastrointestinal tract conveying signals from the gut to the brain following antibiotic administration (Bercik et al., 2011a). However, there is evidence to support a role for the microbiota in control of gut-extrinsic sympathetic activation, a gut-brain circuit involving distal intestine-projecting vagal neurons and brainstem sensory nuclei (Muller et al., 2020).

## The enteric nervous system in microbiota to brain communication

The gastrointestinal tract houses the largest nervous system outside the CNS with an autonomously active enteric nervous system (ENS) in close proximity with its microbiota and capable of contributing across a spectrum of gastrointestinal and CNS disorders (Niesler et al., 2021). Communication between the ENS and the CNS occurs via intestinal neurons to sympathetic ganglia, and extrinsic primary afferent neurons that follow spinal afferent routes potentially facilitate the transmission of sensory information from signals originating in the microbiota (Cryan et al., 2019). In the colon, extrinsic sensory nerves arise from spinal afferent nerves whose cell bodies reside in dorsal root ganglia (Spencer and Hu, 2020). Generally speaking, it is envisaged that the microbiota either directly or indirectly influences ENS signaling via microbial-derived components and microbial metabolites, and by engaging with mucosal enteroendocrine-cell-derived mediators such as serotonin (Hyland and Cryan, 2016). How sensory neurons in the ENS detect and respond to these gut microbiota—derived signals is a topic that has received comparatively less attention than the vagus nerve in a landscape where our knowledge of the molecular mechanisms underpinning extrinsic and intrinsic gut innervation is incomplete (Uesaka et al., 2016). Pattern recognition receptors (PRRs) including toll-like receptors (TLRs) including TLR2, TLR3, TLR4, and TLR5, which are expressed in the ENS, provide another mechanism for the recognition of microbial molecules (Barajon et al., 2009; Fulde et al., 2018; Hyland and Cryan, 2016).

There are recent indications that the aryl hydrocarbon receptor (AHR) might pick up microbial signals from the intestinal lumen to influence motility via the ENS (Hindson, 2020; Obata et al., 2020) with the broader implications of this observation for microbiota to brain communication still to be worked out. Observations in germ-free (GF) animals indicate that the gut microbiota can potentially induce maturation of the adult ENS via the action of 5-HT on 5-HT4 receptors (De Vadder et al., 2018). Early exposure to the gut microbiota is also considered essential for region-specific postnatal development of the ENS with implications for nerve density, the number of neurons per ganglion, and the proportion of myenteric nitrergic neurons (Collins et al., 2014). Moreover, the gut microbiota is essential to facilitate gut-brain signaling via normal intrinsic and extrinsic nerve function (McVey Neufeld et al., 2015), for normal excitability of intrinsic primary afferent neurons (McVey Neufeld et al., 2013) and for homeostasis of glial cells in the intestinal mucosa (Kabouridis et al., 2015).

Postnatal ENS development also occurs in the context of triangular conversations with the assembly trajectory of the gut microbiota and an extensive immune system (Kabouridis and Pachnis, 2015). This is another area where the continued application of single cell transcriptomics, circuit-tracing methods, and functional chemogenetic manipulations is likely to bear fruit as we seek to understand the extent to which the gut microbiota contributions to the luminal ecosystem tunes the normal activity and functional integration of ENS in gut-brain axis neurocircuitry.

Supernatants from *B. longum* NCC3001 were able to inhibit sensory after-hyperpolarizing (AH) neuron excitability, possibly via opening of potassium channels (Khoshdel et al., 2013) while ingestion of *L. reuteri* enhances excitability of colonic AH neurons by inhibiting calcium-dependent potassium channel opening (Kunze et al., 2009). Mice-fed *L. salivarius* UCC118 showed an effect of that strain on neurally driven secretomotor responses (Lomasney et al., 2014). Potential mechanisms at play may relate to gut epithelial action of microvesicles in the case of *L. rhamnosus* JB-1 (Al-Nedawi et al., 2015) while the acute activation of intestinal sensory neurons by *Bacteroides fragilis* is mediated via a capsular exopolysaccharide (Mao et al., 2013). *Escherichia coli* strain Nissle 1917 can package GABA in a lipopeptide capable of crossing the epithelial barrier and subsequently act on the $GABA_B$ receptor on sensory neurons with implications for nociceptor activation and visceral hypersensitivity (Perez-Berezo et al., 2017).

## The immune system in microbiota to brain communication

Development of the immune system is dependent on the gut microbiota, which educates, primes, and supports key features of mucosal, systemic, and CNS immunity (Dinan and Cryan, 2017; Hooper et al., 2012). This is clear from the longstanding knowledge regarding the marked immaturity of lymphoid tissue architecture and defective immune function in GF mice (Zheng et al., 2020). Even after the initial engagement of the microbiota in the maturation of the host immune system, there are continuous and dynamic interactions with the gut microbiota of importance for brain function and behavior (El Aidy et al., 2015; Foster, 2016). Once mature, the gastrointestinal tract is home to the most populous concentration of immune cells in the body providing a variety of communication options with the trillions of microbes that inhabit our gut (Cryan et al., 2019). There is also a division of labor with enterocytes expressing the innate immune receptors linked to the release of cytokines and chemokines and the gut-associated lymphoid tissue recruiting lymphocytes toward the release of immunoglobulins for a more specific immune response (Cryan et al., 2019). One important example is the signals from members of the gut microbiota such as antigens and immunoregulatory small molecules that guide the development and maintenance of intestinal regulatory T cells (Tanoue et al., 2016). Of course, this is also a two-way system and the immune system also plays a key role in sculpting the composition and function of the gut microbiome (Zheng et al., 2020).

The majority of host-microbe interactions occur at the luminal-mucosal interface and enterocytes, secretory cells, chemosensory cells, and gut-associated lymphoid tissue provide the cellular machinery in the gut epithelium to facilitate local host-microbe immune communication (Cryan et al., 2019). Low level exposure of immune cells to bacterial cell wall components such as lipopolysaccharides (LPS) facilitates establishment and maintenance of mucosal homeostasis, including the production of proin-flammatory cytokines by intestinal macrophages and T cells (El Aidy et al., 2015). The innate immune system employs a family of TLRs to recognize specific molecular patterns derived from the surface and intracellular components of microorganisms (Akira and Takeda, 2004). Other examples of bacterial molecules important for host-microbe immune dialog include bacterial polysaccharides (Mazmanian et al., 2005) and peptidoglycans (Arentsen et al., 2017; Gonzalez-Santana and Diaz Heijtz, 2020). Many microbial metabolites including bacteriocins, bile acids, and SCFAs are also immunomodulatory (Cryan et al., 2019). The cytokines produced following stimulation of the gut immune system include those implicated in the low-grade immune activation that is considered a neurobiological feature of stress-related disorders such as depression and irritable bowel syndrome (IBS), and it is plausible that this proinflammatory immune phenotype originates at the gastrointestinal mucosal sites before spreading to distal locations (Raison et al., 2006; Wilmes et al., 2021). It has recently been demonstrated that the gut microbiota influences the priming and recovery of the innate immune system by recruiting immune cell-type-specific changes following an acute stress exposure (van de Wouw et al., 2020).

Much of the recent attention in this area has come to focus on the multiple mechanisms by which the gut microbiota can regulate neuroinflammation (Rea et al., 2016). For example, it has been shown that the gut microbiota plays a key role in the maturation and function of microglia in the CNS (Erny et al., 2015). Microbial regulation of microglial function and neuroinflammation may be facilitated by metabolites such as SCFAs (Erny and Prinz, 2020) and indoles produced from tryptophan (Rothhammer et al., 2016), with potential implications for downstream events such as neurogenesis and functionally associated with memory and learning (Clarke, 2021; Ma et al., 2021). In an experimental stroke model, SCFAs were identified as proregenerative modulators of poststroke neuronal plasticity via a T cell-dependent regulation of microglial activation (Medina-Rodriguez et al., 2020). Targeting the gut microbiome with oligofructose-enriched inulin supplementation, a prebiotic, was also shown to modulate the peripheral immune response and alter neuroinflammation in middle aged mice (Boehme et al., 2020). The aryl hydrocarbon receptor may be an important sensor in the CNS for microbial metabolites in addition to pattern-recognition receptors for bacterial peptidoglycan, and microglial and astroglial TLR signaling (Li et al., 2021a; van Noort and Bsibsi, 2009). It was recently demonstrated that neuroinflammation can arise from the expansion of segmented filamentous bacteria (SFB) in the gut in a mechanism linked to quorum sensing with the accumulation of T helper 17 (Th17) immune cells in the hippocampus conferring an increased susceptibility to depressive-like behavior (Cruz-Pereira and Cryan, 2020; Medina-Rodriguez et al., 2020). Circulating microbiota PRR-derived ligands have also been identified as regulators of the splenic Ly6C$^{high}$ monocyte function, innate immune cells which have been implicated in the development of CNS-related disorders (Kolypetri et al., 2021).

## Neuroendocrine signaling in microbiota to brain communication

The hypothalamic-pituitary-adrenal (HPA) axis represents the core mammalian stress response system and a key pillar of the gut-brain axis. Important aspects and behavioral consequences of microbiota to brain communication involve microbial regulation of neuroendocrine pathways (Cussotto et al., 2018). This contention was initially supported by observations confirming an important role for the gut microbiota in the postnatal development of the HPA stress response in mice (Sudo et al., 2004), a report which has since been replicated numerous times in both mice (Clarke et al., 2013; Lyte et al., 2020) and rats (Crumeyrolle-Arias et al., 2014).

Such reports are consistent with the outcomes of acute stress exposures in clinical populations such as IBS, a disorder of gut-brain axis interactions with sustained HPA axis activity following the Trier Social Stress Test (Kennedy et al., 2014). There is also support from modulation of the acute stress response by *B. longum* 1714 in healthy human subjects (Allen et al., 2016) and the cortisol awakening response by *B. longum* APC1472 in obese individuals (Schellekens et al., 2021). As is the case with many facets of microbiota-gut-brain signaling, the gut microbiota is also responsive to the host experience of stress, be that acute or chronic (Dinan and Cryan, 2012; Foster et al., 2017; Lyte et al., 2020).

A number of behavioral domains associated with microbiota to brain communication in addition to stress, including addiction, learning and memory, sexual behavior, eating behavior, and social interactions, may be mediated by neuroendocrine pathways (Cussotto et al., 2018). This perspective envisages a role not just for glucocorticoids (Clarke et al., 2014) but also hormones secreted by enteroendocrine cells such as cholecystokinin (CCK), peptide YY (PYY), and glucagonlike peptide-1 (GLP-1) (Cussotto et al., 2018). In addition to a role for ghrelin and leptin, control over these hormones is taken to at least partially explain the association between the gut microbiota and host metabolism, adiposity, homeostasis, and energy balance with additional implications for central appetite and food reward signaling in the context of obesity (Torres-Fuentes et al., 2017). Many of these hormones are now considered to have a role beyond hunger and nutritional status such as ghrelin linked to memory via hippocampal neurogenesis (Buntwal et al., 2019), leptin, GLP-1 and PYY to reward, and CCK to anxiety (Skibicka and Dickson, 2013; Xu, 2014). Oxytocin is also an important neuropeptide implicated in social behavior, addiction, and learning and memory (Cussotto et al., 2018; Kingsbury and Bilbo, 2019) and influenced by the gut microbiome (Erdman and Poutahidis, 2016). Elements from four of the six major neuroendocrine pathways (HPA, the hypothalamic—pituitary—thyroid axis (HPT), the hypothalamic—pituitary—gonadal axis (HPG), and the hypothalamic—neurohypophyseal axis (HN)) can potentially be recruited to service microbiota to brain communication (Clarke et al., 2014; Cussotto et al., 2018). Work is ongoing to understand the mechanisms through which these neuroendocrine pathways are influenced by the gut microbiota, but this likely involves microbial metabolites such as SCFAs, the regulation of enteroendocrine cell secretions, immune signaling, and the vagus nerve (Cryan et al., 2019; Fukui et al., 2018).

# Microbial metabolites in microbiota to brain communication

Microbial metabolic pathways (see Fig. 4.2) have come under intense scrutiny in the evaluation of microbiota to brain communication in the context of brain health and disease (Spichak et al., 2021). Important examples include SCFAs (Dalile et al., 2019; Stilling et al., 2016), bile acids (Agus et al., 2021; Long et al., 2017), indole metabolites produced from tryptophan (Agus et al., 2018; Clarke et al., 2021; Gheorghe et al., 2019), and neurotransmitters (Lyte et al., 2011; Patterson et al., 2014). A catalog of gut-brain modules (GBM), microbial metabolic pathways linked to neuroactive compound production or degradation process and thus potentially associated with gut-brain communication has also been created from the analysis of fecal metagenomes (Valles-Colomer et al., 2019). It is also increasingly appreciated that the gut microbiota can produce microbial metabolites that interact with host GPCR receptors, as with bacteria-derived tryptamine acting as a ligand for the gut-epithelium-expressed 5-HT4 receptor to control gastrointestinal motility (Bhattarai et al., 2018; Cryan et al., 2018).

Preclinical evidence supporting a role for oral administration of SCFAs in regulating the host stress response have recently been supported in a

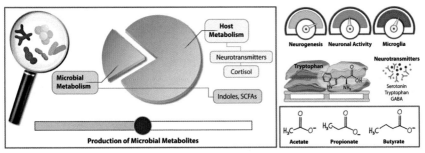

**Figure 4.2** *Microbial metabolites for microbiota to brain communication.* The gut microbiota has the metabolic machinery to process a range of raw materials ingested by the host, and to produce a panel of microbial metabolites from them capable of impacting host physiology, brain function, and behavior. Tryptophan in particular is an important precursor for both the host and our gut microbes, and microbial regulation of the fate of tryptophan encompasses direct and indirect routes. Some microbial metabolites, such as SCFAs, can impact on host physiology and regulate the outputs of the stress response. Bacteria can also produce neurotransmitters which can signal via the ENS or the vagus nerve to the CNS. Taken together, this has implications for neuronal activity, microglia function, and neurogenesis with downstream consequences for neurocognitive performance, mood, and emotion.

clinical study with a colon delivered formulation (Dalile et al., 2020; van de Wouw et al., 2018). SCFAs were also shown to impact ghrelin-mediated signaling through the ghrelin receptor (growth hormone secretagogue receptor (GHSR)-1a) (Torres-Fuentes et al., 2019). Attention has also focused on the capacity of SCFAs to impact on microglia maturation and function (Erny et al., 2015; Erny and Prinz, 2020), most recently in the context of Aβ plaque deposition (Colombo et al., 2021). SCFAs have also been implicated in the functional recovery from spinal cord injury in a mouse model following fecal microbiota transplantation (Jing et al., 2021). It is not entirely clear that these effects are always receptor mediated at the level of the CNS given the importance of SCFAs as an energy source, other physiological actions, and the potential impact on other gut-brain axis signaling pathways via engagement at local gastrointestinal sites (Dalile et al. 2019, 2020). Butyrate, for example, can act via specific receptors (e.g., GPR43/FFAR2; GPR41/FFAR3; GPR109a/HCAR2) and transporters (MCT1/SLC16A1; SMCT1/SLC5A8) and is also an inhibitor of histone deacetylases (HDACs) (Stilling et al., 2016).

A variety of bacteria in the microbiota produce indole from tryptophan, with the parent indole and downstream byproducts representing important examples of interspecies and interkingdom microbial signaling molecules for microbiota to brain communication (Lee et al., 2015; Tomberlin et al., 2017). The aryl hydrocarbon receptor may act as a host indole receptor to link a microbial metabolite to the host immune system (Hubbard et al., 2015; Rothhammer and Quintana, 2019; Stockinger et al., 2021). Supplementation with indole, indoxyl-3-sulfate, indole-3-propionic acid, and indole-3-aldehyde were shown to impact astrocyte activity and neuroinflammation via the aryl hydrocarbon receptor (Rothhammer et al., 2016). Indole activation of the aryl hydrocarbon receptor may also mediate the relationship between gut microbiota and caspase recruitment domain family member 9 (CARD9) to regulate intestinal inflammation (Lamas et al., 2016) and the impact of specific indoles (indole-3-ethanol, indole-3-pyruvate, and indole-3-aldehyde) on gut barrier function (Scott et al., 2020). Indeed the regulation of gut barrier function is a feature often associated with the indoles produced from tryptophan by gut microbes (Bansal et al., 2010; Jennis et al., 2018) with the xenobiotic sensor PXR and TLR4 also implicated (Venkatesh et al., 2014). This function has been assigned to specific peptostreptococcus species in the case of the suppression of inflammation via barrier function by indoleacrylic acid (Wlodarska et al., 2017). Many downstream and often host-processed byproducts of microbial-derived indoles are present in specific brain regions with potential implications for behavior although further research is required to map out the specific mechanisms at play (Clarke et al., 2021; O'Mahony et al., 2015). There is also evidence for a neurodevelopmental role for 3-indoxyl sulfate in the context of fetal thalamocortical axonogenesis (Vuong et al., 2020).

Microbiota-related changes in bile acid metabolism have been linked to gastrointestinal dysfunction in the BTBR mouse model of autism spectrum disorders (Golubeva et al., 2017). Microbially derived bile acids are also elevated at different sites including in the appendix microbiome in Parkinson's Disease (Li et al., 2021b). The immunomodulatory effects of bile acids may also have important implications for microbiota to brain communication via the gut-brain axis (Caspani and Swann, 2019; Ethridge et al., 2021). It remains unclear the extent to which the brain bile acid pool is directly influenced by the action of intestinal bacteria, or indeed how or if these CNS bile acids function mechanistically to impact brain function and behavior (McMillin and DeMorrow, 2016; Monteiro-Cardoso et al., 2021).

## Microbial regulation of tryptophan metabolism in microbiota to brain communication

In addition to direct processing of tryptophan into indoles, gut microbes also play an important role in the fate of this essential amino acid along host metabolic pathways (O'Mahony et al., 2015). This is known from the altered availability and metabolism of tryptophan in GF animals and other rodent models at multiple sites in the gut-brain axis and with relevance for both serotonin and kynurenine pathway metabolites (Clarke et al., 2013; Lukic et al., 2019; Lyte et al., 2020; Marin et al., 2017; Wikoff et al., 2009). In the gastrointestinal tract, it is likely that SCFAs play a role in dictating serotonin biosynthesis from colonic enterochromaffin cells (ECs) (Reigstad et al., 2015; Yano et al., 2015). There is also evidence of bidirectional signaling between the gut microbiota and the host serotonergic system with intestinal serotonin affecting bacterial colonization in the gut (Fung et al., 2019).

The mechanisms at play in microbial regulation of CNS serotonin metabolism are less well understood but may involve impacting on the supply of tryptophan to the CNS (Clarke et al., 2013; Lukic et al., 2019). Serotonin and kynurenine pathway metabolites have important roles in multiple aspects of gastrointestinal and brain function consistent with the known scope of influence of the gut microbiome on behavior and intestinal physiology within the gut-brain axis framework (Clarke et al., 2021; Kennedy et al., 2017; O'Mahony et al., 2015).

## Microbial microvesicles in microbiota to brain communication

Microbial microvesicles or extracellular vesicles (outer membrane vesicles) are produced by gram-negative bacteria and are found in the gut lumen, throughout the GI tract, and in the bloodstream (Stentz et al., 2018). The cargo of these microvesicles can include neurotransmitters, and there are a number of potential mechanisms through which they might be able to influence brain function and behavior including via peripheral nervous system and immunomodulation (Haas-Neill and Forsythe, 2020). The impact on the immune system is likely via immune cells in the lamina propria once they cross the intestinal epithelium (Stentz et al., 2018). The extracellular vesicles of *Akkermansia muciniphila* have been reported to impact on the colonic serotonergic system (Yaghoubfar et al., 2020). Because they have been associated with alterations in brain gene expression and pathological alterations during different stages of neuroinflammation and neurodegeneration, consideration is being given to whether they cross the blood−brain barrier and if they could function as delivery vehicles from a therapeutic perspective (Pirolli et al., 2021; Yang et al., 2018). There is also evidence that, as in the case of *Paenalcaligenes hominis*, these microvesicles can signal via the vagus nerve to impact on cognitive function (Lee et al., 2020).

## Conclusions, perspectives, and future directions

The microbiome-gut-brain axis field has traveled a long way in a relatively short time to chart the scope of influence of the gut microbiome and behavior. Much of the initial efforts concentrated on characterizing the behavioral, structural, and functional impact of the gut microbiome on gut-brain axis signaling. This has yielded a broad range of potential targets, an outline sketch that now requires mechanistic detail added to further understand and take advantage of microbiota to brain communication. Experimental approaches directed to this task have already identified an important role for the vagus nerve, for example, albeit lacking some fine level details of the critical host-microbe points of interaction. Acquiring these details is both a priority and challenge for future research in this area.

Discovery of new molecular mechanisms and pathways from gut to brain can be achieved with a blend of existing experimental systems (see Chapter 1) and state-of-the-art technologies as indicated above such as single cell transcriptomics, circuit-tracing methods, and functional chemogenetic manipulations. Advances in bioinformatic pipelines can also facilitate efforts necessary to expand our fundamental biology knowledge and move from association to causation, and reinvigorate the application of learnings about microbiota to brain communication in clinical practice.

## References

Agus, A., Planchais, J., Sokol, H., 2018. Gut microbiota regulation of tryptophan metabolism in health and disease. Cell Host Microbe 23 (6), 716−724.

Agus, A., Clement, K., Sokol, H., 2021. Gut microbiota-derived metabolites as central regulators in metabolic disorders. Gut 70, 1174−1182.

Akira, S., Takeda, K., 2004. Toll-like receptor signalling. Nat Rev Immunol 4 (7), 499−511.

Al-Nedawi, K., et al., 2015. Gut commensal microvesicles reproduce parent bacterial signals to host immune and enteric nervous systems. FASEB J 29 (2), 684−695.

Allen, A.P., et al., 2016. Bifidobacterium longum 1714 as a translational psychobiotic: modulation of stress, electrophysiology and neurocognition in healthy volunteers. Transl Psychiatry 6 (11), e939.

Arentsen, T., et al., 2017. The bacterial peptidoglycan-sensing molecule Pglyrp2 modulates brain development and behavior. Mol Psychiatr 22 (2), 257−266.

Bansal, T., et al., 2010. The bacterial signal indole increases epithelial-cell tight-junction resistance and attenuates indicators of inflammation. Proc Natl Acad Sci U S A 107 (1), 228−233.

Barajon, I., et al., 2009. Toll-like receptors 3, 4, and 7 are expressed in the enteric nervous system and dorsal root ganglia. J Histochem Cytochem 57 (11), 1013−1023.

Bercik, P., et al., 2011a. The intestinal microbiota affect central levels of brain-derived neurotropic factor and behavior in mice. Gastroenterology 141 (2), 599−609, 609.e1-3.

Bercik, P., et al., 2011b. The anxiolytic effect of Bifidobacterium longum NCC3001 involves vagal pathways for gut-brain communication. Neuro Gastroenterol Motil 23 (12), 1132−1139.

Bhattarai, Y., et al., 2018. Gut microbiota-produced tryptamine activates an epithelial G-protein-coupled receptor to increase colonic secretion. Cell Host Microbe 23 (6), 775−785 e5.

Boehme, M., et al., 2020. Mid-life microbiota crises: middle age is associated with pervasive neuroimmune alterations that are reversed by targeting the gut microbiome. Mol Psychiatr 25 (10), 2567−2583.

Bonaz, B., Sinniger, V., Pellissier, S., 2021. Therapeutic potential of vagus nerve stimulation for inflammatory bowel diseases. Front Neurosci 15, 650971.

Bravo, J.A., et al., 2011. Ingestion of Lactobacillus strain regulates emotional behavior and central GABA receptor expression in a mouse via the vagus nerve. Proc Natl Acad Sci U S A 108 (38), 16050−16055.

Breen, D.P., Halliday, G.M., Lang, A.E., 2019. Gut-brain axis and the spread of alpha-synuclein pathology: vagal highway or dead end? Mov Disord 34 (3), 307–316.

Buntwal, L., et al., 2019. Ghrelin-mediated hippocampal neurogenesis: implications for health and disease. Trends Endocrinol Metab 30 (11), 844–859.

Caspani, G., Swann, J., 2019. Small talk: microbial metabolites involved in the signaling from microbiota to brain. Curr Opin Pharmacol 48, 99–106.

Clarke, G., 2021. Metformin, the gut microbiome and neurogenesis: lessons learned in rebirth of an old drug. Brain Behav Immun 95, 25–26.

Clarke, G., Villalobos-Manriquez, F., Campos Marin, D., 2021. Tryptophan metabolism and the microbiome-gut-brain Axis. In: The Oxford Handbook of the Microbiome-Gut-Brain Axis.

Clarke, G., et al., 2014. Minireview: gut microbiota: the neglected endocrine organ. Mol Endocrinol 28 (8), 1221–1238.

Clarke, G., et al., 2013. The microbiome-gut-brain axis during early life regulates the hippocampal serotonergic system in a sex-dependent manner. Mol Psychiatr 18 (6), 666–673.

Collins, J., et al., 2014. Intestinal microbiota influence the early postnatal development of the enteric nervous system. Neuro Gastroenterol Motil 26 (1), 98–107.

Colombo, A.V., et al., 2021. Microbiota-derived short chain fatty acids modulate microglia and promote Abeta plaque deposition. Elife 10, e59826.

Crumeyrolle-Arias, M., et al., 2014. Absence of the gut microbiota enhances anxiety-like behavior and neuroendocrine response to acute stress in rats. Psychoneuroendocrinology 42, 207–217.

Cruz-Pereira, J.S., Cryan, J.F., 2020. In need of a quorum: from microbes to mood via the immune system. Am J Psychiatr 177 (10), 895–897.

Cryan, J.F., et al., 2018. A microbial drugstore for motility. Cell Host Microbe 23 (6), 691–692.

Cryan, J.F., et al., 2019. The microbiota-gut-brain axis. Physiol Rev 99 (4), 1877–2013.

Cussotto, S., et al., 2018. The neuroendocrinology of the microbiota-gut-brain axis: a behavioural perspective. Front Neuroendocrinol 51, 80–101.

Dalile, B., et al., 2019. The role of short-chain fatty acids in microbiota-gut-brain communication. Nat Rev Gastroenterol Hepatol 16 (8), 461–478.

Dalile, B., et al., 2020. Colon-delivered short-chain fatty acids attenuate the cortisol response to psychosocial stress in healthy men: a randomized, placebo-controlled trial. Neuropsychopharmacology 45 (13), 2257–2266.

De Vadder, F., et al., 2018. Gut microbiota regulates maturation of the adult enteric nervous system via enteric serotonin networks. Proc Natl Acad Sci U S A 115 (25), 6458–6463.

Dinan, T.G., Cryan, J.F., 2012. Regulation of the stress response by the gut microbiota: implications for psychoneuroendocrinology. Psychoneuroendocrinology 37 (9), 1369–1378.

Dinan, T.G., Cryan, J.F., 2017. Microbes, immunity, and behavior: psychoneuroimmunology meets the microbiome. Neuropsychopharmacology 42 (1), 178–192.

El Aidy, S., Dinan, T.G., Cryan, J.F., 2015. Gut microbiota: the conductor in the orchestra of immune-neuroendocrine communication. Clin Therapeut 37 (5), 954–967.

Erdman, S.E., Poutahidis, T., 2016. Microbes and oxytocin: benefits for host physiology and behavior. Int Rev Neurobiol 131, 91–126.

Erny, D., Prinz, M., 2020. How microbiota shape microglial phenotypes and epigenetics. Glia 68 (8), 1655–1672.

Erny, D., et al., 2015. Host microbiota constantly control maturation and function of microglia in the CNS. Nat Neurosci 18 (7), 965–977.

Ethridge, A.D., et al., 2021. Inter-kingdom communication and regulation of mucosal immunity by the microbiome. J Infect Dis 223, S236–S240.

Foster, J.A., 2016. Gut microbiome and behavior: focus on neuroimmune interactions. Int Rev Neurobiol 131, 49–65.

Foster, J.A., Rinaman, L., Cryan, J.F., 2017. Stress & the gut-brain axis: regulation by the microbiome. Neurobiol Stress 7, 124–136.

Fukui, H., Xu, X., Miwa, H., 2018. Role of gut microbiota-gut hormone axis in the pathophysiology of functional gastrointestinal disorders. J Neurogastroenterol Motil 24 (3), 367–386.

Fulde, M., et al., 2018. Neonatal selection by Toll-like receptor 5 influences long-term gut microbiota composition. Nature 560 (7719), 489–493.

Fulling, C., Dinan, T.G., Cryan, J.F., 2019. Gut microbe to brain signaling: what happens in vagus. Neuron 101 (6), 998–1002.

Fung, T.C., et al., 2019. Intestinal serotonin and fluoxetine exposure modulate bacterial colonization in the gut. Nat Microbiol 4 (12), 2064–2073.

Gheorghe, C.E., et al., 2019. Focus on the essentials: tryptophan metabolism and the microbiome-gut-brain axis. Curr Opin Pharmacol 48, 137–145.

Goehler, L.E., et al., 2005. Activation in vagal afferents and central autonomic pathways: early responses to intestinal infection with Campylobacter jejuni. Brain Behav Immun 19 (4), 334–344.

Golubeva, A.V., et al., 2017. Microbiota-related changes in bile acid & tryptophan metabolism are associated with gastrointestinal dysfunction in a mouse model of autism. EBioMedicine 24, 166–178.

Gonzalez-Santana, A., Diaz Heijtz, R., 2020. Bacterial peptidoglycans from microbiota in neurodevelopment and behavior. Trends Mol Med 26 (8), 729–743.

Haas-Neill, S., Forsythe, P., 2020. A budding relationship: bacterial extracellular vesicles in the microbiota-gut-brain axis. Int J Mol Sci 21 (23), 8899.

Han, W., et al., 2018. A neural circuit for gut-induced reward. Cell 175 (3), 665–678 e23.

Hindson, J., 2020. Enteric neuron regulation of gut motility by the microbiota. Nat Rev Gastroenterol Hepatol 17 (4), 194–195.

Hooper, L.V., Littman, D.R., Macpherson, A.J., 2012. Interactions between the microbiota and the immune system. Science 336 (6086), 1268–1273.

Hubbard, T.D., et al., 2015. Adaptation of the human aryl hydrocarbon receptor to sense microbiota-derived indoles. Sci Rep 5, 12689.

Hyland, N.P., Cryan, J.F., 2016. Microbe-host interactions: influence of the gut microbiota on the enteric nervous system. Dev Biol 417 (2), 182–187.

Janik, R., et al., 2016. Magnetic resonance spectroscopy reveals oral Lactobacillus promotion of increases in brain GABA, N-acetyl aspartate and glutamate. Neuroimage 125, 988–995.

Jennis, M., et al., 2018. Microbiota-derived tryptophan indoles increase after gastric bypass surgery and reduce intestinal permeability in vitro and in vivo. Neuro Gastroenterol Motil 30 (2), e13178.

Jing, Y., et al., 2021. Effect of fecal microbiota transplantation on neurological restoration in a spinal cord injury mouse model: involvement of brain-gut axis. Microbiome 9 (1), 59.

Kabouridis, P.S., Pachnis, V., 2015. Emerging roles of gut microbiota and the immune system in the development of the enteric nervous system. J Clin Invest 125 (3), 956–964.

Kabouridis, P.S., et al., 2015. Microbiota controls the homeostasis of glial cells in the gut lamina propria. Neuron 85 (2), 289–295.

Kaelberer, M.M., et al., 2018. A gut-brain neural circuit for nutrient sensory transduction. Science 361 (6408), eaat5236.

Kennedy, P.J., et al., 2017. Kynurenine pathway metabolism and the microbiota-gut-brain axis. Neuropharmacology 112 (Pt B), 399–412.

Kennedy, P.J., et al., 2014. A sustained hypothalamic-pituitary-adrenal axis response to acute psychosocial stress in irritable bowel syndrome. Psychol Med 44 (14), 3123−3134.

Khoshdel, A., et al., 2013. Bifidobacterium longum NCC3001 inhibits AH neuron excitability. Neuro Gastroenterol Motil 25 (7), e478−e484.

Kingsbury, M.A., Bilbo, S.D., 2019. The inflammatory event of birth: how oxytocin signaling may guide the development of the brain and gastrointestinal system. Front Neuroendocrinol 55, 100794.

Kolypetri, P., et al., 2021. Regulation of splenic monocyte homeostasis and function by gut microbial products. iScience 24 (4), 102356.

Kunze, W.A., et al., 2009. Lactobacillus reuteri enhances excitability of colonic AH neurons by inhibiting calcium-dependent potassium channel opening. J Cell Mol Med 13 (8B), 2261−2270.

Lamas, B., et al., 2016. CARD9 impacts colitis by altering gut microbiota metabolism of tryptophan into aryl hydrocarbon receptor ligands. Nat Med 22 (6), 598−605.

Lee, J.H., Wood, T.K., Lee, J., 2015. Roles of indole as an interspecies and interkingdom signaling molecule. Trends Microbiol 23 (11), 707−718.

Lee, K.E., et al., 2020. The extracellular vesicle of gut microbial *Paenalcaligenes hominis* is a risk factor for vagus nerve-mediated cognitive impairment. Microbiome 8 (1), 107.

Li, L., et al., 2021a. Role of astroglial toll-like receptors (TLRs) in central nervous system infections, injury and neurodegenerative diseases. Brain Behav Immun 91, 740−755.

Li, P., et al., 2021b. Gut microbiota dysbiosis is associated with elevated bile acids in Parkinson's disease. Metabolites 11 (1), 29.

Lomasney, K.W., Cryan, J.F., Hyland, N.P., 2014. Converging effects of a Bifidobacterium and Lactobacillus probiotic strain on mouse intestinal physiology. Am J Physiol Gastrointest Liver Physiol 307 (2), G241−G247.

Long, S.L., Gahan, C.G.M., Joyce, S.A., 2017. Interactions between gut bacteria and bile in health and disease. Mol Aspect Med 56, 54−65.

Lukic, I., et al., 2019. Role of tryptophan in microbiota-induced depressive-like behavior: evidence from tryptophan depletion study. Front Behav Neurosci 13, 123.

Lyte, J.M., et al., 2020. Gut-brain axis serotonergic responses to acute stress exposure are microbiome-dependent. Neuro Gastroenterol Motil 32 (11), e13881.

Lyte, M., Vulchanova, L., Brown, D.R., 2011. Stress at the intestinal surface: catecholamines and mucosa-bacteria interactions. Cell Tissue Res 343 (1), 23−32.

Ma, X., et al., 2021. Metformin restores hippocampal neurogenesis and learning and memory via regulating gut microbiota in the obese mouse model. Brain Behav Immun 95, 68−83.

Manaker, S., Zucchi, P.C., 1993. Effects of vagotomy on neurotransmitter receptors in the rat dorsal vagal complex. Neuroscience 52 (2), 427−441.

Mao, Y.K., et al., 2013. Bacteroides fragilis polysaccharide A is necessary and sufficient for acute activation of intestinal sensory neurons. Nat Commun 4, 1465.

Marin, I.A., et al., 2017. Microbiota alteration is associated with the development of stress-induced despair behavior. Sci Rep 7, 43859.

Matteoli, G., Boeckxstaens, G.E., 2013. The vagal innervation of the gut and immune homeostasis. Gut 62 (8), 1214−1222.

Mazmanian, S.K., et al., 2005. An immunomodulatory molecule of symbiotic bacteria directs maturation of the host immune system. Cell 122 (1), 107−118.

McMillin, M., DeMorrow, S., 2016. Effects of bile acids on neurological function and disease. FASEB J 30 (11), 3658−3668.

McVey Neufeld, K.A., et al., 2013. The microbiome is essential for normal gut intrinsic primary afferent neuron excitability in the mouse. Neuro Gastroenterol Motil 25 (2), 183-e88.

McVey Neufeld, K.A., et al., 2015. The gut microbiome restores intrinsic and extrinsic nerve function in germ-free mice accompanied by changes in calbindin. Neuro Gastroenterol Motil 27 (5), 627–636.

Medina-Rodriguez, E.M., et al., 2020. Identification of a signaling mechanism by which the microbiome regulates Th17 cell-mediated depressive-like behaviors in mice. Am J Psychiatr 177 (10), 974–990.

Monteiro-Cardoso, V.F., Corliano, M., Singaraja, R.R., 2021. Bile acids: a communication channel in the gut-brain Axis. NeuroMolecular Med 23 (1), 99–117.

Muller, P.A., et al., 2020. Microbiota modulate sympathetic neurons via a gut-brain circuit. Nature 583 (7816), 441–446.

Murray, K., Reardon, C., 2018. The cholinergic anti-inflammatory pathway revisited. Neuro Gastroenterol Motil 30 (3), e13288.

Niesler, B., et al., 2021. Disorders of the enteric nervous system - a holistic view. Nat Rev Gastroenterol Hepatol 18, 393–410.

O'Leary, O.F., et al., 2018. The vagus nerve modulates BDNF expression and neurogenesis in the hippocampus. Eur Neuropsychopharmacol 28 (2), 307–316.

O'Mahony, S.M., et al., 2015. Serotonin, tryptophan metabolism and the brain-gut-microbiome axis. Behav Brain Res 277, 32–48.

Obata, Y., et al., 2020. Neuronal programming by microbiota regulates intestinal physiology. Nature 578 (7794), 284–289.

Ogbonnaya, E.S., et al., 2015. Adult hippocampal neurogenesis is regulated by the microbiome. Biol Psychiatr 78 (4), e7–e9.

Patterson, E., et al., 2014. Gut microbiota, the pharmabiotics they produce and host health. Proc Nutr Soc 73 (4), 477–489.

Perez-Berezo, T., et al., 2017. Identification of an analgesic lipopeptide produced by the probiotic *Escherichia coli* strain Nissle 1917. Nat Commun 8 (1), 1314.

Perez-Burgos, A., et al., 2013. Psychoactive bacteria *Lactobacillus rhamnosus* (JB-1) elicits rapid frequency facilitation in vagal afferents. Am J Physiol Gastrointest Liver Physiol 304 (2), G211–G220.

Pirolli, N.H., Bentley, W.E., Jay, S.M., 2021. Bacterial extracellular vesicles and the gut-microbiota brain axis: emerging roles in communication and potential as therapeutics. Adv Biol (Weinh) 5, e2000540.

Pu, Y., et al., 2021. A role of the subdiaphragmatic vagus nerve in depression-like phenotypes in mice after fecal microbiota transplantation from Chrna7 knock-out mice with depression-like phenotypes. Brain Behav Immun 94, 318–326.

Raison, C.L., Capuron, L., Miller, A.H., 2006. Cytokines sing the blues: inflammation and the pathogenesis of depression. Trends Immunol 27 (1), 24–31.

Rea, K., Dinan, T.G., Cryan, J.F., 2016. The microbiome: a key regulator of stress and neuroinflammation. Neurobiol Stress 4, 23–33.

Reigstad, C.S., et al., 2015. Gut microbes promote colonic serotonin production through an effect of short-chain fatty acids on enterochromaffin cells. FASEB J 29 (4), 1395–1403.

Rhee, S.H., Pothoulakis, C., Mayer, E.A., 2009. Principles and clinical implications of the brain-gut-enteric microbiota axis. Nat Rev Gastroenterol Hepatol 6 (5), 306–314.

Rothhammer, V., Quintana, F.J., 2019. The aryl hydrocarbon receptor: an environmental sensor integrating immune responses in health and disease. Nat Rev Immunol 19 (3), 184–197.

Rothhammer, V., et al., 2016. Type I interferons and microbial metabolites of tryptophan modulate astrocyte activity and central nervous system inflammation via the aryl hydrocarbon receptor. Nat Med 22 (6), 586–597.

Schellekens, H., et al., 2021. Bifidobacterium longum counters the effects of obesity: partial successful translation from rodent to human. EBioMedicine 63, 103176.

Scott, S.A., Fu, J., Chang, P.V., 2020. Microbial tryptophan metabolites regulate gut barrier function via the aryl hydrocarbon receptor. Proc Natl Acad Sci U S A 117 (32), 19376—19387.

Sgritta, M., et al., 2019. Mechanisms underlying microbial-mediated changes in social behavior in mouse models of autism spectrum disorder. Neuron 101 (2), 246—259 e6.

Siopi, E., et al., 2019. Gut microbiota requires vagus nerve integrity to promote depression. bioRxiv, 547778.

Skibicka, K.P., Dickson, S.L., 2013. Enteroendocrine hormones - central effects on behavior. Curr Opin Pharmacol 13 (6), 977—982.

Spencer, N.J., Hu, H., 2020. Enteric nervous system: sensory transduction, neural circuits and gastrointestinal motility. Nat Rev Gastroenterol Hepatol 17 (6), 338—351.

Spichak, S., et al., 2021. Mining microbes for mental health: determining the role of microbial metabolic pathways in human brain health and disease. Neurosci Biobehav Rev 125, 698—761.

Stentz, R., et al., 2018. Fantastic voyage: the journey of intestinal microbiota-derived microvesicles through the body. Biochem Soc Trans 46 (5), 1021—1027.

Stilling, R.M., et al., 2016. The neuropharmacology of butyrate: the bread and butter of the microbiota-gut-brain axis? Neurochem Int 99, 110—132.

Stockinger, B., Shah, K., Wincent, E., 2021. AHR in the intestinal microenvironment: safeguarding barrier function. Nat Rev Gastroenterol Hepatol 18, 559—570.

Suarez, A.N., et al., 2018. Gut vagal sensory signaling regulates hippocampus function through multi-order pathways. Nat Commun 9 (1), 2181.

Sudo, N., et al., 2004. Postnatal microbial colonization programs the hypothalamic-pituitary-adrenal system for stress response in mice. J Physiol 558 (Pt 1), 263—275.

Szereda-Przestaszewska, M., 1985. Retrograde degeneration within the dorsal motor vagal nucleus following bilateral vagotomy in rabbits. Acta Anat (Basel) 121 (3), 133—139.

Tanoue, T., Atarashi, K., Honda, K., 2016. Development and maintenance of intestinal regulatory T cells. Nat Rev Immunol 16 (5), 295—309.

Tomberlin, J.K., et al., 2017. Indole: an evolutionarily conserved influencer of behavior across kingdoms. Bioessays 39 (2), 1600203.

Torres-Fuentes, C., et al., 2017. The microbiota-gut-brain axis in obesity. Lancet Gastroenterol Hepatol 2 (10), 747—756.

Torres-Fuentes, C., et al., 2019. Short-chain fatty acids and microbiota metabolites attenuate ghrelin receptor signaling. FASEB J 33 (12), 13546—13559.

Uesaka, T., et al., 2016. Development of the intrinsic and extrinsic innervation of the gut. Dev Biol 417 (2), 158—167.

Valles-Colomer, M., et al., 2019. The neuroactive potential of the human gut microbiota in quality of life and depression. Nat Microbiol 4 (4), 623—632.

van de Wouw, M., et al., 2018. Short-chain fatty acids: microbial metabolites that alleviate stress-induced brain-gut axis alterations. J Physiol 596 (20), 4923—4944.

van de Wouw, M., et al., 2020. The role of the microbiota in acute stress-induced myeloid immune cell trafficking. Brain Behav Immun 84, 209—217.

van Noort, J.M., Bsibsi, M., 2009. Toll-like receptors in the CNS: implications for neurodegeneration and repair. Prog Brain Res 175, 139—148.

Venkatesh, M., et al., 2014. Symbiotic bacterial metabolites regulate gastrointestinal barrier function via the xenobiotic sensor PXR and Toll-like receptor 4. Immunity 41 (2), 296—310.

Vuong, H.E., et al., 2020. The maternal microbiome modulates fetal neurodevelopment in mice. Nature 586 (7828), 281—286.

Wikoff, W.R., et al., 2009. Metabolomics analysis reveals large effects of gut microflora on mammalian blood metabolites. Proc Natl Acad Sci U S A 106 (10), 3698—3703.

Wilkinson, J.E., et al., 2021. A framework for microbiome science in public health. Nat Med 27, 766–774.

Wilmes, L., et al., 2021. Of bowels, brain and behavior: a role for the gut microbiota in psychiatric comorbidities in irritable bowel syndrome. Neuro Gastroenterol Motil 33 (3), e14095.

Wlodarska, M., et al., 2017. Indoleacrylic acid produced by commensal peptostreptococcus species suppresses inflammation. Cell Host Microbe 22 (1), 25–37 e6.

Xu, L., 2014. Leptin action in the midbrain: from reward to stress. J Chem Neuroanat 61–62, 256–265.

Yaghoubfar, R., et al., 2020. Modulation of serotonin signaling/metabolism by *Akkermansia muciniphila* and its extracellular vesicles through the gut-brain axis in mice. Sci Rep 10 (1), 22119.

Yang, J., et al., 2018. Microbe-derived extracellular vesicles as a smart drug delivery system. Transl Clin Pharmacol 26 (3), 103–110.

Yano, J.M., et al., 2015. Indigenous bacteria from the gut microbiota regulate host serotonin biosynthesis. Cell 161 (2), 264–276.

Yu, C.D., Xu, Q.J., Chang, R.B., 2020. Vagal sensory neurons and gut-brain signaling. Curr Opin Neurobiol 62, 133–140.

Zheng, D., Liwinski, T., Elinav, E., 2020. Interaction between microbiota and immunity in health and disease. Cell Res 30 (6), 492–506.

# CHAPTER 5

# Microbiota influence behavior—Work in animal models

## Overview

Behavioral neuroscience uses established, ethologically valid, behavioral tests to examine many domains of behavior (Cryan and Holmes, 2005). Several behavioral tests have been employed to demonstrate an association between the microbiome and behavior (Fig. 5.1). Approach avoidance tests including elevated plus maze (EPM), open field (OF), and light—dark tests (LD) measure anxiety-like, exploratory, and locomotor activity. A central feature of these tests is an enclosed "safe" zone and an open "aversive" zone (Bourin and Hascoet, 2003; Prut and Belzung, 2003; Sidor et al., 2010). The time spent and entries into the open, aversive zone (open arm in EPM, center of the OF, light chamber in LD test) is a measure of exploratory and anxiety-like behavior. In addition, locomotor activity is measured by distance traveled and rearing is measure of exploratory behavior. Depressive-like behavior is measured using the forced swim test and the tail suspension test (Can et al., 2012; Cryan et al., 2005; Slattery and Cryan, 2012). In both cases, the mouse or rat attempts to escape the aversive situation, and after a period of time, the rodent becomes immobile, the time spent immobile is an indication of depressive-like behavior. Anhedonia or a reduced ability to experience pleasure is measured using the sucrose preference test (Eagle et al., 2016). Cognitive behavior tests that measure learning and memory include passive avoidance tests, novel object recognition, spatial memory tests, and fear learning (Cryan and Holmes, 2005). Social behavior in rodents is measured through standardized behavioral tests including the social interaction test and the three-chamber test (Moy et al., 2004). Reciprocal social interaction time between two individuals over a short time interval in a Plexiglas arena including close following, touching, and sniffing are quantitated to provide an index of social interaction (Buffington et al., 2016).

*Microbiota Brain Axis*
ISBN 978-0-12-814800-6
https://doi.org/10.1016/B978-0-12-814800-6.00002-9

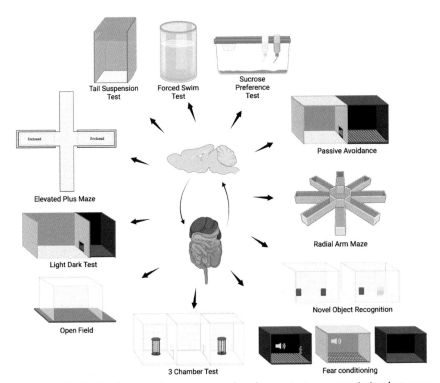

**Figure 5.1** Behavioral neuroscience tests used to demonstrate an association between microbiota and behavior.

Sociability is measured in a three-chamber test based on the amount of time the test mouse spends in the chamber near an unfamiliar conspecific in a wire cage compared to an empty wire cage in a different chamber. Social novelty is measured following sociability, by placing a novel stranger into the empty cage and assessing the social preference of the experimental mouse/rat for the novel stranger compared to the now-familiar one (Moy et al., 2004). Microbiota–brain researchers have utilized behavioral neuroscience tools to demonstrate an association between the microbiome and behavior that is summarized here.

## Behavior in germ-free mice

Much of our understanding about how microbiota influence behavior comes from animal studies that manipulate microbiota, such as experiments using germ-free (GF) animals. The GF mouse/rat has been a useful animal model to help determine the domains of behavior that are influenced by the gut microbiota (Luczynski et al., 2016). The GF mouse model was established in 1957; GF mice are raised in sterile/gnotobiotic environments and have no commensal bacteria (Gustafsson et al., 1957; Gustafsson, 1959). The complete absence of a microbiota results in many host physiological changes including immune system and metabolic alterations (Luczynski et al., 2016; Macpherson and Harris, 2004). A key paper that sparked the interest of neuroscientists showed that GF mice had an exaggerated hypothalamic-pituitary-adrenal (HPA) axis response to stress (Sudo et al., 2004). Next, key studies linked microbiota to stress-related behaviors and revealed that GF mice had reduced anxiety-like behavior, a behavioral phenotype observed in several strains of mice (Clarke et al., 2013; Diaz Heijtz et al., 2011; Neufeld et al., 2011a, 2011b). Different strains of mice have distinct behavioral phenotypes related to anxiety-like and exploratory behavior. Using this natural difference in behavioral phenotype, a proof-of-concept study that colonized GF mice Balb/C (higher anxiety-like behavior) with microbiota from conventionally housed NIH Swiss mice (lower anxiety-like behavior) observed a reduction of anxiety-like behavior in the NIH Swiss colonized Balb/C mice compared to Balb/C mice colonized with microbiota from donor Balb/C mice. Further, the reverse experiment was also effective where NIH Swiss mice colonized with Balb/C microbiota showed increased anxiety-like behavior compared to NIH Swiss mice colonized with microbiota from donor NIH Swiss mice (Bercik et al., 2011). Although GF rats, in contrast to mice, showed increased anxiety-like behavior compared to conventionally housed rats (Crumeyrolle-Arias et al., 2014), additional studies using GF mice have supported the original observations linking the missing microbiota to reduced anxiety-like behaviors (De Palma et al., 2015; Huo et al., 2017; Lu et al., 2018; Luk et al., 2018; Luo et al., 2018; Pan et al., 2019; Qi et al., 2019). Notably, reduced anxiety-like behavior in the light–dark test and open field test as well as reduced depressive-like behavior in the tail suspension test and forced swim test were observed in GF mice compared to conventionally housed mice (De Palma et al., 2015; Luo et al., 2018). In addition to reduced anxiety-like behavior, hyperactivity has been observed in GF mice (Arentsen et al., 2015), and both hyperactivity and reduced anxiety-like behavior have been observed in GF zebrafish (Catron et al., 2019; Davis et al., 2016; Phelps et al., 2017), suggesting that microbiota–brain associations with behavior are generalizable across species.

To determine the critical developmental window during which the microbiota influence brain development and behavior, investigators conventionalized GF mice with specific pathogen-free (SPF) microbiota or with specific profiles of selected bacterial taxa and examined the impact on adult behavior. Interestingly, the reconstitution of microbiota to GF mice early in life was able to normalize EPM behavior and some aspects of light–dark behavior (Clarke et al., 2013; Diaz Heijtz et al., 2011). Further, administration of a cocktail of *Bifidobacterium* spp. starting at postnatal day 1 to GF mice normalized EPM behavior (Luk et al., 2018). In contrast, when GF mice were conventionalized with SPF microbiota in adulthood or around puberty (4–6 weeks), the reduced anxiety-like phenotype observed in the EPM persisted (Neufeld et al., 2011a, 2011b; Pan et al., 2019; Qi et al., 2019).

Additional behavioral domains linked to microbiota through experiments with GF mice include cognition and social behavior. Related to cognition, GF mice showed impaired nonspatial and working memory compared to conventionally housed mice (Gareau et al., 2011). Similarly, GF mice showed reduced passive avoidance memory compared to mice monocolonized with *Lactobacillus* sp. (Mao et al., 2020). Reduced spatial memory was also observed in GF mice compared to conventionally housed mice (Lu et al., 2018). Further support for an association between microbiota and cognition comes from experiments using antibiotic treatment, stress, and fecal transplantation approaches (expanded in sections below). Fear learning, including contextual memory and cued memory, can be assessed in rodents, and GF mice show reduced freezing in cued fear recall and extinction trials compared to conventionally housed mice (Hoban et al., 2018). The conventionalization of GF mice (ex-GF) at weaning (postnatal day 21) normalized fear recall; however, reduced freezing during extinction trials was comparable to GF mice (Hoban et al., 2018), demonstrating the importance of microbiota–host interactions in early postnatal mice to neurodevelopment and behavior.

A role for the microbiome in social behavior has been demonstrated using the three-chamber test (Moy et al., 2004). Although one study found that GF mice displayed increased sociability compared to conventionally housed mice (Arentsen et al., 2015), most studies have shown that GF mice display reduced social interaction (Buffington et al., 2016; Stilling et al., 2018; Wu et al., 2021), reduced sociability (Buffington et al., 2016; Desbonnet et al., 2014; Lu et al., 2018; Stilling et al., 2018), and reduced social novelty (Desbonnet et al., 2014; Stilling et al., 2018). The importance of host—microbiota interactions has been suggested to have influenced the evolution of social behaviors and numerous pathways have been suggested that may mediate microbiota—brain signaling to impact social behavior (Sherwin et al., 2019).

## Antibiotic-related changes in microbiota influence behavior

While the use of GF mice is helpful in identifying host systems that are impacted by microbiota, the complete absence of microbiota from conception to adulthood in GF animals impacts so many host systems that it is difficult to generalize findings in GF mice to normal microbiota—brain interactions. An alternative experimental approach to manipulate the microbiome is the use of antibiotic treatment. This section considers studies that used antibiotic treatment in animal models to provide additional evidence that microbiota and microbiota—brain signaling are important to several domains of behavior.

Antibiotic treatment in adult rodents has been shown to influence behavior (Chu et al., 2019; Frohlich et al., 2016; Glover et al., 2021; Guida et al., 2018; Schmidtner et al., 2019; Stead et al., 2006). Selective breeding of Sprague—Dawley rats generated two groups of adult rats, those that show a high behavioral response to novelty (HR) and those that show a low behavioral response to novelty (LR). LR rats display increased behavioral inhibition, anxiety-like behavior, and reduced stress-coping behavior relative to HR rats (Stead et al., 2006). A recent study examined the microbiota of HR and LR rats and the impact of antibiotic treatment in adult HR and LR rats (Glover et al., 2021). While no microbiota compositional differences were reported between HR and LR rats, both were impacted by antibiotic treatment, and LR rats treated with antibiotics showed increased anxiety-like behavior (Glover et al., 2021). Interestingly, antibiotic treatment increased active coping behavior in the forced swim test in both HR and LR rats compared to control rats (Glover et al., 2021). The absence of a change in microbiota profile between HR and LR rats reminds us of the importance of host genetics in behavior (Ducottet and Belzung, 2005; Moy et al., 2007) and that behavioral phenotypes are wired early in development; as such, treatment with antibiotics in adulthood may have less of an impact on behavior than treatment earlier in life. To note, 16S rRNA sequencing provides compositional and diversity information but does not provide a direct functional readout of the microbiome. In a separate study using Wistar rats selectively bred for high anxiety-like behavior (HAB), administration of minocycline for 3 weeks in adulthood resulted in reduced depressive-like behavior in male HAB rats and also increased sociability in both male and female HAB rats compared to normal anxiety-like behavior (NAB) rats (Schmidtner et al., 2019). Here, trait-related and antibiotic-related differences in Lachnospiraceae and *Clostridiales Family XIII* were observed in the cecal samples (Schmidtner et al., 2019). In addition, Lachnospiraceae abundance was negatively correlated with immobility time in adult male mice treated with an antibiotic cocktail for 2 weeks in a separate study (Guida et al., 2018), suggesting that this family of bacteria may be a key modulator of behavior. Minocycline also inhibits microglial activation, a mechanism of action also proposed to mediate its impact on depression-like behavior, complicating the interpretation of these results (Bassett et al., 2021).

Administration of antibiotics to adult mice also impacted cognitive behavior (Chu et al., 2019; Frohlich et al., 2016; Guida et al., 2018). Administration of a cocktail of five antibiotics to deplete the microbiome in adult male mice resulted in impaired object recognition memory, with no effect on spatial learning (Frohlich et al., 2016). In addition to altered microbiota composition, several microbiota-related metabolites, including short-chain fatty acids (SCFA), were reduced in colonic contents, and the reduction in several circulating metabolites mirrored the changes in the colon (Frohlich et al., 2016). Fear conditioning is an associative learning task that has been reported to be altered in GF mice (Hoban et al., 2018). A direct comparison of antibiotic-treated mice and GF mice strengthens the evidence for this association (Chu et al., 2019). Antibiotic-treated adult male mice showed normal acquisition of fear learning but showed deficits in extinction learning compared to untreated mice (Chu et al., 2019). Impaired extinction learning was also observed in adult male GF mice supporting the role for microbiota in the observed behavioral differences (Chu et al., 2019). To determine if extinction learning deficits in GF mice are reversible, investigators conventionalized GF mice at different time points and demonstrated that exposure to microbiota early in postnatal life is necessary for normal fear extinction learning in adulthood (Chu et al., 2019). Further, extinction learning was associated with microbiota-derived metabolites in the blood and brain (Chu et al., 2019). All but one of the antibiotic studies described above examined adult male mice or rats, so it is important for future work to consider both female and male animals as sex-related differences in microbiota—brain signaling likely contribute to differences in behavior.

The impact of manipulating the microbiome during earlier developmental periods provides further evidence related to critical windows during which microbiota influence behavior. Exposure to antibiotics in utero in mice and rats affected several behavioral domains in offspring (Champagne-Jorgensen et al., 2020; Degroote et al., 2016). Exposure to succinylsulfathiazole in utero in Wistar rats reduced social interaction, increased anxiety-like behavior, and altered sensorimotor gating in offspring (Degroote et al., 2016). Acute exposure to penicillin in the last week of gestation in Balb/C mice reduced anxiety-like behavior in female offspring and reduced sociability in male offspring (Champagne-Jorgensen et al., 2020). Extending the treatment length to include the last week of gestation and the preweaning period resulted in reduced anxiety-like behavior in male offspring, reduced sociability (not separated by sex), and reduced social novelty (not separated by sex) in Balb/C offspring (Leclercq et al., 2017).

Depletion of microbiota at weaning and through adolescence also impacts several behavioral domains (Ceylani et al., 2018; Desbonnet et al., 2015; Zeraati et al., 2019). Balb/C male mice exposed to either ampicillin, cefoperazone, or a mix of the two antibiotics from weaning (P21) for 6 weeks showed reduced locomotor activity and reduced recognition memory. Increased anxiety-like behavior and depressive-like behavior were also observed in the ampicillin-treated group (Ceylani et al., 2018). In a separate study, antibiotic treatment from weaning to adulthood delayed the onset of symptoms in mice with experimental-autoimmune-encephalomyelitis (EAE) and decreased brain cytokine levels in EAE mice. Interestingly, antibiotic treatment increased anxiety-like behavior in both EAE and non-EAE mice (Zeraati et al., 2019).

Overall, these behavioral studies in animals treated with antibiotics provide additional evidence of the importance of microbiota to behavior, and many of the findings corroborate findings in GF rodents. The main action of antibiotics is to kill bacteria, including both pathogenic and commensal bacterial, and have a significant impact on gut microbiota with a general reduction in microbial diversity, composition, and function. Antibiotic treatment impacts microbe—host interactions and microbiota—brain signaling. More research is needed to move beyond associations toward mechanistic insights that help us to better understand specifically how microbiota communicate with the brain.

# The impact of stress on microbiota and behavior

A landmark study showed that GF mice have exaggerated stress reactivity in response to restraint stress with higher stress-related levels of plasma corticosterone levels (CORT) and plasma adrenocorticotrophic hormone levels (ACTH) compared to SPF mice, provided important impetus for neuroscientists to consider the GF mouse model (Sudo et al., 2004). Based on many studies in the past 15 years, we know that the link between stress and microbiota is bidirectional—stress can influence the microbiota and the microbiome can influence the impact of stress (Cryan et al., 2019; Cussotto et al., 2018; Foster et al., 2017). Animal studies have examined the impact of early life stress on behavior using maternal separation during the first 2 weeks of life, as well as housing manipulations such as limited bedding. The use of social isolation and chronic unpredictable mild stress during adolescence impacts behavior. Adult stressors including restraint stress and social stress also impact behavior. This section will present studies that measured microbiome and behavior in animal models of stress.

Social defeat stress is an experimental protocol where experimental mice are exposed to daily bouts of social defeat by an unfamiliar aggressor for 10 days (Berton et al., 2006). Chronic exposure to social defeat results in an enduring reduction in social interaction (Berton et al., 2006), as well as several other persistent changes in activity and stress–related behaviors. More recently, studies have shown that social defeat stress-related changes in social interaction were accompanied by reduced alpha diversity and changes in microbiota composition (Bharwani et al., 2016; McGaughey et al., 2019). A modified protocol of social defeat stress, a milder version, resulted in reduced locomotor activity and reduced exploratory behavior in parallel with increased anxiety-like and depressive-like behavior in adult male C57Bl/6 mice (McGaughey et al., 2019). Several gut bacterial taxa differed in defeated mice, and specific taxa, including *Akkermansia, Oscillospira,* and *Flexispira,* were associated with anxiety-like behavior and anhedonia; whereas, *Ruminococcus* and *Dorea* were associated with anhedonia but not alterations in anxiety-like behavior. While all mice exposed to social defeat stress showed similar changes in anxiety-like behavior and anhedonia, with respect to social behavior, stress-exposed mice segregated into resilient and susceptible groups, the latter of which demonstrated a reduced social approach (McGaughey et al., 2019). A comparison between control, resilient, and susceptible mice revealed an increased abundance of *Prevotella* and *Parabacteroides* in susceptible mice, which mirrors some recent findings observed in individuals with depression (Jiang et al., 2015; Lin et al., 2017; McGaughey et al., 2019). Interestingly, exposure of mice to antibiotics in drinking water for 2 weeks before exposure to chronic defeat stress resulted in a resilient behavior phenotype (Wang et al., 2020). Here, antibiotic-treated mice did not show a stress-related increase in plasma IL-6. Key taxa that distinguished resilient mice included an increased abundance of *Allstipes, Lactobacillus,* and *Candidatus Arthromitus* (Wang et al., 2020). Reduced levels of *Parabacteroides goldsteinii* were suggested to be associated with resilience (Wang et al., 2020). Clearly, these studies demonstrate that stress—microbiome interactions influence behavior, and in addition, they provide insight into key taxa that might be involved. Extending these findings to identify key mechanisms is important; functional readouts of microbiota changes through metagenomics, metatranscriptomics, and metabolomics are needed.

Social isolation is known to have a long-term impact on stress-related behaviors. In rodent models, single-isolation housing, compared to group housing, results in alterations in social and stress-related behaviors (Fone and Porkess, 2008; Sahakian and Robbins, 1977). Recent work has examined how stress-related alterations in microbiota composition may contribute to behavioral changes associated with social isolation (Donovan et al., 2020; Dunphy-Doherty et al., 2018; Huang et al., 2021). Male Lister hooded rats exposed to single isolation housing from weaning to adulthood displayed hyperactivity in the open field and deficits in contextual fear learning (Dunphy-Doherty et al., 2018). These behavioral changes were accompanied by reduced neurogenesis in the hippocampus and reduced levels of hippocampal IL-6 in socially isolated rats, and several differentially abundant taxa were identified between social isolation and group-housed rats. An association of differentially abundant genera in the order Clostridiales with open-field behavior was observed; however, it is difficult to generalize these findings as the correlation analysis included all experimental rats and did not consider housing (Dunphy-Doherty et al., 2018). In male C57Bl/6 mice, social isolation housing from weaning to adulthood resulted in increased anxiety-like behavior and reduced social behavior in adulthood (Huang et al., 2021). An overall reduced alpha diversity in socially isolated mice was observed that was accompanied by a differential relative abundance of bacterial taxa, notably *Lactobacillus murinus and Lactobacillus intestinalis* distinguished the socially isolated group of mice (Huang et al., 2021). A key role for gut microbiota in the behavioral effects was demonstrated as the behavioral phenotype was transferred by fecal microbiota transplantation

(FMT) to antibiotic-treated recipient mice (Huang et al., 2021). Additional molecular analysis and additional proof of concept experiments using GF mice linked microbial-related production of propionic acid to changes in the brain oxytocin receptor signaling and social deficits (Huang et al., 2021). Additional support for the association of social isolation, microbiota, and behavior was provided by exposing the highly social prairie voles, *Microtus ochrogaster*, to social isolation for 6 weeks in adulthood (Donovan et al., 2020). Increased anxiety-like behavior was observed in male and female prairie voles, in parallel with minor alterations in social affiliation, and alterations in gut microbiota composition. Here, correlative associations were observed between stress-related taxa and molecular changes in brain systems (Donovan et al., 2020). Overall, social isolation stress across different animal models (rat, mice, voles) leads to microbiota changes that may contribute to behavioral changes observed.

Maternal separation (MS) is an established animal model of early life stress, where pups are separated from the dam for 3 h daily during the preweaning period and this early life stress model has been shown to increase stress-reactivity, stress-related behaviors, and result in long-term changes in inflammation as well as CNS stress systems (Daniels et al., 2004; Foster et al., 2017; O'Mahony et al., 2009; Plotsky and Meaney, 1993; Slotten et al., 2006). Several studies have also reported alterations in microbiota composition in mice or rats exposed to maternal separation (Bailey and Coe, 1999; Desbonnet et al., 2010; Garcia-Rodenas et al., 2006; Gareau et al., 2007). Recent work has considered MS, microbiota, and behavior revealing a key role for the microbiota in MS-related changes in behavior (De Palma et al., 2015; Moya-Perez et al., 2017). MS for 3 h daily from postnatal day 4 to weaning in mice increased anxiety-like behavior later in life in mice; however, this effect on behavior was not observed in GF mice, despite other stress-related changes in host physiology suggesting a role for microbiota in behavioral changes observed (De Palma et al., 2015). In a recent study, MS-related increased stress reactivity and anxiety-like behavior in male C57Bl/6 mice were accompanied by the increased relative abundance of a select group of taxa, identified at the OTU level (Moya-Perez et al., 2017). Further blast analysis of these OTUs revealed taxa classified into the Bacteroidales family *S24-7*, recently identified as *Muribaculaceae* (Lagkouvardos et al., 2019), known to be involved in the degradation of complex carbohydrates and SCFA production (Lagkouvardos et al., 2016, 2019; Smith et al., 2020; Volk et al., 2019). Concomitant administration of *Bifidobacterium CECT 7765* during early life stress and extended through adolescence reduced the impact of maternal separation including behavioral and microbiota-related changes, as well as inflammation-related changes. Notably, this study highlighted a key role for microbiota-immune signaling in stress-related changes in behavior (Moya-Perez et al., 2017). Importantly, sex differences in prevalence and symptom presentation are well established in anxiety and other stress-related disorders. Evidence linking microbiome changes and inflammatory activation that are sex-specific is emerging (Audet, 2019; Chen et al., 2018) and serves as an important consideration for future studies in this area.

Stress exposure in adulthood leads to increased stress reactivity and increased stress-reactive behaviors. Chronic restraint stress in adult mice altered the microbiota composition in parallel with increased anxiety-like and depressive-like behavior (Guo et al., 2019; Wong et al., 2016). Chronic variable stress in adult rats altered gut microbiota, changes that were associated with an altered fecal metabolite profile (Yu et al., 2017). These gut-related compositional and functional changes were associated with altered hippocampal catecholamine levels, demonstrating a link between microbiota changes and brain changes in response to stress (Yu et al., 2017). Similar findings have been reported in male C57Bl/6 mice, where probiotic treatment with *Lactobacillus helveticus, Lactobacillus plantarum, and Bifidobacterium longum* prevented chronic mild stress-induced increased anxiety-like and depressive-like behavior as well as reversed stress-induced changes in hippocampal cytokines (Li et al., 2018). Other adult stress procedures support the connection between stress, the microbiome, and behavior (Bangsgaard Bendtsen et al., 2012; Langgartner et al., 2018). Chronic psychosocial stress through chronic subordinate colony housing, used as an animal model of posttraumatic stress disorder (Reber et al., 2016), increased anxiety-like behavior in mice with parallel changes in microbiota composition and increased levels of inflammation (Langgartner et al., 2018). Notably, fecal transplantation of microbiota from nonstressed mice to stressed mice reversed the impact of stress and the transfer of microbiota from stressed mice to nonstressed mice resulted in increased inflammation in the recipient mice (Langgartner et al., 2018).

## The impact of diet on microbiota and behavior

Microbes play an important role in host metabolism and diet—microbiota interactions impact the microbiota—brain axis (Ezra-Nevo et al., 2020). Microbiota can ferment fiber that comes from the diet, not only providing essential nutrients to the host but also benefitting the microbial environment as a whole (Deehan et al., 2017; Makki et al., 2018). To note, the presence of fiber within the gut can alter the microenvironment that is optimal for microbial growth through cross-feeding of the substrates produced from the breakdown of fiber to other microbial species (Deehan et al., 2017; Makki et al., 2018). Dietary fibers are important energy sources for microbiota. Some examples of taxa that respond and thrive in the presence of fiber are Actinobacteria and Firmicutes (Deehan et al., 2017; Makki et al., 2018). Importantly, microbes themselves produce metabolites that provide benefits to both the gut and host physiology. An example of these metabolites is SCFA (Baxter et al., 2019; Chambers et al., 2018; Rieder et al., 2017). Gut bacteria found within the gastrointestinal tract can metabolize complex carbohydrates obtained through the diet, which leads to the production of SCFA. The most abundant SCFA are acetate, propionate, and butyrate, which all have important roles in metabolism and in the maintenance of the gut barrier epithelium (Baxter et al., 2019; Chambers et al., 2018; Rieder et al., 2017). In addition, a recent report demonstrated that delivery of SCFA via special diets containing acylated starches altered microbial composition. Further, the delivery of butyrylated or propionylated starches reduced anxiety-like behavior in male BALB/C mice (Kimura-Todani et al., 2020).

The production of SCFA as well as other bioactive molecules by gut microbiota is a prominent mechanism of diet—microbiota—host interactions that can impact the brain (Dalile et al., 2019; O'Riordan et al., 2022). Microbial metabolite communication via peripheral nerves, enteroendocrine cells, immune cells, or directly through blood circulation can impact brain function and behavior (Ezra-Nevo et al., 2020). As such, the nutritional state of an animal can influence behavior through diet—microbiome interactions. This is well demonstrated in experiments that showed that dietary proteins and microbiota interactions influenced food choice in *Drosophila melanogaster* (Henriques et al., 2020; Leitao-Goncalves et al., 2017).

Through their action on the microbiota composition and function, different diets can impact behavior. From an experimental design perspective, commercially supplied diets for animal studies have varied ingredients that could influence physiology and behavior. For example, diet purity has been shown to influence microbiome composition in male C57Bl/6 mice (Toyoda et al., 2018). Diet emulsifiers found in many processed foods were shown to impact both anxiety-like and social behaviors in male and female C57Bl/6 mice (Holder et al., 2019). Understanding the mechanisms of how dietary nutrients and other additives influence microbiota—brain signaling may have important translational value to human diet—microbiome interactions (Ezra-Nevo et al., 2020).

## Fecal matter transplantation alters behavior in recipient

FMT experiments have been used to demonstrate that microbiota contribute to a clinical behavioral phenotype in many cases when the transfer of fecal matter collected from healthy individuals or individuals with specific clinical conditions to recipient animals (mice or rats) results in a similar behavioral phenotype in the recipient (see recent review, recommendations, and guidelines (Gheorghe et al., 2021; Secombe et al., 2021; Settanni et al., 2021)). Transfer of fecal samples from individuals with depression to rodents resulted in an anhedonic and depressive-like phenotype demonstrating an active role for the microbiome in depression and depressive symptoms (Kelly et al., 2016; Zheng et al., 2016). More recently, the transfer of specific taxa, *Mycoacterium neoarum*, isolated from depressed patients resulted in depressive-like behavior in mice (Li et al., 2022). Notably, this was mediated by testosterone degradation demonstrating a potential mechanism that may lead to testosterone deficiencies in patients (Li et al., 2022). FMT from individuals with schizophrenia (SZ) compared to FMT from healthy controls (HC) has been shown to alter behavior and brain metabolites (Liang et al., 2019; Zheng et al., 2019). Using this approach, SZ-FMT mice showed increased activity in the open field, reduced anxiety- and depressive-like behavior, and an increased startle response compared to CON-FMT mice; however, they did not display the expected reduced prepulse inhibition that is observed in animal models of SZ and in patient populations (Zheng et al., 2019). Interestingly, the transfer of fecal microbiota from drug-free individuals with schizophrenia to mice also resulted in altered behavior including in the domains of psychomotor and learning/memory function (Zhu et al., 2020). A substantial body of preclinical and clinical evidence demonstrates a role for the microbiome in autism spectrum disorder (Vuong and Hsiao, 2017). Transfer of fecal microbiota from individuals with autism resulted in autistic-like behaviors in mice, a phenotype not observed following FMT from typical developing individuals (Sharon et al., 2019). Recently, one study used FMT to demonstrate a role for the microbiome in anorexia nervosa (Hata et al., 2019). AN-FMT compared to HC-FMT resulted in weight loss, reduced food intake and food efficiency, and behavioral changes that were associated with microbiota composition in the recipient female mice (Hata et al., 2019). Similarly, FMT from patients with alcoholism or alcohol use disorder to mice compared to FMT from healthy controls resulted in increased anxiety- and depression-like behaviors, decreased social interactions, and resulted in spontaneous alcohol preference, along with molecular changes in the brain (Leclercq et al., 2020; Zhao et al., 2020). The use of this reverse translation experimental approach (FMT) has gained momentum in the past 5 years and provides an opportunity to link the gut microbiota to clinical phenotypes and to examine signaling and molecular mechanisms involved in the microbiota—brain axis. In some cases, researchers have adopted FMT to transfer fecal microbiota between mice as described above. In addition, recent studies showed that FMT from young to aged mice increased SCFA in recipient mice and improved cognitive behavior (Lee et al., 2020) and that improvement in age-related cognitive behavior was associated with peripheral and central immune signaling (Boehme et al., 2021). In this situation, when the behavioral phenotype of the donor mice is recapitulated in the FMT recipient mouse, it provides further evidence of an important role for the microbiome in behavior.

## Future directions

The extended body of research reviewed in this chapter clearly demonstrates that microbiota—brain signaling has an important regulatory role in behavior. The research in this area is expanding quickly to include many more animal models of disease and importantly to utilize non-rodent models to explore microbiota—brain connections (Cusick et al., 2021) and to build mechanistic insights (Nagpal and Cryan, 2021). More attention to identifying the microbiota—brain mechanisms underlying behavior is at the forefront of neuroscience research (Buffington et al., 2016; Needham et al., 2022; Wu et al., 2021). Furthermore, the translation of key findings from preclinical models in clinical populations (Table 5.1) demonstrates that microbiota—brain interactions discovered in animal models can inform and guide clinical neuroscience discovery.

**Table 5.1** Translating behavioral neuroscience to clinical populations.

| Microbiota-associated behaviors | Related clinical finding | References |
|---|---|---|
| Anxiety-like behavior | • Alpha diversity and species richness associated with anxiety severity in inpatient psychiatric population | Madan et al. (2020) |
| | • Altered microbiota composition in individuals with general anxiety disorder | Chen et al. (2019), Jiang et al. (2018) |
| | • Several bacterial taxa associated with anxiety symptoms | Chen et al. (2019) |
| | • Reduced microbiota diversity and taxa differences in individuals with attention deficit hyperactivity disorder | Prehn-Kristensen et al. (2018) |
| Depressive-like behavior | • Alpha diversity and species richness associated with depression severity in inpatient psychiatric population | Madan et al. (2020) |
| | • Differentially abundant taxa between healthy controls and individuals with major depressive disorder and bipolar disorder | See Chapter 8 – Foster and Clarke (2023) |
| | • Association of specific taxa with quality of life and linked to depression | Valles-Colomer et al. (2019) |
| | • Taxa associated with depression severity | Chung et al. (2019), Liu et al. (2020), Zhang et al. (2021) |
| | • Depression-related taxa linked to tryptophan metabolism | Lai et al. (2021), Philippe et al. (2021) |
| | • *Mycobacterium neoaurum* from depressed patients degrades testosterone and leads to depressive-like phenotype in mice | Li et al. (2022) |

*Continued*

**Table 5.1** Translating behavioral neuroscience to clinical populations.—cont'd

| Microbiota-associated behaviors | Related clinical finding | References |
|---|---|---|
| Social behavior | • Fecal matter transplantation to individuals with autism spectrum disorder reduced gastrointestinal symptoms and improved core ASD behaviors | Kang et al. (2017, 2019) |
| | • Increased intestinal permeability associated with sociability in individuals with alcohol use disorder | Leclercq et al. (2020) |
| | • Microbiota diversity and composition in the first year of life associated with social behavior at 3 years old | Laue et al. (2020) |
| Cognitive behavior | • Microbiota diversity at 1 month and 1 year old was associated with nonsocial fear behavior | Carlson et al. (2018, 2021) |

# References

Arentsen, T., Raith, H., Qian, Y., Forssberg, H., Diaz Heijtz, R., 2015. Host microbiota modulates development of social preference in mice. Microb Ecol Health Dis 26, 29719.

Audet, M.C., 2019. Stress-induced disturbances along the gut microbiota-immune-brain axis and implications for mental health: does sex matter? Front Neuroendocrinol 54, 100772.

Bailey, M.T., Coe, C.L., 1999. Maternal separation disrupts the integrity of the intestinal microflora in infant rhesus monkeys. Dev Psychobiol 35, 146—155.

Bangsgaard Bendtsen, K.M., Krych, L., Sorensen, D.B., Pang, W., Nielsen, D.S., Josefsen, K., Hansen, L.H., Sorensen, S.J., Hansen, A.K., 2012. Gut microbiota composition is correlated to grid floor induced stress and behavior in the BALB/c mouse. PLoS One 7, e46231.

Bassett, B., Subramaniyam, S., Fan, Y., Varney, S., Pan, H., Carneiro, A.M.D., Chung, C.Y., 2021. Minocycline alleviates depression-like symptoms by rescuing decrease in neurogenesis in dorsal hippocampus via blocking microglia activation/phagocytosis. Brain Behav Immun 91, 519—530.

Baxter, N.T., Schmidt, A.W., Venkataraman, A., Kim, K.S., Waldron, C., Schmidt, T.M., 2019. Dynamics of human gut microbiota and short-chain fatty acids in response to dietary interventions with three fermentable fibers. mBio 10.

Bercik, P., Denou, E., Collins, J., Jackson, W., Lu, J., Jury, J., Deng, Y., Blennerhassett, P., Macri, J., McCoy, K.D., Verdu, E.F., Collins, S.M., 2011. The intestinal microbiota affect central levels of brain-derived neurotropic factor and behavior in mice. Gastroenterology 141, 599—609.

Berton, O., McClung, C.A., Dileone, R.J., Krishnan, V., Renthal, W., Russo, S.J., Graham, D., Tsankova, N.M., Bolanos, C.A., Rios, M., Monteggia, L.M., Self, D.W., Nestler, E.J., 2006. Essential role of BDNF in the mesolimbic dopamine pathway in social defeat stress. Science 311, 864—868.

Bharwani, A., Mian, M.F., Foster, J.A., Surette, M.G., Bienenstock, J., Forsythe, P., 2016. Structural & functional consequences of chronic psychosocial stress on the microbiome & host. Psychoneuroendocrinology 63, 217—227.

Boehme, M., Guzzetta, K.E., Bastiaanssen, T.F.S., van de Wouw, M., Moloney, G.M., Gual-Grau, A., Spichak, S., Olavarría-Ramírez, L., Fitzgerald, P., Morillas, E., Ritz, N.L., Jaggar, M., Cowan, C.S.M., Crispie, F., Donoso, F., Halitzki, E., Neto, M.C., Sichetti, M., Golubeva, A.V., Fitzgerald, R.S., Claesson, M.J., Cotter, P.D., O'Leary, O.F., Dinan, T.G., Cryan, J.F., 2021. Microbiota from young mice counteracts selective age-associated behavioral deficits. Nature Aging 1, 666—676.

Bourin, M., Hascoet, M., 2003. The mouse light/dark box test. Eur J Pharmacol 463, 55—65.

Buffington, S.A., Di Prisco, G.V., Auchtung, T.A., Ajami, N.J., Petrosino, J.F., Costa-Mattioli, M., 2016. Microbial reconstitution reverses maternal diet-induced social and synaptic deficits in offspring. Cell 165, 1762—1775.

Can, A., Dao, D.T., Arad, M., Terrillion, C.E., Piantadosi, S.C., Gould, T.D., 2012. The mouse forced swim test. J Vis Exp, e3638.

Carlson, A.L., Xia, K., Azcarate-Peril, M.A., Goldman, B.D., Ahn, M., Styner, M.A., Thompson, A.L., Geng, X., Gilmore, J.H., Knickmeyer, R.C., 2018. Infant gut microbiome associated with cognitive development. Biol Psychiatry 83, 148—159.

Carlson, A.L., Xia, K., Azcarate-Peril, M.A., Rosin, S.P., Fine, J.P., Mu, W., Zopp, J.B., Kimmel, M.C., Styner, M.A., Thompson, A.L., Propper, C.B., Knickmeyer, R.C., 2021. Infant gut microbiome composition is associated with non-social fear behavior in a pilot study. Nat Commun 12, 3294.

Catron, T.R., Swank, A., Wehmas, L.C., Phelps, D., Keely, S.P., Brinkman, N.E., McCord, J., Singh, R., Sobus, J., Wood, C.E., Strynar, M., Wheaton, E., Tal, T., 2019. Microbiota alter metabolism and mediate neurodevelopmental toxicity of 17beta-estradiol. Sci Rep 9, 7064.

Ceylani, T., Jakubowska-Dogru, E., Gurbanov, R., Teker, H.T., Gozen, A.G., 2018. The effects of repeated antibiotic administration to juvenile BALB/c mice on the microbiota status and animal behavior at the adult age. Heliyon 4, e00644.

Chambers, E.S., Preston, T., Frost, G., Morrison, D.J., 2018. Role of gut microbiota-generated short-chain fatty acids in metabolic and cardiovascular health. Curr Nutr Rep 7, 198—206.

Champagne-Jorgensen, K., Mian, M.F., Kay, S., Hanani, H., Ziv, O., McVey Neufeld, K.A., Koren, O., Bienenstock, J., 2020. Prenatal low-dose penicillin results in long-term sex-specific changes to murine behaviour, immune regulation, and gut microbiota. Brain Behav Immun 84, 154—163.

Chen, J.J., Zheng, P., Liu, Y.Y., Zhong, X.G., Wang, H.Y., Guo, Y.J., Xie, P., 2018. Sex differences in gut microbiota in patients with major depressive disorder. Neuropsychiatr Dis Treat 14, 647—655.

Chen, Y.H., Bai, J., Wu, D., Yu, S.F., Qiang, X.L., Bai, H., Wang, H.N., Peng, Z.W., 2019. Association between fecal microbiota and generalized anxiety disorder: severity and early treatment response. J Affect Disord 259, 56—66.

Chu, C., Murdock, M.H., Jing, D., Won, T.H., Chung, H., Kressel, A.M., Tsaava, T., Addorisio, M.E., Putzel, G.G., Zhou, L., Bessman, N.J., Yang, R., Moriyama, S., Parkhurst, C.N., Li, A., Meyer, H.C., Teng, F., Chavan, S.S., Tracey, K.J., Regev, A., Schroeder, F.C., Lee, F.S., Liston, C., Artis, D., 2019. The microbiota regulate neuronal function and fear extinction learning. Nature 574, 543—548.

Chung, Y.E., Chen, H.C., Chou, H.L., Chen, I.M., Lee, M.S., Chuang, L.C., Liu, Y.W., Lu, M.L., Chen, C.H., Wu, C.S., Huang, M.C., Liao, S.C., Ni, Y.H., Lai, M.S., Shih, W.L., Kuo, P.H., 2019. Exploration of microbiota targets for major depressive disorder and mood related traits. J Psychiatr Res 111, 74–82.

Clarke, G., Grenham, S., Scully, P., Fitzgerald, P., Moloney, R.D., Shanahan, F., Dinan, T.G., Cryan, J.F., 2013. The microbiome-gut-brain axis during early life regulates the hippocampal serotonergic system in a sex-dependent manner. Mol Psychiatr 18, 666–673.

Crumeyrolle-Arias, M., Jaglin, M., Bruneau, A., Vancassel, S., Cardon, A., Dauge, V., Naudon, L., Rabot, S., 2014. Absence of the gut microbiota enhances anxiety-like behavior and neuroendocrine response to acute stres in rats. Psychoneuroendocrinology 42, 207–271. https://doi.org/10.1016/j.psyneuen.2014.01.014.

Cryan, J.F., Holmes, A., 2005. The ascent of mouse: advances in modelling human depression and anxiety. Nat Rev 4, 775–790.

Cryan, J.F., Mombereau, C., Vassout, A., 2005. The tail suspension test as a model for assessing antidepressant activity: review of pharmacological and genetic studies in mice. Neurosci Biobehav Rev 29, 571–625.

Cryan, J.F., O'Riordan, K.J., Cowan, C.S.M., Sandhu, K.V., Bastiaanssen, T.F.S., Boehme, M., Codagnone, M.G., Cussotto, S., Fulling, C., Golubeva, A.V., Guzzetta, K.E., Jaggar, M., Long-Smith, C.M., Lyte, J.M., Martin, J.A., Molinero-Perez, A., Moloney, G., Morelli, E., Morillas, E., O'Connor, R., Cruz-Pereira, J.S., Peterson, V.L., Rea, K., Ritz, N.L., Sherwin, E., Spichak, S., Teichman, E.M., van de Wouw, M., Ventura-Silva, A.P., Wallace-Fitzsimons, S.E., Hyland, N., Clarke, G., Dinan, T.G., 2019. The microbiota-gut-brain Axis. Physiol Rev 99, 1877–2013.

Cusick, J.A., Wellman, C.L., Demas, G.E., 2021. The call of the wild: using non-model systems to investigate microbiome-behaviour relationships. J Exp Biol 224.

Cussotto, S., Sandhu, K.V., Dinan, T.G., Cryan, J.F., 2018. The neuroendocrinology of the microbiota-gut-brain axis: a behavioural perspective. Front Neuroendocrinol 51, 80–101.

Dalile, B., Van Oudenhove, L., Vervliet, B., Verbeke, K., 2019. The role of short-chain fatty acids in microbiota-gut-brain communication. Nat Rev Gastroenterol Hepatol 16, 461–478.

Daniels, W.M., Pietersen, C.Y., Carstens, M.E., Stein, D.J., 2004. Maternal separation in rats leads to anxiety-like behavior and a blunted ACTH response and altered neurotransmitter levels in response to a subsequent stressor. Metab Brain Dis 19, 3–14.

Davis, D.J., Bryda, E.C., Gillespie, C.H., Ericsson, A.C., 2016. Microbial modulation of behavior and stress responses in zebrafish larvae. Behav Brain Res 311, 219–227.

De Palma, G., Blennerhassett, P., Lu, J., Deng, Y., Park, A.J., Green, W., Denou, E., Silva, M.A., Santacruz, A., Sanz, Y., Surette, M.G., Verdu, E.F., Collins, S.M., Bercik, P., 2015. Microbiota and host determinants of behavioural phenotype in maternally separated mice. Nat Commun 6, 7735.

Deehan, E.C., Duar, R.M., Armet, A.M., Perez-Munoz, M.E., Jin, M., Walter, J., 2017. Modulation of the gastrointestinal microbiome with nondigestible fermentable carbohydrates to improve human health. Microbiol Spectr 5.

Degroote, S., Hunting, D.J., Baccarelli, A.A., Takser, L., 2016. Maternal gut and fetal brain connection: increased anxiety and reduced social interactions in Wistar rat offspring following peri-conceptional antibiotic exposure. Prog Neuro-Psychopharmacol Biol Psychiatry 71, 76–82.

Desbonnet, L., Clarke, G., Shanahan, F., Dinan, T.G., Cryan, J.F., 2014. Microbiota is essential for social development in the mouse. Mol Psychiatr 19, 146–148.

Desbonnet, L., Clarke, G., Traplin, A., O'Sullivan, O., Crispie, F., Moloney, R.D., Cotter, P.D., Dinan, T.G., Cryan, J.F., 2015. Gut microbiota depletion from early adolescence in mice: implications for brain and behaviour. Brain Behav Immun 48, 165–173.

Desbonnet, L., Garrett, L., Clarke, G., Kiely, B., Cryan, J.F., Dinan, T.G., 2010. Effects of the probiotic Bifidobacterium infantis in the maternal separation model of depression. Neuroscience 170, 1179—1188.

Diaz Heijtz, R., Wang, S., Anuar, F., Qian, Y., Bjorkholm, B., Samuelsson, A., Hibberd, M.L., Forssberg, H., Pettersson, S., 2011. Normal gut microbiota modulates brain development and behavior. Proc Natl Acad Sci U S A 108, 3047—3052.

Donovan, M., Mackey, C.S., Platt, G.N., Rounds, J., Brown, A.N., Trickey, D.J., Liu, Y., Jones, K.M., Wang, Z., 2020. Social isolation alters behavior, the gut-immune-brain axis, and neurochemical circuits in male and female prairie voles. Neurobiol Stress 13, 100278.

Ducottet, C., Belzung, C., 2005. Correlations between behaviours in the elevated plus-maze and sensitivity to unpredictable subchronic mild stress: evidence from inbred strains of mice. Behav Brain Res 156, 153—162.

Dunphy-Doherty, F., O'Mahony, S.M., Peterson, V.L., O'Sullivan, O., Crispie, F., Cotter, P.D., Wigmore, P., King, M.V., Cryan, J.F., Fone, K.C.F., 2018. Post-weaning social isolation of rats leads to long-term disruption of the gut microbiota-immune-brain axis. Brain Behav Immun 68, 261—273.

Eagle, A., Mazei-Robison, M., Robison, A., 2016. Sucrose preference test to measure stress-induced anhedonia. Bio-Protocol 6.

Ezra-Nevo, G., Henriques, S.F., Ribeiro, C., 2020. The diet-microbiome tango: how nutrients lead the gut brain axis. Curr Opin Neurobiol 62, 122—132.

Fone, K.C., Porkess, M.V., 2008. Behavioural and neurochemical effects of post-weaning social isolation in rodents-relevance to developmental neuropsychiatric disorders. Neurosci Biobehav Rev 32, 1087—1102.

Foster, J.A., Clarke, G., 2023. Microbiota Brain Axis. Elsevier.

Foster, J.A., Rinaman, L., Cryan, J.F., 2017. Stress & the gut-brain axis: regulation by the microbiome. Neurobiol Stress 7, 124—136.

Frohlich, E.E., Farzi, A., Mayerhofer, R., Reichmann, F., Jacan, A., Wagner, B., Zinser, E., Bordag, N., Magnes, C., Frohlich, E., Kashofer, K., Gorkiewicz, G., Holzer, P., 2016. Cognitive impairment by antibiotic-induced gut dysbiosis: analysis of gut microbiota-brain communication. Brain Behav Immun 56, 140—155.

Garcia-Rodenas, C.L., Bergonzelli, G.E., Nutten, S., Schumann, A., Cherbut, C., Turini, M., Ornstein, K., Rochat, F., Corthesy-Theulaz, I., 2006. Nutritional approach to restore impaired intestinal barrier function and growth after neonatal stress in rats. J Pediatr Gastroenterol Nutr 43, 16—24.

Gareau, M.G., Jury, J., MacQueen, G., Sherman, P.M., Perdue, M.H., 2007. Probiotic treatment of rat pups normalises corticosterone release and ameliorates colonic dysfunction induced by maternal separation. Gut 56, 1522—1528.

Gareau, M.G., Wine, E., Rodrigues, D.M., Cho, J.H., Whary, M.T., Philpott, D.J., Macqueen, G., Sherman, P.M., 2011. Bacterial infection causes stress-induced memory dysfunction in mice. Gut 60, 307—317.

Gheorghe, C.E., Ritz, N.L., Martin, J.A., Wardill, H.R., Cryan, J.F., Clarke, G., 2021. Investigating causality with fecal microbiota transplantation in rodents: applications, recommendations and pitfalls. Gut Microb 13, 1941711.

Glover, M.E., Cohen, J.L., Singer, J.R., Sabbagh, M.N., Rainville, J.R., Hyland, M.T., Morrow, C.D., Weaver, C.T., Hodes, G.E., Kerman, I.A., Clinton, S.M., 2021. Examining the role of microbiota in emotional behavior: antibiotic treatment exacerbates anxiety in high anxiety-prone male rats. Neuroscience 459, 179—197.

Guida, F., Turco, F., Iannotta, M., De Gregorio, D., Palumbo, I., Sarnelli, G., Furiano, A., Napolitano, F., Boccella, S., Luongo, L., Mazzitelli, M., Usiello, A., De Filippis, F., Iannotti, F.A., Piscitelli, F., Ercolini, D., de Novellis, V., Di Marzo, V., Cuomo, R., Maione, S., 2018. Antibiotic-induced microbiota perturbation causes gut endocannabinoidome changes, hippocampal neuroglial reorganization and depression in mice. Brain Behav Immun 67, 230—245.

Guo, Y., Xie, J.P., Deng, K., Li, X., Yuan, Y., Xuan, Q., Xie, J., He, X.M., Wang, Q., Li, J.J., Luo, H.R., 2019. Prophylactic effects of Bifidobacterium adolescentis on anxiety and depression-like phenotypes after chronic stress: a role of the gut microbiota-inflammation axis. Front Behav Neurosci 13, 126.

Gustafsson, B., Kahlson, G., Rosengren, E., 1957. Biogenesis of histamine studied by its distribution and urinary excretion in germ free reared and not germ free rats fed a histamine free diet. Acta Physiol Scand 41, 217−228.

Gustafsson, B.E., 1959. Lightweight stainless steel systems for rearing germfree animals. Ann N Y Acad Sci 78, 17−28.

Hata, T., Miyata, N., Takakura, S., Yoshihara, K., Asano, Y., Kimura-Todani, T., Yamashita, M., Zhang, X.T., Watanabe, N., Mikami, K., Koga, Y., Sudo, N., 2019. The gut microbiome derived from anorexia nervosa patients impairs weight gain and behavioral performance in female mice. Endocrinology 160, 2441−2452.

Henriques, S.F., Dhakan, D.B., Serra, L., Francisco, A.P., Carvalho-Santos, Z., Baltazar, C., Elias, A.P., Anjos, M., Zhang, T., Maddocks, O.D.K., Ribeiro, C., 2020. Metabolic cross-feeding in imbalanced diets allows gut microbes to improve reproduction and alter host behaviour. Nat Commun 11, 4236.

Hoban, A.E., Stilling, R.M., Moloney, G., Shanahan, F., Dinan, T.G., Clarke, G., Cryan, J.F., 2018. The microbiome regulates amygdala-dependent fear recall. Mol Psychiatr 23, 1134−1144.

Holder, M.K., Peters, N.V., Whylings, J., Fields, C.T., Gewirtz, A.T., Chassaing, B., de Vries, G.J., 2019. Dietary emulsifiers consumption alters anxiety-like and social-related behaviors in mice in a sex-dependent manner. Sci Rep 9, 172.

Huang, L., Duan, C., Xia, X., Wang, H., Wang, Y., Zhong, Z., Wang, B., Ding, W., Yang, Y., 2021. Commensal microbe-derived propionic acid mediates juvenile social isolation-induced social deficits and anxiety-like behaviors. Brain Res Bull 166, 161−171.

Huo, R., Zeng, B., Zeng, L., Cheng, K., Li, B., Luo, Y., Wang, H., Zhou, C., Fang, L., Li, W., Niu, R., Wei, H., Xie, P., 2017. Microbiota modulate anxiety-like behavior and endocrine abnormalities in hypothalamic-pituitary-adrenal axis. Front Cell Infect Microbiol 7, 489.

Jiang, H., Ling, Z., Zhang, Y., Mao, H., Ma, Z., Yin, Y., Wang, W., Tang, W., Tan, Z., Shi, J., Li, L., Ruan, B., 2015. Altered fecal microbiota composition in patients with major depressive disorder. Brain Behav Immun 48, 186−194.

Jiang, H.Y., Zhang, X., Yu, Z.H., Zhang, Z., Deng, M., Zhao, J.H., Ruan, B., 2018. Altered gut microbiota profile in patients with generalized anxiety disorder. J Psychiatr Res 104, 130−136.

Kang, D.W., Adams, J.B., Coleman, D.M., Pollard, E.L., Maldonado, J., McDonough-Means, S., Caporaso, J.G., Krajmalnik-Brown, R., 2019. Long-term benefit of Microbiota Transfer Therapy on autism symptoms and gut microbiota. Sci Rep 9, 5821.

Kang, D.W., Adams, J.B., Gregory, A.C., Borody, T., Chittick, L., Fasano, A., Khoruts, A., Geis, E., Maldonado, J., McDonough-Means, S., Pollard, E.L., Roux, S., Sadowsky, M.J., Lipson, K.S., Sullivan, M.B., Caporaso, J.G., Krajmalnik-Brown, R., 2017. Microbiota Transfer Therapy alters gut ecosystem and improves gastrointestinal and autism symptoms: an open-label study. Microbiome 5, 10.

Kelly, J.R., Borre, Y., Ciaran, O.B., Patterson, E., El Aidy, S., Deane, J., Kennedy, P.J., Beers, S., Scott, K., Moloney, G., Hoban, A.E., Scott, L., Fitzgerald, P., Ross, P., Stanton, C., Clarke, G., Cryan, J.F., Dinan, T.G., 2016. Transferring the blues: depression-associated gut microbiota induces neurobehavioural changes in the rat. J Psychiatr Res 82, 109−118.

Kimura-Todani, T., Hata, T., Miyata, N., Takakura, S., Yoshihara, K., Zhang, X.T., Asano, Y., Altaisaikhan, A., Tsukahara, T., Sudo, N., 2020. Dietary delivery of acetate to the colon using acylated starches as a carrier exerts anxiolytic effects in mice. Physiol Behav 223, 113004.

Lagkouvardos, I., Lesker, T.R., Hitch, T.C.A., Galvez, E.J.C., Smit, N., Neuhaus, K., Wang, J., Baines, J.F., Abt, B., Stecher, B., Overmann, J., Strowig, T., Clavel, T., 2019. Sequence and cultivation study of Muribaculaceae reveals novel species, host preference, and functional potential of this yet undescribed family. Microbiome 7, 28.

Lagkouvardos, I., Pukall, R., Abt, B., Foesel, B.U., Meier-Kolthoff, J.P., Kumar, N., Bresciani, A., Martinez, I., Just, S., Ziegler, C., Brugiroux, S., Garzetti, D., Wenning, M., Bui, T.P., Wang, J., Hugenholtz, F., Plugge, C.M., Peterson, D.A., Hornef, M.W., Baines, J.F., Smidt, H., Walter, J., Kristiansen, K., Nielsen, H.B., Haller, D., Overmann, J., Stecher, B., Clavel, T., 2016. The Mouse Intestinal Bacterial Collection (miBC) provides host-specific insight into cultured diversity and functional potential of the gut microbiota. Nat Microbiol 1, 16131.

Lai, W.T., Zhao, J., Xu, S.X., Deng, W.F., Xu, D., Wang, M.B., He, F.S., Liu, Y.H., Guo, Y.Y., Ye, S.W., Yang, Q.F., Zhang, Y.L., Wang, S., Li, M.Z., Yang, Y.J., Liu, T.B., Tan, Z.M., Xie, X.H., Rong, H., 2021. Shotgun metagenomics reveals both taxonomic and tryptophan pathway differences of gut microbiota in bipolar disorder with current major depressive episode patients. J Affect Disord 278, 311−319.

Langgartner, D., Vaihinger, C.A., Haffner-Luntzer, M., Kunze, J.F., Weiss, A.J., Foertsch, S., Bergdolt, S., Ignatius, A., Reber, S.O., 2018. The role of the intestinal microbiome in chronic psychosocial stress-induced pathologies in male mice. Front Behav Neurosci 12, 252.

Laue, H.E., Korrick, S.A., Baker, E.R., Karagas, M.R., Madan, J.C., 2020. Prospective associations of the infant gut microbiome and microbial function with social behaviors related to autism at age 3 years. Sci Rep 10, 15515.

Leclercq, S., Le Roy, T., Furgiuele, S., Coste, V., Bindels, L.B., Leyrolle, Q., Neyrinck, A.M., Quoilin, C., Amadieu, C., Petit, G., Dricot, L., Tagliatti, V., Cani, P.D., Verbeke, K., Colet, J.M., Starkel, P., de Timary, P., Delzenne, N.M., 2020. Gut microbiota-induced changes in beta-hydroxybutyrate metabolism are linked to altered sociability and depression in alcohol use disorder. Cell Rep 33, 108238.

Leclercq, S., Mian, F.M., Stanisz, A.M., Bindels, L.B., Cambier, E., Ben-Amram, H., Koren, O., Forsythe, P., Bienenstock, J., 2017. Low-dose penicillin in early life induces long-term changes in murine gut microbiota, brain cytokines and behavior. Nat Commun 8, 15062.

Lee, J., Venna, V.R., Durgan, D.J., Shi, H., Hudobenko, J., Putluri, N., Petrosino, J., McCullough, L.D., Bryan, R.M., 2020. Young versus aged microbiota transplants to germ-free mice: increased short-chain fatty acids and improved cognitive performance. Gut Microb 12, 1−14.

Leitao-Goncalves, R., Carvalho-Santos, Z., Francisco, A.P., Fioreze, G.T., Anjos, M., Baltazar, C., Elias, A.P., Itskov, P.M., Piper, M.D.W., Ribeiro, C., 2017. Commensal bacteria and essential amino acids control food choice behavior and reproduction. PLoS Biol 15, e2000862.

Li, D., Liu, R., Wang, M., Peng, R., Fu, S., Fu, A., Le, J., Yao, Q., Yuan, T., Chi, H., Mu, X., Sun, T., Liu, H., Yan, P., Wang, S., Cheng, S., Deng, Z., Liu, Z., Wang, G., Li, Y., Liu, T., 2022. 3beta-Hydroxysteroid dehydrogenase expressed by gut microbes degrades testosterone and is linked to depression in males. Cell Host Microbe 30, 329−339.

Li, N., Wang, Q., Wang, Y., Sun, A., Lin, Y., Jin, Y., Li, X., 2018. Oral probiotics ameliorate the behavioral deficits induced by chronic mild stress in mice via the gut microbiota-inflammation axis. Front Behav Neurosci 12, 266.

Liang, W., Huang, Y., Tan, X., Wu, J., Duan, J., Zhang, H., Yin, B., Li, Y., Zheng, P., Wei, H., Xie, P., 2019. Alterations of glycerophospholipid and fatty acyl metabolism in multiple brain regions of schizophrenia microbiota recipient mice. Neuropsychiatr Dis Treat 15, 3219−3229.

Lin, P., Ding, B., Feng, C., Yin, S., Zhang, T., Qi, X., Lv, H., Guo, X., Dong, K., Zhu, Y., Li, Q., 2017. Prevotella and Klebsiella proportions in fecal microbial communities are potential characteristic parameters for patients with major depressive disorder. J Affect Disord 207, 300–304.

Liu, R.T., Rowan-Nash, A.D., Sheehan, A.E., Walsh, R.F.L., Sanzari, C.M., Korry, B.J., Belenky, P., 2020. Reductions in anti-inflammatory gut bacteria are associated with depression in a sample of young adults. Brain Behav Immun 88, 308–324.

Lu, J., Synowiec, S., Lu, L., Yu, Y., Bretherick, T., Takada, S., Yarnykh, V., Caplan, J., Caplan, M., Claud, E.C., Drobyshevsky, A., 2018. Microbiota influence the development of the brain and behaviors in C57BL/6J mice. PLoS One 13, e0201829.

Luczynski, P., McVey Neufeld, K.A., Oriach, C.S., Clarke, G., Dinan, T.G., Cryan, J.F., 2016. Growing up in a bubble: using germ-free animals to assess the influence of the gut microbiota on brain and behavior. Int J Neuropsychopharmacol 19.

Luk, B., Veeraragavan, S., Engevik, M., Balderas, M., Major, A., Runge, J., Luna, R.A., Versalovic, J., 2018. Postnatal colonization with human "infant-type" Bifidobacterium species alters behavior of adult gnotobiotic mice. PLoS One 13, e0196510.

Luo, Y., Zeng, B., Zeng, L., Du, X., Li, B., Huo, R., Liu, L., Wang, H., Dong, M., Pan, J., Zheng, P., Zhou, C., Wei, H., Xie, P., 2018. Gut microbiota regulates mouse behaviors through glucocorticoid receptor pathway genes in the hippocampus. Transl Psychiatry 8, 187.

Macpherson, A.J., Harris, N.L., 2004. Interactions between commensal intestinal bacteria and the immune system. Nat Rev Immunol 4, 478–485.

Madan, A., Thompson, D., Fowler, J.C., Ajami, N.J., Salas, R., Frueh, B.C., Bradshaw, M.R., Weinstein, B.L., Oldham, J.M., Petrosino, J.F., 2020. The gut microbiota is associated with psychiatric symptom severity and treatment outcome among individuals with serious mental illness. J Affect Disord 264, 98–106.

Makki, K., Deehan, E.C., Walter, J., Backhed, F., 2018. The impact of dietary fiber on gut microbiota in host health and disease. Cell Host Microbe 23, 705–715.

Mao, J.H., Kim, Y.M., Zhou, Y.X., Hu, D., Zhong, C., Chang, H., Brislawn, C.J., Fansler, S., Langley, S., Wang, Y., Peisl, B.Y.L., Celniker, S.E., Threadgill, D.W., Wilmes, P., Orr, G., Metz, T.O., Jansson, J.K., Snijders, A.M., 2020. Genetic and metabolic links between the murine microbiome and memory. Microbiome 8, 53.

McGaughey, K.D., Yilmaz-Swenson, T., Elsayed, N.M., Cruz, D.A., Rodriguiz, R.M., Kritzer, M.D., Peterchev, A.V., Roach, J., Wetsel, W.C., Williamson, D.E., 2019. Relative abundance of Akkermansia spp. and other bacterial phylotypes correlates with anxiety- and depressive-like behavior following social defeat in mice. Sci Rep 9, 3281.

Moy, S.S., Nadler, J.J., Perez, A., Barbaro, R.P., Johns, J.M., Magnuson, T.R., Piven, J., Crawley, J.N., 2004. Sociability and preference for social novelty in five inbred strains: an approach to assess autistic-like behavior in mice. Genes Brain Behav 3, 287–302.

Moy, S.S., Nadler, J.J., Young, N.B., Perez, A., Holloway, L.P., Barbaro, R.P., Barbaro, J.R., Wilson, L.M., Threadgill, D.W., Lauder, J.M., Magnuson, T.R., Crawley, J.N., 2007. Mouse behavioral tasks relevant to autism: phenotypes of 10 inbred strains. Behav Brain Res 176, 4–20.

Moya-Perez, A., Perez-Villalba, A., Benitez-Paez, A., Campillo, I., Sanz, Y., 2017. Bifidobacterium CECT 7765 modulates early stress-induced immune, neuroendocrine and behavioral alterations in mice. Brain Behav Immun 65, 43–56.

Nagpal, J., Cryan, J.F., 2021. Microbiota-brain interactions: moving toward mechanisms in model organisms. Neuron 109, 3930–3953.

Needham, B.D., Funabashi, M., Adame, M.D., Wang, Z., Boktor, J.C., Haney, J., Wu, W.L., Rabut, C., Ladinsky, M.S., Hwang, S.J., Guo, Y., Zhu, Q., Griffiths, J.A., Knight, R., Bjorkman, P.J., Shapiro, M.G., Geschwind, D.H., Holschneider, D.P., Fischbach, M.A., Mazmanian, S.K., 2022. A gut-derived metabolite alters brain activity and anxiety behaviour in mice. Nature 602, 647–653.

Neufeld, K.A., Kang, N., Bienenstock, J., Foster, J.A., 2011a. Effects of intestinal microbiota on anxiety-like behavior. Commun Integr Biol 4, 492–494.

Neufeld, K.M., Kang, N., Bienenstock, J., Foster, J.A., 2011b. Reduced anxiety-like behavior and central neurochemical change in germ-free mice. Neuro Gastroenterol Motil 23, 255–264.

O'Mahony, S.M., Marchesi, J.R., Scully, P., Codling, C., Ceolho, A.M., Quigley, E.M., Cryan, J.F., Dinan, T.G., 2009. Early life stress alters behavior, immunity, and microbiota in rats: implications for irritable bowel syndrome and psychiatric illnesses. Biol Psychiatry 65, 263–267.

O'Riordan, K.J., Collins, M.K., Moloney, G.M., Knox, E.G., Aburto, M.R., Fulling, C., Morley, S.J., Clarke, G., Schellekens, H., Cryan, J.F., 2022. Short chain fatty acids: microbial metabolites for gut-brain axis signalling. Mol Cell Endocrinol 546, 111572.

Pan, J.X., Deng, F.L., Zeng, B.H., Zheng, P., Liang, W.W., Yin, B.M., Wu, J., Dong, M.X., Luo, Y.Y., Wang, H.Y., Wei, H., Xie, P., 2019. Absence of gut microbiota during early life affects anxiolytic Behaviors and monoamine neurotransmitters system in the hippocampal of mice. J Neurol Sci 400, 160–168.

Phelps, D., Brinkman, N.E., Keely, S.P., Anneken, E.M., Catron, T.R., Betancourt, D., Wood, C.E., Espenschied, S.T., Rawls, J.F., Tal, T., 2017. Microbial colonization is required for normal neurobehavioral development in zebrafish. Sci Rep 7, 11244.

Philippe, C., Szabo de Edelenyi, F., Naudon, L., Druesne-Pecollo, N., Hercberg, S., Kesse-Guyot, E., Latino-Martel, P., Galan, P., Rabot, S., 2021. Relation between mood and the host-microbiome Co-metabolite 3-indoxylsulfate: results from the observational prospective NutriNet-sante study. Microorganisms 9.

Plotsky, P.M., Meaney, M.J., 1993. Early, postnatal experience alters hypothalamic corticotropin-releasing factor (CRF) mRNA, median eminence CRF content and stress-induced release in adult rats. Brain Res Mol Brain Res 18, 195–200.

Prehn-Kristensen, A., Zimmermann, A., Tittmann, L., Lieb, W., Schreiber, S., Baving, L., Fischer, A., 2018. Reduced microbiome alpha diversity in young patients with ADHD. PLoS One 13, e0200728.

Prut, L., Belzung, C., 2003. The open field as a paradigm to measure the effects of drugs on anxiety-like behaviors: a review. Eur J Pharmacol 463, 3–33.

Qi, X., Wang, G., Zhong, X., Zeng, B., Chen, J., Zeng, L., Bai, S., Xu, S., Wang, W., Cheng, K., Chen, Z., Wei, H., Xie, P., 2019. Sema3A - mediated modulation of NR1D1 expression may be involved in the regulation of axonal guidance signaling by the microbiota. Life Sci 223, 54–61.

Reber, S.O., Siebler, P.H., Donner, N.C., Morton, J.T., Smith, D.G., Kopelman, J.M., Lowe, K.R., Wheeler, K.J., Fox, J.H., Hassell Jr., J.E., Greenwood, B.N., Jansch, C., Lechner, A., Schmidt, D., Uschold-Schmidt, N., Fuchsl, A.M., Langgartner, D., Walker, F.R., Hale, M.W., Lopez Perez, G., Van Treuren, W., Gonzalez, A., Halweg-Edwards, A.L., Fleshner, M., Raison, C.L., Rook, G.A., Peddada, S.D., Knight, R., Lowry, C.A., 2016. Immunization with a heat-killed preparation of the environmental bacterium Mycobacterium vaccae promotes stress resilience in mice. Proc Natl Acad Sci U S A 113, E3130–E3139.

Rieder, R., Wisniewski, P.J., Alderman, B.L., Campbell, S.C., 2017. Microbes and mental health: a review. Brain Behav Immun 66, 9–17.

Sahakian, B.J., Robbins, T.W., 1977. Isolation-rearing enhances tail pinch-induced oral behavior in rats. Physiol Behav 18, 53–58.

Schmidtner, A.K., Slattery, D.A., Glasner, J., Hiergeist, A., Gryksa, K., Malik, V.A., Hellmann-Regen, J., Heuser, I., Baghai, T.C., Gessner, A., Rupprecht, R., Di Benedetto, B., Neumann, I.D., 2019. Minocycline alters behavior, microglia and the gut microbiome in a trait-anxiety-dependent manner. Transl Psychiatry 9, 223.

Secombe, K.R., Al-Qadami, G.H., Subramaniam, C.B., Bowen, J.M., Scott, J., Van Sebille, Y.Z.A., Snelson, M., Cowan, C., Clarke, G., Gheorghe, C.E., Cryan, J.F., Wardill, H.R., 2021. Guidelines for reporting on animal fecal transplantation (GRAFT) studies: recommendations from a systematic review of murine transplantation protocols. Gut Microb 13, 1979878.

Settanni, C.R., Ianiro, G., Bibbo, S., Cammarota, G., Gasbarrini, A., 2021. Gut microbiota alteration and modulation in psychiatric disorders: current evidence on fecal microbiota transplantation. Prog Neuro-Psychopharmacol Biol Psychiatry 109, 110258.

Sharon, G., Cruz, N.J., Kang, D.W., Gandal, M.J., Wang, B., Kim, Y.M., Zink, E.M., Casey, C.P., Taylor, B.C., Lane, C.J., Bramer, L.M., Isern, N.G., Hoyt, D.W., Noecker, C., Sweredoski, M.J., Moradian, A., Borenstein, E., Jansson, J.K., Knight, R., Metz, T.O., Lois, C., Geschwind, D.H., Krajmalnik-Brown, R., Mazmanian, S.K., 2019. Human gut microbiota from autism spectrum disorder promote behavioral symptoms in mice. Cell 177, 1600—1618.

Sherwin, E., Bordenstein, S.R., Quinn, J.L., Dinan, T.G., Cryan, J.F., 2019. Microbiota and the social brain. Science 366.

Sidor, M.M., Rilett, K., Foster, J.A., 2010. Validation of an automated system for measuring anxiety-related behaviours in the elevated plus maze. J Neurosci Methods 188, 7—13.

Slattery, D.A., Cryan, J.F., 2012. Using the rat forced swim test to assess antidepressant-like activity in rodents. Nat Protoc 7, 1009—1014.

Slotten, H.A., Kalinichev, M., Hagan, J.J., Marsden, C.A., Fone, K.C., 2006. Long-lasting changes in behavioural and neuroendocrine indices in the rat following neonatal maternal separation: gender-dependent effects. Brain Res 1097, 123—132.

Smith, B.J., Miller, R.A., Schmidt, T.M., 2020. Muribaculaceae genomes assembled from metagenomes suggest genetic drivers of differential response to acarbose treatment in mice. bioRxiv. https://doi.org/10.1101/2020.07.01.183202.

Stead, J.D., Clinton, S., Neal, C., Schneider, J., Jama, A., Miller, S., Vazquez, D.M., Watson, S.J., Akil, H., 2006. Selective breeding for divergence in novelty-seeking traits: heritability and enrichment in spontaneous anxiety-related behaviors. Behav Genet 36, 697—712.

Stilling, R.M., Moloney, G.M., Ryan, F.J., Hoban, A.E., Bastiaanssen, T.F., Shanahan, F., Clarke, G., Claesson, M.J., Dinan, T.G., Cryan, J.F., 2018. Social interaction-induced activation of RNA splicing in the amygdala of microbiome-deficient mice. Elife 7.

Sudo, N., Chida, Y., Aiba, Y., Sonoda, J., Oyama, N., Yu, X.N., Kubo, C., Koga, Y., 2004. Postnatal microbial colonization programs the hypothalamic-pituitary-adrenal system for stress response in mice. J Physiol 558, 263—275.

Toyoda, A., Shimonishi, H., Sato, M., Usuda, K., Ohsawa, N., Nagaoka, K., 2018. Effects of non-purified and semi-purified commercial diets on behaviors, plasma corticosterone levels, and cecum microbiome in C57BL/6J mice. Neurosci Lett 670, 36—40.

Valles-Colomer, M., Falony, G., Darzi, Y., Tigchelaar, E.F., Wang, J., Tito, R.Y., Schiweck, C., Kurilshikov, A., Joossens, M., Wijmenga, C., Claes, S., Van Oudenhove, L., Zhernakova, A., Vieira-Silva, S., Raes, J., 2019. The neuroactive potential of the human gut microbiota in quality of life and depression. Nat Microbiol 4, 623—632.

Volk, J.K., Nystrom, E.E.L., van der Post, S., Abad, B.M., Schroeder, B.O., Johansson, A., Svensson, F., Javerfelt, S., Johansson, M.E.V., Hansson, G.C., Birchenough, G.M.H., 2019. The Nlrp6 inflammasome is not required for baseline colonic inner mucus layer formation or function. J Exp Med 216, 2602—2618.

Vuong, H.E., Hsiao, E.Y., 2017. Emerging roles for the gut microbiome in autism spectrum disorder. Biol Psychiatry 81, 411—423.

Wang, S., Qu, Y., Chang, L., Pu, Y., Zhang, K., Hashimoto, K., 2020. Antibiotic-induced microbiome depletion is associated with resilience in mice after chronic social defeat stress. J Affect Disord 260, 448—457.

Wong, M.L., Inserra, A., Lewis, M.D., Mastronardi, C.A., Leong, L., Choo, J., Kentish, S., Xie, P., Morrison, M., Wesselingh, S.L., Rogers, G.B., Licinio, J., 2016. Inflammasome signaling affects anxiety- and depressive-like behavior and gut microbiome composition. Mol Psychiatr 21, 797—805.

Wu, W.L., Adame, M.D., Liou, C.W., Barlow, J.T., Lai, T.T., Sharon, G., Schretter, C.E., Needham, B.D., Wang, M.I., Tang, W., Ousey, J., Lin, Y.Y., Yao, T.H., Abdel-Haq, R., Beadle, K., Gradinaru, V., Ismagilov, R.F., Mazmanian, S.K., 2021. Microbiota regulate social behaviour via stress response neurons in the brain. Nature 595, 409—414.

Yu, M., Jia, H., Zhou, C., Yang, Y., Zhao, Y., Yang, M., Zou, Z., 2017. Variations in gut microbiota and fecal metabolic phenotype associated with depression by 16S rRNA gene sequencing and LC/MS-based metabolomics. J Pharm Biomed Anal 138, 231—239.

Zeraati, M., Enayati, M., Kafami, L., Shahidi, S.H., Salari, A.A., 2019. Gut microbiota depletion from early adolescence alters adult immunological and neurobehavioral responses in a mouse model of multiple sclerosis. Neuropharmacology 157, 107685.

Zhang, Q., Yun, Y., An, H., Zhao, W., Ma, T., Wang, Z., Yang, F., 2021. Gut microbiome composition associated with major depressive disorder and sleep quality. Front Psychiatr 12, 645045.

Zhao, W., Hu, Y., Li, C., Li, N., Zhu, S., Tan, X., Li, M., Zhang, Y., Xu, Z., Ding, Z., Hu, L., Liu, Z., Sun, J., 2020. Transplantation of fecal microbiota from patients with alcoholism induces anxiety/depression behaviors and decreases brain mGluR1/PKC epsilon levels in mouse. Biofactors 46, 38—54.

Zheng, P., Zeng, B., Liu, M., Chen, J., Pan, J., Han, Y., Liu, Y., Cheng, K., Zhou, C., Wang, H., Zhou, X., Gui, S., Perry, S.W., Wong, M.L., Licinio, J., Wei, H., Xie, P., 2019. The gut microbiome from patients with schizophrenia modulates the glutamate-glutamine-GABA cycle and schizophrenia-relevant behaviors in mice. Sci Adv 5, eaau8317.

Zheng, P., Zeng, B., Zhou, C., Liu, M., Fang, Z., Xu, X., Zeng, L., Chen, J., Fan, S., Du, X., Zhang, X., Yang, D., Yang, Y., Meng, H., Li, W., Melgiri, N.D., Licinio, J., Wei, H., Xie, P., 2016. Gut microbiome remodeling induces depressive-like behaviors through a pathway mediated by the host's metabolism. Mol Psychiatr 21, 786—796.

Zhu, F., Guo, R., Wang, W., Ju, Y., Wang, Q., Ma, Q., Sun, Q., Fan, Y., Xie, Y., Yang, Z., Jie, Z., Zhao, B., Xiao, L., Yang, L., Zhang, T., Liu, B., Guo, L., He, X., Chen, Y., Chen, C., Gao, C., Xu, X., Yang, H., Wang, J., Dang, Y., Madsen, L., Brix, S., Kristiansen, K., Jia, H., Ma, X., 2020. Transplantation of microbiota from drug-free patients with schizophrenia causes schizophrenia-like abnormal behaviors and dysregulated kynurenine metabolism in mice. Mol Psychiatr 25, 2905—2918.

# CHAPTER 6

# Microbiota influence brain systems—Work in animal models

## Introduction

Over the past 15 years, the field of microbiota—brain research has leveraged animal models to examine the impact of the microbiome on the brain (Cryan et al., 2019). The gut microbiota impacts biological and physiological systems that influence neurodevelopment and healthy brain function as well as psychiatric and neurological disorders. Animal models have been paramount in linking the microbiome to fundamental neural processes and mechanisms. Moreover, observations in animal studies pave the way for ongoing and future human studies that translate these observations to healthy and clinical populations (Chin Fatt et al., 2023; Cryan and Mazmanian, 2022; Foster, 2022; Needham et al., 2022; Stewart Campbell et al., 2022). This chapter will put key research findings into a broader contextual domain, including neuroimmune signaling, neurotransmission, and neuroplasticity. The focus here is on biological mechanisms and not functional or behavioral outcomes. It aims to provide insight into how recent evidence can guide researchers to go beyond association to better inform mechanisms of action. The systems considered, and the related mechanisms, are as important to healthy brain development and function, as they are in response to stress and other environmental triggers, and in disease states.

## Neuroimmune signaling

The field of neuroimmunology was ignited in 1987 by the discovery that peripheral administration of the proinflammatory cytokine interleukin-1β (IL-1β) could activate the hypothalamic pituitary adrenal (HPA) axis by a central nervous system (CNS)-mediated event, namely the increased production of corticotropin-releasing hormone (CRH) by parvocellular neurons in the paraventricular nucleus of the hypothalamus (Berkenbosch et al., 1987; Sapolsky et al., 1987). Following this landmark discovery, researchers focused on the signal transduction mechanisms involved in

*Microbiota Brain Axis*
ISBN 978-0-12-814800-6
https://doi.org/10.1016/B978-0-12-814800-6.00010-8

processing immune signals generated by inflammation or injury originating in the periphery or in the brain. Molecules originally thought to be unique to the immune system are present and functional in the brain. In fact, the brain and immune system share many molecules including neurotransmitters and their receptors, cytokines and chemokines and their receptors, integrins, and cell adhesion molecules. The discovery that immune molecules may have non-immune roles in the brain changed the way that we think about the immune system and stimulated new research directions aimed at identifying novel roles for immune molecules in normal brain function.

Several communication pathways by which peripheral immune signals communicate with the brain have been identified (Turnbull and Rivier, 1999). The vagus nerve (Maier et al., 1998), circumventricular organs (Blatteis et al., 1983; Katsuura et al., 1990), and blood—brain barrier cells (Cao et al., 1999; Quan et al., 1998) are known to relay peripheral immune signals to the brain. These neuroimmune pathways are also essential pathways in the microbiota—brain axis in health and in disease. The integration of 25 years of neuroimmune research with more than a decade of microbiome—brain research has led to recent discoveries that have moved the conversation from the identification of microbiota—immune—brain pathways to direct mechanisms of action (Blacher et al., 2019; Cryan and Mazmanian, 2022; Foster, 2022; Needham et al., 2022). In this section, the healthy, homeostatic role of neuroimmune signaling will be considered in the context of how microglia, astrocytes, oligodendrocytes, and peripheral immune cells influence microbiota—brain signaling.

Microglia are the resident immune cells in the brain (Kreutzberg, 1995, 1996) and in that role respond to a wide variety of local, peripheral, and environmental signals (Thion et al., 2018). Beyond their classical immune role, microglia are a key regulator of brain homeostasis over the lifespan (Matcovitch-Natan et al., 2016; Nimmerjahn et al., 2005; Thion et al., 2018). Microglial ontogeny, from myeloid progenitors in the yolk sac to the brain, occurs in early embryonic life and continues through postnatal development and has been reviewed elsewhere (Thion et al., 2018). During postnatal development, microglia play an essential role in brain wiring and are active in synaptic pruning (Pasciuto et al., 2020; Schafer et al., 2012). Peripheral signals, including those originating with the microbiome, influence the development of the microglial regulatory network (Matcovitch-Natan et al., 2016; Pasciuto et al., 2020; Salter and Stevens, 2017). Recent attention to the embryonic and postnatal maturation of microglia using transcriptomics and epigenetics identified three distinct developmental phases of microglial development, including early microglia (up to embryonic day 14), premicroglia (E14 to a few weeks after birth), and adult microglia (Matcovitch-Natan et al., 2016).

The potential role for the microbiome in microglial development was suggested from the observation that morphological and gene expression differences in microglia derived from germ-free (GF) mice compared to conventionally housed/colonized mice (Erny et al., 2015, 2017). It is well established that the microbiome influences the maturation, function, and homeostatic balance of the mucosal and systemic immune system during development and over the lifespan (Hansen et al., 2012; Macpherson and Harris, 2004; Maynard et al., 2012). Moreover, the observation of increased expression of inflammation pathways in microglia in adult GF mice suggests that microbiota—immune signaling contributes to brain homeostasis over the lifespan (Matcovitch-Natan et al., 2016). Interestingly, when the trajectory of transcriptomic and epigenetic changes observed in healthy mice was compared to that of GF mice, genes that were dysregulated in the postnatal period in GF mice were associated with adult microglial genes that have a role in homeostasis (Matcovitch-Natan et al., 2016), suggesting that the absence of gut bacteria altered the trajectory of brain development. To complement this observation using a different animal model, researchers observed altered microglial development in offspring exposed to a maternal immune activation (MIA). This highlights the importance of peripheral immune—brain pathways to microglial development (Matcovitch-Natan et al., 2016). In contrast to the observation noted above in GF mice, MIA impacted the premicroglia transcriptional profile resulting in a transcriptional shift in developmental gene expression where there was a significant overlap between the genes expressed in postnatal offspring exposed to MIA with genes normally expressed in adult microglia (Matcovitch-Natan et al., 2016). Both the maternal microbiome and the offspring microbiome are impacted by maternal exposure to immune activation, and these observations further highlight the importance of microbiota—immune brain pathways in brain development. Together, these studies demonstrate the need for balanced regulation of immune responses during microglial and brain development and that disruptions by genetic or environmental perturbations that lead to an imbalance in microbiota—immune—brain signaling may contribute to dysregulation of microglia that may contribute to neurodevelopmental outcomes and neurodevelopmental disorders (Matcovitch-Natan et al., 2016).

Recent work has drawn attention to the neonatal microbiome and its developmental trajectory throughout early postnatal life, which parallels the ontogenesis of host immunity and proper wiring of neural circuitry in the brain (Jasarevic and Bale, 2019). There is extensive evidence that activation of the peripheral immune system during pregnancy and early in life impacts brain development, including microglial development. Understanding which aspects of the peripheral immune system and the related mechanisms that contribute to brain development is an active area of research. An extended body of neuroimmunology research has considered the importance of the adaptive immune system, and in particular, T lymphocytes or T cells, in normal neurodevelopment and brain homeostasis, notably in the absence of infection or immune activation (Ellwardt et al., 2016). In mice, the adaptive immune system undergoes expansion and maturation in the first 3 weeks of postnatal life that is directly regulated by gut bacteria (Maynard et al., 2012). Bacterial taxa and their metabolites are known to promote the proliferation of distinct T-cell subsets (Atarashi et al., 2008; Ivanov et al., 2008). The exponential expansion of lymphocyte pools within the first week of life occurs in tandem with hierarchical wiring of neural circuitry in the neonate brain (Tanabe and Yamashita, 2018b). During the first few weeks of postnatal development, T-cells populate the neonate brain and maintain stable, self-renewing pools in the meningeal space (Tanabe and Yamashita, 2018a). Modulatory effects of T cells are present before birth, as maternal immune activation (MIA) has been shown to impact synaptic development in the offspring via secretion of IL-17a from maternal Th17 lymphocytes (Choi et al., 2016). A recent study demonstrated that γδ T-cells, the first T-cell subsets to seed peripheral tissues, secrete IL17a to regulate anxiety-like behavior (Alves de Lima et al., 2020). Interestingly, this experiment also demonstrated that secretion of IL17a was partially dependent on commensal-derived signals, suggesting an intermediary role of the adaptive immune system in transmitting microbial signals to the brain. T-cell—derived IFN-γ, which displays significantly reduced levels in germ-free mice, has been shown to alter the balance between excitatory and inhibitory synapse formation in favor of inhibitory signaling (Filiano et al., 2016).

Recent work using T-cell—deficient mice, as a result of double knock-out of the β and δ chains of the T-cell receptor (*TCR β-/-δ-/-*), showed that in the absence of T-cells, the developmental trajectory of the gut microbiota was altered (Caspani et al., 2022; Francella et al., 2022). By examining both the microbiota (cecal and fecal) and the metabolome (cecal, fecal, gut, and brain) at several time points over postnatal development, key T-cell—related taxa were identified that may contribute to microbiota—immune—brain signaling on neurodevelopment. Notably, the absence of T-cells resulted in a reorganization of the microbial community that was accompanied by reduced alpha diversity, reduced levels of the short-chain fatty acid (SCFA) butyrate, and higher levels of glucose-6-phosphate and 5-aminovalerate, suggesting a modification in the biochemical signaling between microbe and host in the immunocompromised situation (Caspani et al., 2022). In addition, a potential role for *Muribaculaceae* taxa in modulating the cecal availability of metabolites during early postnatal life was evident (Caspani et al., 2022). A novel finding of this work was that T-cell—related changes in the gut microbiome affected neuroactive metabolites in the brain, with elevated levels of GABA and glycine in the brains of T-cell—deficient mice, reinforcing the importance of microbe—host interactions at the gut barrier on postnatal brain development (Caspani et al., 2022).

While less studied, other peripheral immune cells have been considered in the context of microbiota—immune brain communication. A recent study examined microbiota—immune mechanisms underlying the impact of prebiotic supplementation on microglial activation and demonstrated a differential impact of this diet supplement on microbiota profile that was age dependent (Boehme et al., 2020). Notably, microbiota differences were associated with altered age-related infiltration of Ly-6C[hi] monocytes and microglial activation (Boehme et al., 2020). The discovery of lymphatic vasculature in the CNS, followed by the emerging evidence that demonstrates the functional importance of meningeal immune cells, particularly T cells, during development and over the life span, has provided new potential mechanisms of how peripheral immune cells communicate with brain (Castellani et al., 2023; Kipnis, 2016; Louveau et al., 2015, 2017). Interestingly, the production of IFNγ by meningeal natural killer cells, which are influenced by microbiota, influences the homeostatic anti-inflammatory function of astrocytes and can limit CNS inflammation (Sanmarco et al., 2021). Combined with the extensive work noted above on T cells, this work highlights the importance of the microbiome to neuroimmune influences on brain development and emphasizes the importance of considering both the peripheral immune system and the microbiome to gain a wholistic mechanistic understanding of the factors that are important to brain development.

Beyond microglia, studies have also demonstrated an influence of the gut microbiome on other glia, that is, astrocytes and oligodendrocytes. These influences are important to consider as neuroimmune cross-talk between neurons and glia plays a pivotal role in brain homeostasis. Astrocytes are a key contributor to neuronal–glial communication and respond to local signals in the brain, but also have been shown to be modulated by gut microbiota (Zhao et al., 2021). For example, gut microbiota were shown to impact gene expression that is important to the metabolic coupling between neurons and astrocytes in mice (Margineanu et al., 2020). In addition, microbiome-targeted treatments such as soluble fiber intake have been shown to influence astrocyte activation in the context of improved cognitive function in an animal model of Alzheimer's disease (Cuervo-Zanatta et al., 2023). Further, this influence was modulated by a concomitant reduction in the SCFA propionate and higher levels of butyrate (Cuervo-Zanatta et al., 2023).

A role for the microbiome in homeostatic regulation of myelination has been demonstrated using unbiased genome-wide RNA sequencing in the prefrontal cortex (PFC) of control, GF, and colonized GF mice (Hoban et al., 2016). Interestingly, in the absence of microbiota, an upregulation of genes linked to myelination was accompanied by hypermyelinated axons in the PFC of GF mice (Hoban et al., 2016). The transcriptional observations were reversed when GF mice were colonized at weaning suggesting that some of these effects may be reversible (Hoban et al., 2016). Antibiotic treatment from postnatal day 7 (P7) to weaning at P23, altered gut microbiota and also led to increased expression of myelin-related genes and transcription factors in adult mice (Keogh et al., 2021). Antibiotic treatment of adult mice followed by lysolecithin-induced demyelination impaired myelin clearance and impacted remyelination, which is known to be important in diseases such as multiple sclerosis (McMurran et al., 2019). In mice, increased levels of the gut metabolite 4-ethylphenyl-sulfate (4EPS) were linked to increased anxiety-like behavior, in parallel with reduced myelination (Needham et al., 2022) suggesting that alterations in oligodendrocyte function and myelination may contribute to behavioral deficits or behavioral alterations. Overall, the cross-talk between the microbiome and the peripheral immune system is an important component of gut–brain signaling and has an impact on neurodevelopment, through its action on neuroimmune signaling in the brain including microglia, astrocytes, and oligodendrocytes.

## Neurotransmission

Before reviewing the link between the microbiome and neurotransmission, this section will first provide an overview of the neuroanatomical studies that consider a role for the microbiome. As in other areas of study, studies have utilized GF mice to identify neuroanatomical brain systems that are influenced by the microbiome. Neuroanatomical differences observed in GF mice, compared to SPF mice, included increased amygdala and hippocampus volumes, changes that were associated with increased dendritic arborization and spine density in the same regions (Luczynski et al., 2016). An increased number of cortical microglia was observed in GF mice in parallel with an immature morphological microglial phenotype, including longer processes with more branch points, compared to control mice (Erny et al., 2015). Using magnetic resonance imaging (MRI) in adult GF mice compared to controls, widespread regional volume differences were observed, including decreased myelination in white matter regions and fiber tracts of these GF mice, suggesting less mature myelination patterns in the absence of microbiota (Lu et al., 2018). These studies demonstrate links between microbiota and neuroanatomy, but lack detail on key members of the microbiota or the microbial factors that may influence the connection.

A recent study in healthy adult Sprague—Dawley rats examined the white matter structure in the context of diet-related changes in microbiota composition (Ong et al., 2018). White matter tracts were visualized using diffusion tensor imaging (DTI), and diet-related (standard, high fat, high fiber, high protein/low carb) changes in white matter structural integrity were observed. Moreover, an abundance of diet-related microbiota was associated with changes in brain structure (Ong et al., 2018). This study strengthens the observations shown previously in GF mice and provides a foundation for similar studies that target specific mechanisms that may underly how microbiota influence brain structure. To this end, a recent study colonized mice with bacterial strains bioengineered to produce 4-ethylphenol (4-EP) that can be sulfonated by the host to 4-EPS and showed elevated levels of 4-EPS in the serum and brain tissue of 4-EP positive mice (Needham et al., 2022). The functional impact of this microbial-metabolite signal to the brain included lower structural connectivity, visualized using functional ultrasound imaging, in hippocampus, thalamus, amygdala, hypothalamus, and cortex—demonstrating a direct microbe—metabolite—brain pathway that can impact brain connectivity (Needham et al., 2022).

Discovery and targeted analysis of how the microbiome influences gene expression, including both RNA and microRNAs, demonstrates that neurotransmitter systems are influenced by microbiota—brain communication (Hoban et al., 2017; Philip et al., 2021; Zhou et al., 2020). A targeted analysis of gene expression, related to excitation—inhibition balance, using GF and SPF Swiss Webster mice revealed differential expression of key glutamatergic and GABAergic mRNAs in hippocampus, amygdala, and prefrontal cortex (Philip et al., 2021). Colonization of GF mice normalized some of the differential expression observed. Interestingly, these investigators compared gene expression profiles in GF mice to regional gene expression profiles generated from postmortem tissue from depressed individuals and demonstrated a significant correlation between the gene expression profiles (Philip et al., 2021). Differential expression of mono-amine neurotransmitter-related genes in the hippocampus from GF mice compared to SPF mice has also been reported (Pan et al., 2019). Interestingly, in a separate study, whole-cell patch-clamp electrophysiology was used to examine the impact of probiotic administration, *Lactobacillus rhamnosus* and *Bifidobacterium animalis*, on the intrinsic excitability of CA1 hippocampal neurons and demonstrated an ability of these probiotics to increase intrinsic excitability, whereas this impact was not observed in GF mice treated with the same probiotic mixture (Kim et al., 2022). This is an important observation as several studies have demonstrated that reconstitution of gut microbiota later in life did not normalize the deficits observed in GF mice and therefore highlights the importance of appropriate controls for these experiments and care in interpretation of results in these studies.

The potential of gut bacteria to produce neuroactive metabolites including neurotransmitters such as dopamine, norepinephrine, serotonin, and gamma-aminobutyric acid (GABA) has garnered a lot of attention (Strandwitz, 2018; Strandwitz et al., 2019). Microbial-derived molecules, including neurotransmitters, short-chain fatty acids, indoles, bile acids, choline metabolites, lactate, and vitamins, exert local effects in the gastrointestinal environment and are essential to maintaining barrier integrity, immune regulation, as well as other roles (see also Chapter 4) (Caspani et al., 2019). When and how gut-derived neuroactive molecules impact the brain is less well understood. Evidence that gut microbiota influence brain neurotransmitters stems from animal studies. To demonstrate the importance of *Lactobacillus* sp. on memory, investigators colonized GF mice with *L. reuteri* F275, *L. plantarum* BDGP2, and *L. brevis* BDGP6 and demonstrated improved memory compared to GF mice. This behavioral change was accompanied by increased levels of lactate in the stool of *Lactobacillus*-colonized mice and increased levels of hippocampal GABA in these mice (Mao et al., 2020). Whether gut microbe—derived GABA can act directly on the CNS has not been demonstrated. Notably, monocolonization of GF mice with *Bifidobacterium dentium*, a bacterial taxa that they demonstrated can produce GABA in vitro, resulted in increased colonic levels of GABA, glutamate, and glutamine but no parallel changes in these neurotransmitters were observed in the brain (Luck et al., 2021). In contrast, a few recent studies provided evidence that microbe—immune cross-talk, particularly involving the adaptive immune system, can influence brain neurotransmitter systems. In mice lacking all functional T cells, the developmental trajectory of the gut microbiome and its metabolome was altered (Caspani et al., 2022). In parallel with reduced alpha diversity, T-cell—deficient mice showed reduced cecal and fecal levels of short-chain fatty acid (SCFA) butyrate (Caspani et al., 2022). These gut-related changes impacted the brain metabolome, with a higher abundance of GABA, glycine, aspartate, acetate, and glycerophosphocholine measured in the brains of T-cell—deficient mice across the lifespan (Caspani et al., 2022). Using a different immunocompromised mouse model, that is mice homozygous for the severe combined immune deficiency spontaneous mutation $Prkdc_{scid}$, commonly referred to as SCID, researchers demonstrated that T-cell—derived cytokine interferon-$\gamma$ (IFN-$\gamma$) increased GABAergic neurotransmission and was associated with hyperconnectivity of frontocortical brain regions, as well as social behavior deficits (Filiano et al., 2015). Overall, these studies demonstrate the influence of microbiota—brain signaling on central neurotransmitter systems.

## Neural plasticity

Studies have demonstrated the importance of the microbiota—brain signaling to key neural plastic functions including neurogenesis and synaptic plasticity using GF mice as well as antibiotic treatment in rodents. In the context of ongoing neuroimmune work, these studies point to microbiota—immune pathways as key regulators of neural plasticity. Neurogenesis refers to the production of new neural cells (neurons and glia) in key brain regions, such as the hippocampus, and is an important neural plastic process that has been shown to influence cognitive function, learning, memory, and related behaviors (Gage et al., 1998; Garcia-Verdugo et al., 1998; Gould and Gross, 2002; Kee et al., 2007; van

Praag et al., 2002). Neurogenesis assessment includes measurement of cell proliferation and maturation. Moreover, 5-bromo-2′deoxyuridine (BrdU) is a thymidine analog, which is incorporated into DNA during the DNA synthetic (S-phase) of the cell cycle. BrdU is widely used to assess the proliferation and maturation of defined populations of cells in the brain (Miller and Nowakowski, 1988). Doublecortin (DCX) is a protein that is expressed in proliferating cells and has also been used as a proxy for neurogenesis in animal studies (Lucassen et al., 2009).

An initial study using GF mice showed increased neurogenesis in Swiss Webster GF adult mice that was not altered by postweaning microbial colonization of GF mice (Ogbonnaya et al., 2015). More recently, subtle differences in neurogenesis at 4 and 8 weeks old, visualized as the number of DCX-positive cells in the dentate gyrus of the hippocampus, were observed in GF mice (C57Bl/6) compared to control mice demonstrating a potential role for the microbiome in this process (Scott et al., 2020). Notably, the changes in DCX-positive cell counts were distinct in male and female GF mice; whereas, a gradual decrease in DCX-positive cells over development was similar in male and female control mice (Scott et al., 2020). Differential hippocampal cell proliferation, BrDU-positive cells, paralleled the observations using DCX (Scott et al., 2020). In line with these findings, reduced neurogenesis (reduced DCX-positive cells) in adult C57BL/6 GF mice has been reported (Wei et al., 2021). However, this study went further to investigate the microbiota—brain mechanism that mediates adult neurogenesis. They demonstrated a role for indole, a microbial metabolite produced from dietary tryptophan, in mediating neurogenesis. First, GF mice colonized with a genetically modified *Escherichia coli*, that was unable to produce indole, showed reduced neurogenesis similar to GF mice. Second, systemic administration of indole increased neurogenesis in control mice (Wei et al., 2021). The aryl hydrocarbon receptor (AhR) is a transcription factor that is expressed in both the gut and the brain and is a target for tryptophan metabolites including indoles (Dong and Perdew, 2020). In this study, systemic indole administration did not result in increased neurogenesis in *AhR*-knock out mice. Further, the activation of indole-AhR signaling resulted in gene expression changes that are linked to adult neurogenesis (Wei et al., 2021). This work demonstrates an important microbe—metabolite pathway that influences neuroplasticity.

The impact of stress on neurogenesis is well established in animal models (Sahay and Hen, 2007), and a recent study demonstrated a role for the microbiome in stress-related changes in behavior and neurogenesis (Chevalier et al., 2020). First, they demonstrated a direct action for the microbiome as fecal microbiota transplantation from stressed mice to unstressed mice resulted in behavioral changes and reduced neurogenesis. Second, they demonstrated that oral administration of *Lactobaccilus plantarum* normalized the behavioral effects of stress and partially normalized the neurogenesis (Chevalier et al., 2020). Treatment of adult mice with broad-spectrum antibiotics, known to deplete gut microbiota, also resulted in decreased hippocampal neurogenesis (Mohle et al., 2016). By costaining with neuronal-specific antibody NeuN, it was shown that both neuronal progenitors (BrdU, DCX, NeuN-positive) and mature neurons (BrdU, NeuN-positive) were reduced following 7 weeks of antibiotic treatment (Mohle et al., 2016). Reduced neurogenesis was associated with reduced numbers of Ly6C$^{hi}$ monocytes in the brain of antibiotic-treated mice, reinforcing the importance of microbiota-immune signaling in neuro-plasticity as was detailed above for microglial development and function (Mohle et al., 2016). Several studies have shown that exposure to antibiotics during early life leads to long-term changes in microbiome composition and function (Champagne-Jorgensen et al., 2020; Cho et al., 2012; Cox et al., 2014; Lebovitz et al., 2019; Leclercq et al., 2017). A recent study demonstrated that exposure to antibiotics for the first 4 weeks of postnatal life reduced adult neurogenesis, reduced relative hippocampal neuron spine density, and impaired long-term potentiation (LTP) of synaptic transmission (Liu et al., 2022). Reconstitution with normal gut microbiota normalized adult neurogenesis, as well as the behavioral deficits observed (Liu et al., 2022). Mechanistically, this study also considered microbiota—metabolite signaling to the brain and demonstrated that elevated levels of circulating 4-methylphenol, produced by gut microbiota, mediated the impact of antibiotic-induced changes in hippocampal plasticity (Liu et al., 2022).

The above-noted studies link gut microbiota to brain systems and neuroplasticity; however, it is important to also note that gut microbiota influence neuroplasticity of the enteric nervous system (ENS). Studies using GF mice have demonstrated the importance of microbiota to ENS function. An examination of the electrophysiological properties of neurons in the myenteric plexus of the ENS in GF mice showed reduced excitability in myenteric neurons measured as a lower resting membrane potential and a prolonged after hyperpolarization in GF mice compared to controls (McVey Neufeld et al., 2013). These myenteric plexus neurons have nerve fiber endings near gut epithelial cells lining the lumen, and as such, may respond to bacteria and their metabolites to mediate local neural function and to communicate with the central nervous system. Using immunohistochemistry, a recent report showed decreased levels of pan-neuronal marker PGP9.5 in the mucosal and myenteric plexus in GF mice compared to control (Aktar et al., 2020). This effect was reversed when GF mice were monocolonized with *Bacteroides thetaiotamicron*, a dominant commensal bacteria. Further, the importance of *B. thetaiotamicron* on the neurochemical profile of enteric neurons was reported, as well as an impact on gene expression related to barrier integrity, and gut motility (Aktar et al., 2020). These studies provide insight into how gut microbiota influence ENS function and provide evidence for specific commensal bacteria. Additional work is needed to better understand how these ENS functional changes related to the microbiome may influence brain health.

## Future directions

Neuroscience research efforts focused on understanding the links between the microbiome and the brain has continued to expand over the past decade. The amassed body of work has allowed us to better understand how and when the microbiome interacts with the brain. Moreover, mechanistic studies mapping the signaling pathways including neural, humoral, metabolic, and cellular are starting to emerge—and yet, this area of research needs more attention in the future (Cryan and Mazmanian, 2022). Just as important is the microbiome-related translational work ongoing and in development that leverages preclinical findings to advance microbiome research in clinical neuroscience including psychology, psychiatry, and neurology (Cryan and Mazmanian, 2022; Foster, 2022). Remarkably, all of the research studies reviewed here focused on gut bacteria, used rodent models, and did not consider other body sites or other biomes. As sequencing information and analytical tools continue to expand, the potential to consider other components of the microbiome including the virome, mycobiome, archaeome, and parasitome will become more important and integrate into mainstream neuroscience research. Such advances, if translated to the clinic, have the potential to advance precision health approaches through microbiome-informed or microbiome-targeted therapeutics.

# References

Aktar, R., Parkar, N., Stentz, R., Baumard, L., Parker, A., Goldson, A., Brion, A., Carding, S., Blackshaw, A., Peiris, M., 2020. Human resident gut microbe Bacteroides thetaiotaomicron regulates colonic neuronal innervation and neurogenic function. Gut Microb 11, 1745—1757.

Alves de Lima, K., Rustenhoven, J., Da Mesquita, S., Wall, M., Salvador, A.F., Smirnov, I., Martelossi Cebinelli, G., Mamuladze, T., Baker, W., Papadopoulos, Z., Lopes, M.B., Cao, W.S., Xie, X.S., Herz, J., Kipnis, J., 2020. Meningeal γδ T cells regulate anxiety-like behavior via IL-17a signaling in neurons. Nat Immunol 21, 1421—1429.

Atarashi, K., Nishimura, J., Shima, T., Umesaki, Y., Yamamoto, M., Onoue, M., Yagita, H., Ishii, N., Evans, R., Honda, K., Takeda, K., 2008. ATP drives lamina propria T(H)17 cell differentiation. Nature 455, 808—812.

Berkenbosch, F., van Oers, J., del Rey, A., Tilders, F., Besedovsky, H., 1987. Corticotropin-releasing factor-producing neurons in the rat activated by interleukin-1. Science 238, 524—526.

Blacher, E., Bashiardes, S., Shapiro, H., Rothschild, D., Mor, U., Dori-Bachash, M., Kleimeyer, C., Moresi, C., Harnik, Y., Zur, M., Zabari, M., Brik, R.B., Kviatcovsky, D., Zmora, N., Cohen, Y., Bar, N., Levi, I., Amar, N., Mehlman, T., Brandis, A., Biton, I., Kuperman, Y., Tsoory, M., Alfahel, L., Harmelin, A., Schwartz, M., Israelson, A., Arike, L., Johansson, M.E.V., Hansson, G.C., Gotkine, M., Segal, E., Elinav, E., 2019. Potential roles of gut microbiome and metabolites in modulating ALS in mice. Nature 572, 474—480.

Blatteis, C.M., Bealer, S.L., Hunter, W.S., Llanos, Q.J., Ahokas, R.A., Mashburn Jr., T.A., 1983. Suppression of fever after lesions of the anteroventral third ventricle in Guinea pigs. Brain Res Bull 11, 519—526.

Boehme, M., van de Wouw, M., Bastiaanssen, T.F.S., Olavarria-Ramirez, L., Lyons, K., Fouhy, F., Golubeva, A.V., Moloney, G.M., Minuto, C., Sandhu, K.V., Scott, K.A., Clarke, G., Stanton, C., Dinan, T.G., Schellekens, H., Cryan, J.F., 2020. Mid-life microbiota crises: middle age is associated with pervasive neuroimmune alterations that are reversed by targeting the gut microbiome. Mol Psychiatr 25, 2567—2583.

Cao, C., Matsumura, K., Ozaki, M., Watanabe, Y., 1999. Lipopolysaccharide injected into the cerebral ventricle evokes fever through induction of cyclooxygenase-2 in brain endothelial cells. J Neurosci 19, 716—725.

Caspani, G., Green, M., Swann, J.R., Foster, J.A., 2022. Microbe-immune crosstalk: evidence that T cells influence the development of the brain metabolome. Int J Mol Sci 23, 3259.

Caspani, G., Kennedy, S., Foster, J.A., Swann, J., 2019. Gut microbial metabolites in depression: understanding the biochemical mechanisms. Microb Cell 6, 454—481.

Castellani, G., Croese, T., Peralta Ramos, J.M., Schwartz, M., 2023. Transforming the understanding of brain immunity. Science 380, eabo7649.

Champagne-Jorgensen, K., Mian, M.F., Kay, S., Hanani, H., Ziv, O., McVey Neufeld, K.A., Koren, O., Bienenstock, J., 2020. Prenatal low-dose penicillin results in long-term sex-specific changes to murine behaviour, immune regulation, and gut microbiota. Brain Behav Immun 84, 154—163.

Chevalier, G., Siopi, E., Guenin-Mace, L., Pascal, M., Laval, T., Rifflet, A., Boneca, I.G., Demangel, C., Colsch, B., Pruvost, A., Chu-Van, E., Messager, A., Leulier, F., Lepousez, G., Eberl, G., Lledo, P.M., 2020. Effect of gut microbiota on depressive-like behaviors in mice is mediated by the endocannabinoid system. Nat Commun 11, 6363.

Chin Fatt, C.R., Asbury, S., Jha, M.K., Minhajuddin, A., Sethuram, S., Mayes, T., Kennedy, S.H., Foster, J.A., Trivedi, M.H., 2023. Leveraging the microbiome to understand clinical heterogeneity in depression: findings from the T-RAD study. Transl Psychiatry 13, 139.

Cho, I., Yamanishi, S., Cox, L., Methe, B.A., Zavadil, J., Li, K., Gao, Z., Mahana, D., Raju, K., Teitler, I., Li, H., Alekseyenko, A.V., Blaser, M.J., 2012. Antibiotics in early life alter the murine colonic microbiome and adiposity. Nature 488, 621−626.

Choi, B., Yim Yeong, S., Wong, H., Kim, S., Kim, H., Kim Sangwon, V., Hoeffer Charles, A., Littman Dan, R., Huh Jun, R., 2016. The maternal interleukin-17a pathway in mice promotes autism-like phenotypes in offspring. Science 351, 933−939.

Cox, L.M., Yamanishi, S., Sohn, J., Alekseyenko, A.V., Leung, J.M., Cho, I., Kim, S.G., Li, H., Gao, Z., Mahana, D., Zarate Rodriguez, J.G., Rogers, A.B., Robine, N., Loke, P., Blaser, M.J., 2014. Altering the intestinal microbiota during a critical developmental window has lasting metabolic consequences. Cell 158, 705−721.

Cryan, J.F., Mazmanian, S.K., 2022. Microbiota-brain axis: context and causality. Science 376, 938−939.

Cryan, J.F., O'Riordan, K.J., Cowan, C.S.M., Sandhu, K.V., Bastiaanssen, T.F.S., Boehme, M., Codagnone, M.G., Cussotto, S., Fulling, C., Golubeva, A.V., Guzzetta, K.E., Jaggar, M., Long-Smith, C.M., Lyte, J.M., Martin, J.A., Molinero-Perez, A., Moloney, G., Morelli, E., Morillas, E., O'Connor, R., Cruz-Pereira, J.S., Peterson, V.L., Rea, K., Ritz, N.L., Sherwin, E., Spichak, S., Teichman, E.M., van de Wouw, M., Ventura-Silva, A.P., Wallace-Fitzsimons, S.E., Hyland, N., Clarke, G., Dinan, T.G., 2019. The microbiota-gut-brain axis. Physiol Rev 99, 1877−2013.

Cuervo-Zanatta, D., Syeda, T., Sanchez-Valle, V., Irene-Fierro, M., Torres-Aguilar, P., Torres-Ramos, M.A., Shibayama-Salas, M., Silva-Olivares, A., Noriega, L.G., Torres, N., Tovar, A.R., Ruminot, I., Barros, L.F., Garcia-Mena, J., Perez-Cruz, C., 2023. Dietary fiber modulates the release of gut bacterial products preventing cognitive decline in an Alzheimer's mouse model. Cell Mol Neurobiol 43, 1595−1618.

Dong, F., Perdew, G.H., 2020. The aryl hydrocarbon receptor as a mediator of host-microbiota interplay. Gut Microb 12, 1859812.

Ellwardt, E., Walsh, J.T., Kipnis, J., Zipp, F., 2016. Understanding the role of T cells in CNS homeostasis. Trends Immunol 37, 154−165.

Erny, D., Hrabe de Angelis, A.L., Jaitin, D., Wieghofer, P., Staszewski, O., David, E., Keren-Shaul, H., Mahlakoiv, T., Jakobshagen, K., Buch, T., Schwierzeck, V., Utermohlen, O., Chun, E., Garrett, W.S., McCoy, K.D., Diefenbach, A., Staeheli, P., Stecher, B., Amit, I., Prinz, M., 2015. Host microbiota constantly control maturation and function of microglia in the CNS. Nat Neurosci 18, 965−977.

Erny, D., Hrabe de Angelis, A.L., Prinz, M., 2017. Communicating systems in the body: how microbiota and microglia cooperate. Immunology 150, 7−15.

Filiano, A.J., Gadani, S.P., Kipnis, J., 2015. Interactions of innate and adaptive immunity in brain development and function. Brain Res 1617, 18−27.

Filiano, A.J., Xu, Y., Tustison, N.J., Marsh, R.L., Baker, W., Smirnov, I., Overall, C.C., Gadani, S.P., Turner, S.D., Weng, Z., Peerzade, S.N., Chen, H., Lee, K.S., Scott, M.M., Beenhakker, M.P., Litvak, V., Kipnis, J., 2016. Unexpected role of interferon-γ in regulating neuronal connectivity and social behaviour. Nature 535, 425−429.

Foster, J.A., 2022. Modulating brain function with microbiota. Science 376, 936−937.

Francella, C., Green, M., Caspani, G., Lai, J.K.Y., Rilett, K.C., Foster, J.A., 2022. Microbe-immune-stress interactions impact behaviour during postnatal development. Int J Mol Sci 23, 15064.

Gage, F.H., Kempermann, G., Palmer, T.D., Peterson, D.A., Ray, J., 1998. Multipotent progenitor cells in the adult dentate gyrus. J Neurobiol 36, 249−266.

Garcia-Verdugo, J.M., Doetsch, F., Wichterle, H., Lim, D.A., Alvarez-Buylla, A., 1998. Architecture and cell types of the adult subventricular zone: in search of the stem cells. J Neurobiol 36, 234−248.

Gould, E., Gross, C.G., 2002. Neurogenesis in adult mammals: some progress and problems. J Neurosci 22, 619−623.

Hansen, C.H., Nielsen, D.S., Kverka, M., Zakostelska, Z., Klimesova, K., Hudcovic, T., Tlaskalova-Hogenova, H., Hansen, A.K., 2012. Patterns of early gut colonization shape future immune responses of the host. PLoS One 7, e34043.

Hoban, A.E., Stilling, R.M., Moloney, G.M., Moloney, R.D., Shanahan, F., Dinan, T.G., Cryan, J.F., Clarke, G., 2017. Microbial regulation of microRNA expression in the amygdala and prefrontal cortex. Microbiome 5, 102.

Hoban, A.E., Stilling, R.M., Ryan, F.J., Shanahan, F., Dinan, T.G., Claesson, M.J., Clarke, G., Cryan, J.F., 2016. Regulation of prefrontal cortex myelination by the microbiota. Transl Psychiatry 6, e774.

Ivanov, I.I., Frutos Rde, L., Manel, N., Yoshinaga, K., Rifkin, D.B., Sartor, R.B., Finlay, B.B., Littman, D.R., 2008. Specific microbiota direct the differentiation of IL-17-producing T-helper cells in the mucosa of the small intestine. Cell Host Microbe 4, 337−349.

Jasarevic, E., Bale, T.L., 2019. Prenatal and postnatal contributions of the maternal microbiome on offspring programming. Front Neuroendocrinol 55, 100797.

Katsuura, G., Arimura, A., Koves, K., Gottschall, P.E., 1990. Involvement of organum vasculosum of lamina terminalis and preoptic area in interleukin 1 beta-induced ACTH release. Am J Physiol 258, E163−E171.

Kee, N., Teixeira, C.M., Wang, A.H., Frankland, P.W., 2007. Preferential incorporation of adult-generated granule cells into spatial memory networks in the dentate gyrus. Nat Neurosci 10, 355−362.

Keogh, C.E., Kim, D.H.J., Pusceddu, M.M., Knotts, T.A., Rabasa, G., Sladek, J.A., Hsieh, M.T., Honeycutt, M., Brust-Mascher, I., Barboza, M., Gareau, M.G., 2021. Myelin as a regulator of development of the microbiota-gut-brain axis. Brain Behav Immun 91, 437−450.

Kim, J., Kim, D.W., Lee, A., Mason, M., Jouroukhin, Y., Woo, H., Yolken, R.H., Pletnikov, M.V., 2022. Homeostatic regulation of neuronal excitability by probiotics in male germ-free mice. J Neurosci Res 100, 444−460.

Kipnis, J., 2016. Multifaceted interactions between adaptive immunity and the central nervous system. Science 353, 766−771.

Kreutzberg, G.W., 1995. Microglia, the first line of defence in brain pathologies. Arznei-mittelforschung 45, 357−360.

Kreutzberg, G.W., 1996. Microglia: a sensor for pathological events in the CNS. Trends Neurosci 19, 312−318.

Lebovitz, Y., Kowalski, E.A., Wang, X., Kelly, C., Lee, M., McDonald, V., Ward, R., Creasey, M., Mills, W., Gudenschwager Basso, E.K., Hazy, A., Hrubec, T., Theus, M.H., 2019. Lactobacillus rescues postnatal neurobehavioral and microglial dysfunction in a model of maternal microbiome dysbiosis. Brain Behav Immun 81, 617−629.

Leclercq, S., Mian, F.M., Stanisz, A.M., Bindels, L.B., Cambier, E., Ben-Amram, H., Koren, O., Forsythe, P., Bienenstock, J., 2017. Low-dose penicillin in early life induces long-term changes in murine gut microbiota, brain cytokines and behavior. Nat Commun 8, 15062.

Liu, G., Yu, Q., Tan, B., Ke, X., Zhang, C., Li, H., Zhang, T., Lu, Y., 2022. Gut dysbiosis impairs hippocampal plasticity and behaviors by remodeling serum metabolome. Gut Microb 14, 2104089.

Louveau, A., Harris, T.H., Kipnis, J., 2015. Revisiting the mechanisms of CNS immune privilege. Trends Immunol 36, 569—577.

Louveau, A., Plog, B.A., Antila, S., Alitalo, K., Nedergaard, M., Kipnis, J., 2017. Understanding the functions and relationships of the glymphatic system and meningeal lymphatics. J Clin Invest 127, 3210—3219.

Lu, J., Lu, L., Yu, Y., Cluette-Brown, J., Martin, C.R., Claud, E.C., 2018. Effects of intestinal microbiota on brain development in humanized gnotobiotic mice. Sci Rep 8, 5443.

Lucassen, P.J., Bosch, O.J., Jousma, E., Kromer, S.A., Andrew, R., Seckl, J.R., Neumann, I.D., 2009. Prenatal stress reduces postnatal neurogenesis in rats selectively bred for high, but not low, anxiety: possible key role of placental 11beta-hydroxysteroid dehydrogenase type 2. Eur J Neurosci 29, 97—103.

Luck, B., Horvath, T.D., Engevik, K.A., Ruan, W., Haidacher, S.J., Hoch, K.M., Oezguen, N., Spinler, J.K., Haag, A.M., Versalovic, J., Engevik, M.A., 2021. Neurotransmitter profiles are altered in the gut and brain of mice mono-associated with Bifidobacterium dentium. Biomolecules 11, 1091.

Luczynski, P., McVey Neufeld, K.A., Oriach, C.S., Clarke, G., Dinan, T.G., Cryan, J.F., 2016. Growing up in a bubble: using germ-free animals to assess the influence of the gut microbiota on brain and behavior. Int J Neuropsychopharmacol 19, pyw020.

Macpherson, A.J., Harris, N.L., 2004. Interactions between commensal intestinal bacteria and the immune system. Nat Rev Immunol 4, 478—485.

Maier, S.F., Goehler, L.E., Fleshner, M., Watkins, L.R., 1998. The role of the vagus nerve in cytokine-to-brain communication. Ann N Y Acad Sci 840, 289—300.

Mao, J.H., Kim, Y.M., Zhou, Y.X., Hu, D., Zhong, C., Chang, H., Brislawn, C.J., Fansler, S., Langley, S., Wang, Y., Peisl, B.Y.L., Celniker, S.E., Threadgill, D.W., Wilmes, P., Orr, G., Metz, T.O., Jansson, J.K., Snijders, A.M., 2020. Genetic and metabolic links between the murine microbiome and memory. Microbiome 8, 53.

Margineanu, M.B., Sherwin, E., Golubeva, A., Peterson, V., Hoban, A., Fiumelli, H., Rea, K., Cryan, J.F., Magistretti, P.J., 2020. Gut microbiota modulates expression of genes involved in the astrocyte-neuron lactate shuttle in the hippocampus. Eur Neuropsychopharmacol 41, 152—159.

Matcovitch-Natan, O., Winter, D.R., Giladi, A., Vargas Aguilar, S., Spinrad, A., Sarrazin, S., Ben-Yehuda, H., David, E., Zelada Gonzalez, F., Perrin, P., Keren-Shaul, H., Gury, M., Lara-Astaiso, D., Thaiss, C.A., Cohen, M., Bahar Halpern, K., Baruch, K., Deczkowska, A., Lorenzo-Vivas, E., Itzkovitz, S., Elinav, E., Sieweke, M.H., Schwartz, M., Amit, I., 2016. Microglia development follows a stepwise program to regulate brain homeostasis. Science 353, aad8670.

Maynard, C.L., Elson, C.O., Hatton, R.D., Weaver, C.T., 2012. Reciprocal interactions of the intestinal microbiota and immune system. Nature 489, 231—241.

McMurran, C.E., Guzman de la Fuente, A., Penalva, R., Ben Menachem-Zidon, O., Dombrowski, Y., Falconer, J., Gonzalez, G.A., Zhao, C., Krause, F.N., Young, A.M.H., Griffin, J.L., Jones, C.A., Hollins, C., Heimesaat, M.M., Fitzgerald, D.C., Franklin, R.J.M., 2019. The microbiota regulates murine inflammatory responses to toxin-induced CNS demyelination but has minimal impact on remyelination. Proc Natl Acad Sci U S A 116, 25311—25321.

McVey Neufeld, K.A., Mao, Y.K., Bienenstock, J., Foster, J.A., Kunze, W.A., 2013. The microbiome is essential for normal gut intrinsic primary afferent neuron excitability in the mouse. Neuro Gastroenterol Motil 25, 183—e188.

Miller, M.W., Nowakowski, R.S., 1988. Use of bromodeoxyuridine-immunohistochemistry to examine the proliferation, migration and time of origin of cells in the central nervous system. Brain Res 457, 44—52.

Mohle, L., Mattei, D., Heimesaat, M.M., Bereswill, S., Fischer, A., Alutis, M., French, T., Hambardzumyan, D., Matzinger, P., Dunay, I.R., Wolf, S.A., 2016. Ly6C(hi) monocytes provide a link between antibiotic-induced changes in gut microbiota and adult hippocampal neurogenesis. Cell Rep 15, 1945–1956.

Needham, B.D., Funabashi, M., Adame, M.D., Wang, Z., Boktor, J.C., Haney, J., Wu, W.L., Rabut, C., Ladinsky, M.S., Hwang, S.J., Guo, Y., Zhu, Q., Griffiths, J.A., Knight, R., Bjorkman, P.J., Shapiro, M.G., Geschwind, D.H., Holschneider, D.P., Fischbach, M.A., Mazmanian, S.K., 2022. A gut-derived metabolite alters brain activity and anxiety behaviour in mice. Nature 602, 647–653.

Nimmerjahn, A., Kirchhoff, F., Helmchen, F., 2005. Resting microglial cells are highly dynamic surveillants of brain parenchyma in vivo. Science 308, 1314–1318.

Ogbonnaya, E.S., Clarke, G., Shanahan, F., Dinan, T.G., Cryan, J.F., O'Leary, O.F., 2015. Adult hippocampal neurogenesis is regulated by the microbiome. Biol Psychiatr 78, e7–e9.

Ong, I.M., Gonzalez, J.G., McIlwain, S.J., Sawin, E.A., Schoen, A.J., Adluru, N., Alexander, A.L., Yu, J.J., 2018. Gut microbiome populations are associated with structure-specific changes in white matter architecture. Transl Psychiatry 8, 6.

Pan, J.X., Deng, F.L., Zeng, B.H., Zheng, P., Liang, W.W., Yin, B.M., Wu, J., Dong, M.X., Luo, Y.Y., Wang, H.Y., Wei, H., Xie, P., 2019. Absence of gut microbiota during early life affects anxiolytic Behaviors and monoamine neurotransmitters system in the hippocampal of mice. J Neurol Sci 400, 160–168.

Pasciuto, E., Burton, O.T., Roca, C.P., Lagou, V., Rajan, W.D., Theys, T., Mancuso, R., Tito, R.Y., Kouser, L., Callaerts-Vegh, Z., de la Fuente, A.G., Prezzemolo, T., Mascali, L.G., Brajic, A., Whyte, C.E., Yshii, L., Martinez-Muriana, A., Naughton, M., Young, A., Moudra, A., Lemaitre, P., Poovathingal, S., Raes, J., De Strooper, B., Fitzgerald, D.C., Dooley, J., Liston, A., 2020. Microglia require CD4 T cells to complete the fetal-to-adult transition. Cell 182, 625–640 e624.

Philip, V., Newton, D.F., Oh, H., Collins, S.M., Bercik, P., Sibille, E., 2021. Transcriptional markers of excitation-inhibition balance in germ-free mice show region-specific dysregulation and rescue after bacterial colonization. J Psychiatr Res 135, 248–255.

Quan, N., Whiteside, M., Herkenham, M., 1998. Cyclooxygenase 2 mRNA expression in rat brain after peripheral injection of lipopolysaccharide. Brain Res 802, 189–197.

Sahay, A., Hen, R., 2007. Adult hippocampal neurogenesis in depression. Nat Neurosci 10, 1110–1115.

Salter, M.W., Stevens, B., 2017. Microglia emerge as central players in brain disease. Nat Med 23, 1018–1027.

Sanmarco, L.M., Wheeler, M.A., Gutierrez-Vazquez, C., Polonio, C.M., Linnerbauer, M., Pinho-Ribeiro, F.A., Li, Z., Giovannoni, F., Batterman, K.V., Scalisi, G., Zandee, S.E.J., Heck, E.S., Alsuwailm, M., Rosene, D.L., Becher, B., Chiu, I.M., Prat, A., Quintana, F.J., 2021. Gut-licensed IFNgamma(+) NK cells drive LAMP1(+) TRAIL(+) anti-inflammatory astrocytes. Nature 590, 473–479.

Sapolsky, R., Rivier, C., Yamamoto, G., Plotsky, P., Vale, W., 1987. Interleukin-1 stimulates the secretion of hypothalamic corticotropin-releasing factor. Science 238, 522–524.

Schafer, D.P., Lehrman, E.K., Kautzman, A.G., Koyama, R., Mardinly, A.R., Yamasaki, R., Ransohoff, R.M., Greenberg, M.E., Barres, B.A., Stevens, B., 2012. Microglia sculpt postnatal neural circuits in an activity and complement-dependent manner. Neuron 74, 691–705.

Scott, G.A., Terstege, D.J., Vu, A.P., Law, S., Evans, A., Epp, J.R., 2020. Disrupted neurogenesis in germ-free mice: effects of age and sex. Front Cell Dev Biol 8, 407.

Stewart Campbell, A., Needham, B.D., Meyer, C.R., Tan, J., Conrad, M., Preston, G.M., Bolognani, F., Rao, S.G., Heussler, H., Griffith, R., Guastella, A.J., Janes, A.C., Frederick, B., Donabedian, D.H., Mazmanian, S.K., 2022. Safety and target engagement of an oral small-molecule sequestrant in adolescents with autism spectrum disorder: an open-label phase 1b/2a trial. Nat Med 28, 528—534.

Strandwitz, P., 2018. Neurotransmitter modulation by the gut microbiota. Brain Res 1693, 128—133.

Strandwitz, P., Kim, K.H., Terekhova, D., Liu, J.K., Sharma, A., Levering, J., McDonald, D., Dietrich, D., Ramadhar, T.R., Lekbua, A., Mroue, N., Liston, C., Stewart, E.J., Dubin, M.J., Zengler, K., Knight, R., Gilbert, J.A., Clardy, J., Lewis, K., 2019. GABA-modulating bacteria of the human gut microbiota. Nat Microbiol 4, 396—403.

Tanabe, S., Yamashita, T., 2018a. B-1a lymphocytes promote oligodendrogenesis during brain development. Nat Neurosci 21, 506—516.

Tanabe, S., Yamashita, T., 2018b. The role of immune cells in brain development and neurodevelopmental diseases. Int Immunol 30, 437—444.

Thion, M.S., Ginhoux, F., Garel, S., 2018. Microglia and early brain development: an intimate journey. Science 362, 185—189.

Turnbull, A.V., Rivier, C.L., 1999. Regulation of the hypothalamic-pituitary-adrenal axis by cytokines: actions and mechanisms of action. Physiol Rev 79, 1—71.

van Praag, H., Schinder, A.F., Christie, B.R., Toni, N., Palmer, T.D., Gage, F.H., 2002. Functional neurogenesis in the adult hippocampus. Nature 415, 1030—1034.

Wei, G.Z., Martin, K.A., Xing, P.Y., Agrawal, R., Whiley, L., Wood, T.K., Hejndorf, S., Ng, Y.Z., Low, J.Z.Y., Rossant, J., Nechanitzky, R., Holmes, E., Nicholson, J.K., Tan, E.K., Matthews, P.M., Pettersson, S., 2021. Tryptophan-metabolizing gut microbes regulate adult neurogenesis via the aryl hydrocarbon receptor. Proc Natl Acad Sci U S A 118, e2021091118.

Zhao, Y.F., Wei, D.N., Tang, Y., 2021. Gut microbiota regulate astrocytic functions in the brain: possible therapeutic consequences. Curr Neuropharmacol 19, 1354—1366.

Zhou, T., Yang, M., Zhang, G., Kang, L., Yang, L., Guan, H., 2020. Long non-coding RNA nuclear paraspeckle assembly transcript 1 protects human lens epithelial cells against $H(2)O(2)$ stimuli through the nuclear factor kappa b/p65 and p38/mitogen-activated protein kinase axis. Ann Transl Med 8, 1653.

# CHAPTER 7

# Microbiota in neurodevelopmental disorders

## Introduction

The development, composition, stability, and diversity of the infant gut microbiota and fundamental aspects of gut–brain axis development are intertwined with implications for neurodevelopment (Clarke et al., 2014). Deviations from the optimal assembly of the infant gut microbiome during critical time windows are likely to impact the central nervous system (CNS), a suggestion that is now supported by substantial preclinical literature (Codagnone et al., 2019b). The clinical data have started to catch up, and, in this chapter, we outline these translational efforts and evaluate the evidence for microbiome alterations in neurodevelopment disorders with a focus on autism spectrum disorders (ASDs). We consider the importance of microbiota to brain communication in neurodevelopment as divined from studies evaluating the maternal microbiota and brain development from conception to birth, the microbiota in healthy postnatal brain development, and factors influencing the assembly of the microbiota in early life such as mode of delivery and diet (Fig. 7.1).

## Microbiota to brain communication in neurodevelopment

Much of the clinical literature to date has focused on gut microbiota composition during different stages of infancy and subsequent behavioral outcomes distal to that assessment. These studies have often screened for cognitive outcomes and communication deficits, an emphasis supported by the preclinical literature (Codagnone et al., 2019b). Higher gut microbiota alpha diversity at 1 year old was associated with lower scores on the overall composite score on Mullen Scales of Early Learning and on the visual

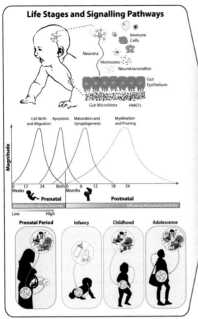
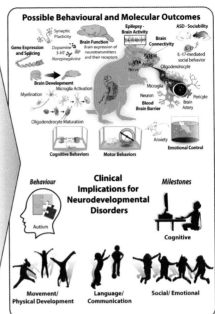

**Figure 7.1** *The gut microbiome and brain development.* There is considerable overlap between the assembly of the infant gut microbiota across multiple life stages and fundamental aspects of brain development. Suboptimal trajectories of gut microbiota assembly can lead to aberrant microbiota—gut—brain axis signaling to manifest as a range of behavioral and molecular outcomes, with substantial support from the preclinical literature including animal models of neurodevelopmental disorders. Research efforts are now focused on understanding the clinical implications in terms of associations between gut microbiota composition and function, and markers of cognitive, social, emotional, communication and physical development. The data from gut microbiota assessments in neurodevelopmental disorders such as ASD also support associations with symptom expression and severity.

reception scale, and expressive language scale at 2 years old (Carlson et al., 2018). The neuroimaging data from this study did not indicate a major role for gut microbiota composition at 1 year old on regional brain volumes at 1 and 2 years old. A decreased relative abundance of *Prevotella* in fecal samples collected at 12 months old was associated with increased internalizing problem scores at 2 years (Loughman et al., 2020). Using the Bayley Scale of Infant Development, a *Bacteroidetes*-dominated gut microbiota at 12 months was associated with higher scores for cognition and language at age 2, a feature most prominent in male infants (Tamana et al., 2021).

The analysis of stool samples collected from 2.5-month-old infants, when compared to maternal reports of Infant Behavior Questionnaire-

Revised (IBQ-R) at the age of 6 months, revealed the presence of microbiota clusters associated with temperament in a sex-dependent manner (Aatsinki et al., 2019). There was also an association between a lower abundance of *Bifidobacterium* and a higher abundance of *Clostridium* at 2.5 months with increased attention toward fearful versus happy/neutral faces at 8 months (Aatsinki et al., 2020a). Cortisol stress responsiveness after a mild stressor was weakly associated with a diverse fecal microbiota, both assessed at the age of 2.5 months (Keskitalo et al., 2021). Prosocial-self-regulating temperament, termed "Effortful Control" and derived from maternal ratings of temperament, was associated with the relative abundance of *Akkermansia* in the gut microbiome of typically developing children assessed at 18−27 months old (Delgadillo et al., 2022). Associations between internalizing behaviors and gut microbiota alpha diversity in preschool children (3−5 years old) have also been reported, a study notable for the suggested links between short-chain fatty acids (SCFAs), and behavioral problems and somatic complaints (van de Wouw et al., 2022). Many of these initial studies relied on taxa associations with behavioral outcomes in the absence of a functional microbiota readout, and the need for reproducibility in larger sample sizes is clear.

While these studies represent an important first step toward understanding the importance of microbiota to brain communication in neurodevelopment, open questions remain. In particular, we are lacking knowledge of "the how and the when," in terms of which mechanisms are most important during which critical time window (Brodin, 2022). An increased focus on longitudinal studies is paramount to better understand the temporal relationships between gut microbiota composition and function, and neurodevelopmental outcomes in early life. The benefits of such an approach can be seen in a recent study that assessed the fecal microbiota composition of participants at 1, 3, and 4 months, and 6 and 10 years old, identifying microbiota clusters associated with externalizing behavior at age 10 (Ou et al., 2022). These clusters were defined according to varying relative abundances of *Bifidobacterium*, *Streptococcus*, *Enterococcus*, and *Prevotella_9* but await replication and validation in other cohorts (Ou et al., 2022). Sex-specific relationships between the infant microbiome and early-childhood behavioral outcomes relating to anxiety, depression, hyperactivity, and social behaviors have also been reported with higher diversity at 6 weeks associated with internalizing problems in boys but not girls at 3 years old (Laue et al., 2021). This study also evaluated samples harvested at 6 weeks, 1 year, and 2 years old, indicating that the relative abundance of *Bifidobacterium* at 6 weeks was also associated with adaptive skills scores in boys but not girls at 3 years old (Laue et al., 2021).

## Microbiota during pregnancy

There is a marked remodeling of the gut microbiota during pregnancy (Koren et al., 2012), a feature that has seen the maternal microbiota considered in the context of prenatal factors that contribute to the developmental origins of adult disease. This is a risk factor often framed as inappropriate remodeling of the maternal microbiota during pregnancy and subsequent vertical transmission of a suboptimal microbiota at birth (Codagnone et al., 2019b). Maternal stress has most frequently been considered as a modifier of the maternal microbiota, impacting optimal microbial transmission at birth and the most appropriate subsequent bacterial assembly trajectory of the neonate gut microbiota, via either the stress-induced modifications of the vaginal or gut microbiota (Jasarevic et al., 2015a, 2017).

Preclinical studies in a mouse model of early prenatal stress have implicated the maternal stress—related altered microbiota composition in the neonate gut with metabolic reprogramming of the offspring gut and brain (Jasarevic et al., 2015b). Prenatal stress was also associated with alterations in the fetal intestinal transcriptome and with the pattern of hypothalamic gene expression following the impact of chronic stress exposure in adulthood (Jasarevic et al., 2018). Adult male offspring exposed to prenatal stress exhibited decreases in *Bacteroides* and *Parabacteroides* and a reduction in social behavior in adulthood coupled with increased corticosterone release following social interaction (Gur et al., 2019). Meanwhile, it was reported that adult female offspring exposed to prenatal stress exhibited alterations in their gut microbiota as well as increased anxiety-like behavior and alterations in cognition (Gur et al., 2017).

A number of potential mechanisms (see Fig. 7.2) have been considered at the interface between prenatal stress, the maternal microbiome, and neurodevelopment. Maternal stress during pregnancy in mice revealed reductions in microbial strains, genes, and metabolic pathways associated with proinflammatory function (Antonson et al., 2020). The altered expression of genes involved in tryptophan and serotonin metabolism as a consequence of prenatal stress exposures has also been implicated in aberrant fetal neurodevelopment (Galley et al., 2021). This is consistent with neuroinflammation and decreased serotonin metabolism in the cortex in

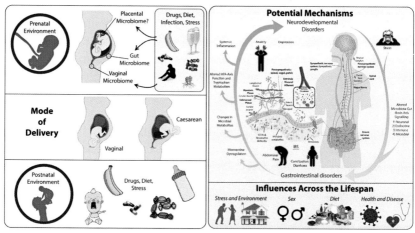

**Figure 7.2** *Potential mechanisms at the interface between the gut microbiome, prenatal and postnatal factors, and neurodevelopment.* A variety of prenatal and postnatal factors are thought to contribute to neurodevelopmental outcomes. Potential mechanisms at the interface between these factors, the gut microbiome, and neurodevelopment can be viewed through the lens of the microbiome—gut—brain axis. The assembly of the infant gut microbiome follows a number of paths. For example, the gut and vaginal microbiome of the mother can be altered by diet, drugs, infection, and stress before birth. If the birth occurs through vaginal delivery, the maternal vaginal microbiota first colonizes the newborn's gut. If Cesarean delivery occurs, the maternal skin microbiome first colonizes the infant's gut. Early in life, diet, drugs, and stress can also affect the infant microbiota composition, factors which can also exert an influence across the lifespan. Deviations from the optimal assembly can manifest as alterations in gut—brain axis signaling in key pathways such as tryptophan metabolism, immune function, and HPA axis outputs with implications for brain and gastrointestinal function in health and disease.

adulthood of male offspring exposed to prenatal stress (Gur et al., 2019). A role for gut microbiota-derived metabolites produced during pregnancy in the formation of neural circuits during fetal growth has also been proposed (Jasarevic and Bale, 2019). This possibility was supported by the report that the maternal microbiome supports fetal thalamocortical axonogenesis, likely via microbial metabolites (Vuong et al., 2020).

The gut microbiome has also been assessed in prenatal murine models of neurodevelopmental disorders such as ASDs. Mice exposed in utero to valproic acid and subsequently displaying an ASD-like phenotype exhibited increased levels of cecal butyrate as well as alterations with the main phyla of *Bacteroidetes* and *Firmicutes* (de Theije et al., 2014). Maternal immune activation (MIA) via infection during pregnancy is also known to impact neurodevelopment and increase the risk for ASD (Minakova and Warner, 2018).

Animal models of MIA include treating pregnant mice with the viral mimetic polyinosinic-polycytidylic acid (Poly I:C) during key neuro-developmental time windows (Lammert and Lukens, 2019). The MIA protocol induces social deficits that have been associated with altered endocrine responses to stress and gut permeability deficits consistent with an impact on gut–brain axis signaling (Morais et al., 2018) as well as gut microbiota alterations (Hsiao et al., 2013). Pregnant mice were more likely to produce offspring with MIA-associated abnormalities if they had been colonized with mouse commensal segmented filamentous bacteria or human commensal bacteria that induce intestinal TH17 cells (Kim et al., 2017). The induction of MIA using lipopolysaccharide (LPS) resulted in an ASD-like microbiota profile in the offspring in addition to the expected deficits in social, anxiety, and repetitive behaviors (Lee et al., 2021).

Poor quality diet and obesity impair brain function and cognition across the lifespan (Kendig et al., 2021). Maternal obesity during pregnancy has been associated with an increased risk of neurodevelopmental disorders and it has been reported in the preclinical literature that a maternal high-fat diet (MHFD) induces a shift in both the maternal and offspring gut microbiota that negatively impacts offspring social behavior (Buffington et al., 2016). It has also been demonstrated that a maternal diet deficient in n-3 PUFAs worsens MIA-induced early-life alterations in offspring gut microbiota composition and local inflammation, deficits that correlate with effects on transcriptional and behavioral levels in adulthood (Leyrolle et al., 2021).

Transmission of a suboptimal maternal microbiota at birth and an adverse impact of the maternal microbiota and/or its metabolites during fetal development have thus been linked in the preclinical literature to the behavioral consequences of maternal exposures to stress, infection, and poor diet (Jasarevic and Bale, 2019). There are relatively few studies in human populations that have evaluated the interface between these prenatal ex-posures, neurodevelopment, and the gut microbiota. A recent study investigating the relationship between the prenatal fecal microbiota, pre-natal diet, and childhood behavior reported that alpha diversity of the maternal fecal microbiota during the third trimester of pregnancy predicted child internalizing behavior at 2 years old (Dawson et al., 2021). The au-thors also reported that taxa from butyrate-producing families, a key SCFA for gastrointestinal (GI) function also implicated in signaling cascades important for neurodevelopment (O'Riordan et al., 2022), were more abundant in mothers of children with normative behavior and that a healthy prenatal diet was associated with a higher alpha diversity of maternal fecal microbiota (Dawson et al., 2021).

Higher levels of prenatal maternal stress, depression, and anxiety have also been associated with GI permeability and systemic inflammation, indicative of altered gut—brain axis signaling, although the implications for the offspring were not evaluated (Keane et al., 2021). The offspring of mothers with high cumulative self-reported and physiological stress loads during pregnancy had significantly higher relative abundances of Proteobacterial groups and lower relative abundances of lactic acid bacteria (e.g., *Lactobacillus*) and *Bifidobacteria* (Zijlmans et al., 2015). Subsequently, an association between maternal general anxiety, but not psychosocial stress variables, and microbial composition was reported (Hechler et al., 2019). Chronic maternal prenatal psychological distress and hair cortisol concentration were associated with infant fecal microbiota composition assessed at 2.5 months old (Aatsinki et al., 2020b). Maternal depressive symptoms have also been linked to reduced fecal immunoglobulin A concentrations, important for infant gut mucosal immunity, in the offspring (Kang et al., 2018b). Tryptophan metabolites, primary and secondary bile acids, and SCFAs have also been associated with maternal depression and anxiety during pregnancy (Kimmel et al., 2022).

## Stress and the postnatal microbiota

Adult mental health problems are frequently attributed to adverse childhood experiences (Kessler et al., 2010). In the preclinical literature, stress during the early postnatal period has most often been investigated using the maternal separation model of brain—gut axis dysfunction (O'Mahony et al., 2011). This early-life stress exposure leads in adult rats to behavioral alterations reminiscent of both depression and irritable bowel syndrome (IBS), accompanied by gut microbiota alterations (O'Mahony et al., 2009). In mice, it has been demonstrated that aspects of the aberrant behavioral phenotype require the presence of the gut microbiota in early life (De Palma et al., 2015).

In humans, early-life adversity has been associated in adulthood with four gut-regulated metabolites within the glutamate pathway (5-oxoproline, malate, urate, and glutamate gamma methyl ester) and alterations in functional brain connectivity linked to scores for perceived stress, anxiety, and depression (Coley et al., 2021). In physically and psychiatrically healthy adult pregnant women, the experience of multiple childhood adversities was associated with changes in gut microbiota composition during pregnancy which the authors linked to altered inflammatory status and glucocorticoid responses to an acute stressor (Hantsoo et al., 2019). In healthy 5–7-year-old children, associations between socioeconomic risk, behavioral dysregulation, caregiver behavior, and the gut microbiome were evaluated. Interesting associations were reported between the relative abundance of *B. fragilis* and lower incidents of family turmoil and fewer events on the Life Events Checklist, and reduced levels of aggressive behavior, emotional reactivity, externalizing behavior, sadness, and impulsivity (Hantsoo and Zemel, 2021). A higher relative abundance of the *Prevotella*, *Bacteroides*, *Coprococcus*, *Streptococcus*, and *Escherichia* genera has been recorded in adolescents who had recently been in institutional care, used as a proxy for adverse early-life experiences, compared to noninstitutionalized adolescents (Reid et al., 2021). Previous adverse care experiences were also associated with an increased incidence of GI symptoms, associated with concurrent and future anxiety (Callaghan et al., 2020). Adversity in this study was also associated with both indices of both alpha and beta diversity of microbial communities while the relative abundance of adversity-associated and adversity-independent bacteria was correlated with prefrontal cortex activation to emotional faces (Callaghan et al., 2020).

## C-section

Mode of birth results in a marked divergence in the initial assembly of the infant gut microbiota between Cesarean section (C-section) and vaginally delivered infants before converging again by weaning at 6–9 months old (Dominguez-Bello et al., 2010; Shao et al., 2019). The stunted maturation of the gut microbiota has been investigated for links to the health complications associated with C-section-delivered infants including childhood atopic diseases such as asthma (Fujimura et al., 2016; Renz and Skevaki, 2021; Stokholm et al., 2018, 2020). The neurodevelopmental consequences of C-section delivery have also come under scrutiny although the epidemiological evidence for this association is less clear than for other health complications, probably due to familial confounding by genetic and/ or environmental factors (Curran et al., 2015a,b; Zhang et al., 2019, 2021). Meanwhile, the preclinical literature supports the presence of enduring behavioral effects (Morais et al., 2020) and metabolic dysfunction (Martinez et al., 2017) following C-section delivery in the mouse.

# Preterm

Characteristic patterns of microbial colonization associated with preterm infant gut have also been reported (Aguilar-Lopez et al., 2021), with an increased relative abundance of *Klebsiella* spp. identified as a possible risk factor for necrotizing enterocolitis (Rao et al., 2021). A significant effect of gestational age at birth endures up to 4 years old with the preterm delivery microbiota best discriminated by the relative abundance of *Lactobacillus* (year 1), *Streptococcus* (year 2), and *Carnobacterium* (year 4) (Fouhy et al., 2019). The literature evaluating the associations between the preterm gut microbiota and markers of neurodevelopment includes a report documenting a reduced relative abundance of *Bacteroidota* and *Lachnospiraceae* in preterm infants with suboptimal head circumference growth trajectories (Oliphant et al., 2021). A study evaluating the impact of antibiotic use in symptomatic preterm neonates in the first 48 h after birth revealed the enduring impact of antibiotics on the relative abundance of *Veillonella*, a gram-negative gut bacteria associated with fecal GABA concentrations (Russell et al., 2021).

A study evaluating the gut microbiota, immune system, and neurophysiological development of 60 extremely premature infants reported a *Klebsiella*-dominated gut microbiota as a compositional signature of infants with brain pathologies (Seki et al., 2021). The delayed onset of electrocortical maturation and the transient suppression of neurophysiological development reported in this study suggest a role for aberrant microbiota—gut—brain axis signaling in perinatal white matter injury (Clarke et al., 2021; Seki et al., 2021). This is in line with the preclinical literature linking microbiota depletion to hypermyelination in the prefrontal cortex (Gacias et al., 2016; Hoban et al., 2016; Keogh et al., 2021; Lynch et al., 2021). Further insights are required to understand how this aberrant assembly trajectory of the microbiota can be beneficially reshaped in preterm infants toward supporting healthy brain development and to parse the potentially confounding influence of birth mode and antibiotic use (Clarke et al., 2021; Healy et al., 2022).

# Antibiotics

Preclinical research indicates that antibiotic use (see also Chapter 10) is an important feature of the postnatal period with neonatal exposure to antibiotics in rodents impacting the microbiota—gut—brain axis, visible as alterations in the microbiota, brain inflammation, and behavior

(Leclercq et al., 2017). The potential implications of this study were further substantiated when newborn mice exposed to low-dose penicillin resulted in alterations in intestinal microbiota population structure and composition concurrently with transcriptomic alterations in pathways in the frontal cortex and amygdala implicated in neurodevelopmental and neuropsychiatric disorders (Volkova et al., 2021). The effects of antibiotic exposure on the developing infant gut microbiome have also been explored in clinical samples to reveal decreased microbiome stability and diversity and transiently increased transcription of antibiotic-resistant genes (Yassour et al., 2016). The impact of commonly prescribed antibiotics in early life on brain health later in life in the context of the microbiota−gut−brain axis requires further elaboration in clinical populations, although separating the effects of the antibiotics from the impact of the reason for their deployment remains a challenge.

## Mode of feeding

The assembly trajectory of the early postnatal gut microbiota community is strongly influenced by diet as demonstrated by the microbiota differences reported between breastfed and formula-fed infants (Penders et al., 2006). There may be significant long-term neurodevelopmental benefits of breastfeeding (Bar et al., 2016) but the role of the gut microbiota as a mediator of these benefits is currently unclear. This is likely to be an important focus as future studies seek to establish the relationship between gut microbiota and neurodevelopmental outcomes.

## General

The preclinical literature continues to outline the possible interventions targeting the infant gut microbiota to support neurodevelopment and this includes options such as prebiotics, probiotics, and diet (Codagnone et al., 2019b). These approaches have not yet been evaluated systematically in clinical populations although it is notable that daily administration of a combination of *Bifidobacterium bifidum* and *Lactobacillus acidophilus* was not sufficient to prevent the *Klebsiella* bloom or the associated perinatal white matter injury in preterm infants with brain injury (Seki et al., 2021). This is consistent with the outcome of a meta-analysis of randomized controlled trials (RCTs) in preterm infants indicating that prebiotic and/or probiotic supplementation did not alter neurodevelopmental outcomes

(Upadhyay et al., 2020). Nutritional interventions targeting the microbiota—gut—brain axis in early life are warranted in less severe pathologies where the neurodevelopmental alterations may be more tractable. Advances in infant formula science, based on refining nutritional content to better mimic the functionality of breast milk, are geared toward optimal programming of the developing microbiome and immune system (Ahern et al., 2019).

## Microbiota in neurodevelopmental disorders

The preclinical research linking the gut microbiome to brain development is supported by clinical studies focused on the association between the gut microbiota and various developmental milestones linked to emotion, social behavior, communication, and cognition (Codagnone et al., 2019a). A further line of evidence comes from the evaluation of the gut microbiota in relevant clinical neurodevelopmental disorders. Of these, the majority have focused on the role of the gut microbiome in ASD. ASD is a multifaceted neurodevelopmental disorder with a panoply of alterations across domains like social communication, anxiety, repetitive stereotypical behaviors, immune dysregulation, and GI issues (Lord et al., 2020).

The presence of autistic symptoms has been associated with lower relative abundances of the genera *Prevotella*, *Coprococcus*, and unclassified Veillonellaceae in an early pyrosequencing study conducted in 20 neurotypical and 20 autistic children (Kang et al., 2013). There are now many studies looking at the gut microbiome community in ASD and although there is an absence of a consensus signature, most have reported a difference in the composition of the gut bacteria between ASD cases and their controls (Vellingiri et al., 2022). These compositional assessments have given way to more functional microbiota readouts with differences in fecal microbial metabolites such as higher isopropanol and p-cresol concentrations and lower GABA concentrations reported (Kang et al., 2018a, 2020).

An open-label study using a fecal microbiota transplantation (FMT) demonstrated improvements in GI symptoms, ASD symptoms, and the microbiome (Kang et al., 2017), and these improvements were sustained postintervention suggesting a potential long-term benefit associated with this option (Kang et al., 2019). In addition to modulation of the ASD-associated gut microbiome via engraftment of the donor microbiota, there is also evidence that the alterations in fecal and plasma metabolites in children with ASD are modulated following FMT, including metabolic features related to nicotinate/nicotinamide and purine metabolism (Kang et al., 2020).

Despite the accumulation of evidence from clinical studies supporting alteration in gut microbiota function, there has been ongoing debate about how or if these observations are causally related to ASD. Indeed, it has been difficult to reconcile the microbiome literature with the fact that there is a large (host) genetic contribution to ASD (Lord et al., 2020). Concerns that the gut microbiome alterations reported in ASD were diet-related received support recently from a study that found *Romboutsia timonensis* was the only taxa associated with an autism diagnosis (Yap et al., 2021). This dataset instead supported less-diverse dietary preferences in ASD as the driver of reduced microbial taxonomic diversity and looser stool consistency (Yap et al., 2021).

Whatever the driver of the reported gut microbiota alterations in ASD, it remains clear that the microbiota can still have a role in symptom modification. This view is supported by studies where transplanted gut microbiota from human donors with ASD-induced hallmark autistic behaviors in the recipient animals as well as alternative splicing of ASD-relevant genes in the CNS, which was linked to specific bacterial taxa and their metabolites (Sharon et al., 2019). Mechanistic options include the microbial metabolite 4-ethylphenyl sulfate (4EPS), which can reach the brain and is associated with oligodendrocyte function and myelination patterns of neuronal axons (Needham et al., 2022). This is aligned with a study that illustrated that microbial metabolites such as imidazole propionate or those produced following additional host processing such as 3-indoxyl-sulfate, trimethylamine-N-oxide, and phenylacetylglycine are present in the CNS of mice in the neonatal period and into adulthood (Swann et al., 2020). Of note, it was reported that a circulating marker of intestinal epithelial damage, I-FABP, was associated with more severe behavioral phenotypes in very young children with ASD (Teskey et al., 2021). Alternative mechanisms suggested from animal models of ASD-like phenotypes include a role for the vagus nerve in social behavior (Sgritta et al., 2019) and alterations in tryptophan metabolism and bile acids in GI dysfunction (Golubeva et al., 2017).

Schizophrenia is a complex heterogenous psychiatric disorder that involves abnormal neurodevelopment likely arising from the complex interplay between genetic and environmental risk factors (Kahn et al., 2015). The symptom profile associated with schizophrenia includes reduced social, cognitive, and emotional function in addition to the characteristic psychotic features. All these behavioral domains have been associated with disrupted microbiota—gut—brain axis signaling, informing research efforts to evaluate these signaling pathways in the development of schizophrenia (Kelly et al., 2021). A number of studies have evaluated the gut microbiome composition in adulthood and recent systematic reviews and meta-analyses have attempted to create a consensus profile of compositional alterations. While indices of alpha diversity were not consistently altered in schizophrenia, multiple studies reported similar compositional alterations with an increased relative abundance of the genus *Eggerthella* and decreased relative abundances of the genera *Faecalibacterium* and *Coprococcus* (McGuinness et al., 2022; Nikolova et al., 2021). These compositional alterations were particularly notable as they were also similarly enriched and depleted in other psychiatric disorders including major depressive disorder, bipolar disorder, psychosis, and anxiety. Moreover, most studies reported β-diversity differences between schizophrenia cases and controls (McGuinness et al., 2022; Nikolova et al., 2021).

Additional analyses are underway to understand better the functional implications of these compositional alterations (Murray et al., 2021). Reports implicate alterations in butyrate production as well as bacterial glutamate and GABA metabolism (McGuinness et al., 2022; Nikolova et al., 2021), important microbiota—gut—brain axis signaling pathways, which fit conceptually with this hypothesis (Kelly et al., 2021). Indeed, transplantation of the schizophrenia-associated gut microbiome to mice results in relevant behavioral alterations in addition to modulation of kynurenine pathway metabolism (Zhu et al., 2020) and the glutamate—glutamine—GABA cycle (Zheng et al., 2019).

Our picture of the gut microbiota alterations in schizophrenia is somewhat blurred by the knowledge that the antipsychotic drugs, which are the mainstay of treatment, are associated with antimicrobial effects (Maier et al., 2018). This might explain the preclinical literature indicating that olanzapine treatment remodels gut microbiota composition (Davey et al., 2012), an impact associated with drug-induced weight gain (Davey et al., 2013). Nevertheless, there are some studies that have been able to assess microbiota composition in drug naïve patients, with lower numbers of fecal *Bifidobacterium* responsive to subsequent risperidone treatment (Yuan et al., 2018). It has also been reported that the increased relative abundance of *Lactobacillus* in first-episode psychosis is associated with symptom severity (Schwarz et al., 2018) as was the relative abundance of

*Succinvibrio* and *Corynebacterium* in patients with a schizophrenia diagnosis (Li et al., 2020).

## Conclusions, perspectives, and future directions

Considerable evidence from preclinical studies links the gut microbiota to brain development. Support for these observations from the clinical literature is steadily accumulating, both in terms of the overlapping trajectory of microbiota assembly and brain function and in neurodevelopmental disorders. Open questions remain in relation to the contribution of diet and medication use. Atypical neurodevelopment gives rise to complex and heterogenous clinical disorders, and it is now clear that the gut microbiota represents an important component of these complex diseases, with implications for the varying symptom profiles on display. It remains to be determined if these associations are causal in nature with additional mechanistic insights also required to help understand in more granular detail what microbiota—gut—brain axis signaling pathways are involved and when they are most important. This includes during adolescence, a time of enormous developmental change that has thus far been largely neglected in this field. Such information will be essential to expedite the development of successful interventions that target the microbiota and can be applied during the most appropriate critical time window in early life to prevent the emergence or combat the expression of neurodevelopmental deficits.

## References

Aatsinki, A.K., et al., 2020a. Infant fecal microbiota composition and attention to emotional faces. Emotion 22, 1159.

Aatsinki, A.K., et al., 2019. Gut microbiota composition is associated with temperament traits in infants. Brain Behav Immun 80, 849—858.

Aatsinki, A.K., et al., 2020b. Maternal prenatal psychological distress and hair cortisol levels associate with infant fecal microbiota composition at 2.5 months of age. Psychoneuroendocrinology 119, 104754.

Aguilar-Lopez, M., et al., 2021. A systematic review of the factors influencing microbial colonization of the preterm infant gut. Gut Microb 13 (1), 1—33.

Ahern, G.J., et al., 2019. Advances in infant formula science. Annu Rev Food Sci Technol 10, 75—102.

Antonson, A.M., et al., 2020. Unique maternal immune and functional microbial profiles during prenatal stress. Sci Rep 10 (1), 20288.

Bar, S., Milanaik, R., Adesman, A., 2016. Long-term neurodevelopmental benefits of breastfeeding. Curr Opin Pediatr 28 (4), 559—566.

Brodin, P., 2022. Immune-microbe interactions early in life: a determinant of health and disease long term. Science 376 (6596), 945—950.

Buffington, S.A., et al., 2016. Microbial reconstitution reverses maternal diet-induced social and synaptic deficits in offspring. Cell 165 (7), 1762—1775.

Callaghan, B.L., et al., 2020. Mind and gut: associations between mood and gastrointestinal distress in children exposed to adversity. Dev Psychopathol 32 (1), 309—328.

Carlson, A.L., et al., 2018. Infant gut microbiome associated with cognitive development. Biol Psychiatr 83 (2), 148—159.

Clarke, G., Aatsinki, A., O'Mahony, S.M., 2021. Brain development in premature infants: a bug in the programming system? Cell Host Microbe 29 (10), 1477—1479.

Clarke, G., et al., 2014. Priming for health: gut microbiota acquired in early life regulates physiology, brain and behaviour. Acta Paediatr 103 (8), 812—819.

Codagnone, M.G., et al., 2019a. Microbiota and neurodevelopmental trajectories: role of maternal and early-life nutrition. Ann Nutr Metab 74 (Suppl. 2), 16—27.

Codagnone, M.G., et al., 2019b. Programming bugs: microbiota and the developmental origins of brain health and disease. Biol Psychiatr 85 (2), 150—163.

Coley, E.J.L., et al., 2021. Early life adversity predicts brain-gut alterations associated with increased stress and mood. Neurobiol Stress 15, 100348.

Curran, E.A., et al., 2015a. Association between obstetric mode of delivery and autism spectrum disorder: a population-based sibling design study. JAMA Psychiatr 72 (9), 935—942.

Curran, E.A., et al., 2015b. Research review: birth by caesarean section and development of autism spectrum disorder and attention-deficit/hyperactivity disorder: a systematic review and meta-analysis. J Child Psychol Psychiatry 56 (5), 500—508.

Davey, K.J., et al., 2013. Antipsychotics and the gut microbiome: olanzapine-induced metabolic dysfunction is attenuated by antibiotic administration in the rat. Transl Psychiatry 3, e309.

Davey, K.J., et al., 2012. Gender-dependent consequences of chronic olanzapine in the rat: effects on body weight, inflammatory, metabolic and microbiota parameters. Psychopharmacology (Berl) 221 (1), 155—169.

Dawson, S.L., et al., 2021. Maternal prenatal gut microbiota composition predicts child behaviour. EBioMedicine 68, 103400.

De Palma, G., et al., 2015. Microbiota and host determinants of behavioural phenotype in maternally separated mice. Nat Commun 6, 7735.

de Theije, C.G., et al., 2014. Altered gut microbiota and activity in a murine model of autism spectrum disorders. Brain Behav Immun 37, 197—206.

Delgadillo, D., et al., 2022. Associations between gut microbes and social behavior in healthy 2-year-old children. Psychosom Med 84, 749—756.

Dominguez-Bello, M.G., et al., 2010. Delivery mode shapes the acquisition and structure of the initial microbiota across multiple body habitats in newborns. Proc Natl Acad Sci U S A 107 (26), 11971—11975.

Fouhy, F., et al., 2019. Perinatal factors affect the gut microbiota up to four years after birth. Nat Commun 10 (1), 1517.

Fujimura, K.E., et al., 2016. Neonatal gut microbiota associates with childhood multi-sensitized atopy and T cell differentiation. Nat Med 22 (10), 1187—1191.

Gacias, M., et al., 2016. Microbiota-driven transcriptional changes in prefrontal cortex override genetic differences in social behavior. Elife 5, e13442.

Galley, J.D., et al., 2021. Prenatal stress-induced disruptions in microbial and host tryptophan metabolism and transport. Behav Brain Res 414, 113471.

Golubeva, A.V., et al., 2017. Microbiota-related changes in bile acid & tryptophan metabolism are associated with gastrointestinal dysfunction in a mouse model of autism. EBioMedicine 24, 166—178.

Gur, T.L., et al., 2017. Prenatal stress affects placental cytokines and neurotrophins, commensal microbes, and anxiety-like behavior in adult female offspring. Brain Behav Immun 64, 50−58.

Gur, T.L., et al., 2019. Prenatal stress disrupts social behavior, cortical neurobiology and commensal microbes in adult male offspring. Behav Brain Res 359, 886−894.

Hantsoo, L., Zemel, B.S., 2021. Stress gets into the belly: early life stress and the gut microbiome. Behav Brain Res 414, 113474.

Hantsoo, L., et al., 2019. Childhood adversity impact on gut microbiota and inflammatory response to stress during pregnancy. Brain Behav Immun 75, 240−250.

Healy, D.B., et al., 2022. Clinical implications of preterm infant gut microbiome development. Nat Microbiol 7 (1), 22−33.

Hechler, C., et al., 2019. Association between psychosocial stress and fecal microbiota in pregnant women. Sci Rep 9 (1), 4463.

Hoban, A.E., et al., 2016. Regulation of prefrontal cortex myelination by the microbiota. Transl Psychiatry 6, e774.

Hsiao, E.Y., et al., 2013. Microbiota modulate behavioral and physiological abnormalities associated with neurodevelopmental disorders. Cell 155 (7), 1451−1463.

Jasarevic, E., Bale, T.L., 2019. Prenatal and postnatal contributions of the maternal microbiome on offspring programming. Front Neuroendocrinol 55, 100797.

Jasarevic, E., Rodgers, A.B., Bale, T.L., 2015a. A novel role for maternal stress and microbial transmission in early life programming and neurodevelopment. Neurobiol Stress 1, 81−88.

Jasarevic, E., et al., 2015b. Alterations in the vaginal microbiome by maternal stress are associated with metabolic reprogramming of the offspring gut and brain. Endocrinology 156 (9), 3265−3276.

Jasarevic, E., et al., 2017. Stress during pregnancy alters temporal and spatial dynamics of the maternal and offspring microbiome in a sex-specific manner. Sci Rep 7, 44182.

Jasarevic, E., et al., 2018. The maternal vaginal microbiome partially mediates the effects of prenatal stress on offspring gut and hypothalamus. Nat Neurosci 21 (8), 1061−1071.

Kahn, R.S., et al., 2015. Schizophrenia. Nat Rev Dis Prim 1, 15067.

Kang, D.W., et al., 2020. Distinct fecal and plasma metabolites in children with autism spectrum disorders and their modulation after microbiota transfer therapy. mSphere 5 (5), 10−1128.

Kang, D.W., et al., 2013. Reduced incidence of Prevotella and other fermenters in intestinal microflora of autistic children. PLoS One 8 (7), e68322.

Kang, D.W., et al., 2019. Long-term benefit of Microbiota Transfer Therapy on autism symptoms and gut microbiota. Sci Rep 9 (1), 5821.

Kang, D.W., et al., 2018a. Differences in fecal microbial metabolites and microbiota of children with autism spectrum disorders. Anaerobe 49, 121−131.

Kang, D.W., et al., 2017. Microbiota Transfer Therapy alters gut ecosystem and improves gastrointestinal and autism symptoms: an open-label study. Microbiome 5 (1), 10.

Kang, L.J., et al., 2018b. Maternal depressive symptoms linked to reduced fecal immunoglobulin A concentrations in infants. Brain Behav Immun 68, 123−131.

Keane, J.M., et al., 2021. Identifying a biological signature of prenatal maternal stress. JCI Insight 6 (2), e143007.

Kelly, J.R., et al., 2021. The role of the gut microbiome in the development of schizophrenia. Schizophr Res 234, 4−23.

Kendig, M.D., Leigh, S.J., Morris, M.J., 2021. Unravelling the impacts of western-style diets on brain, gut microbiota and cognition. Neurosci Biobehav Rev 128, 233−243.

Keogh, C.E., et al., 2021. Myelin as a regulator of development of the microbiota-gut-brain axis. Brain Behav Immun 91, 437−450.

Keskitalo, A., et al., 2021. Gut microbiota diversity but not composition is related to saliva cortisol stress response at the age of 2.5 months. Stress 24 (5), 551—560.

Kessler, R.C., et al., 2010. Childhood adversities and adult psychopathology in the WHO world mental health surveys. Br J Psychiatry 197 (5), 378—385.

Kim, S., et al., 2017. Maternal gut bacteria promote neurodevelopmental abnormalities in mouse offspring. Nature 549 (7673), 528—532.

Kimmel, M., et al., 2022. Metabolite trajectories across the perinatal period and mental health: a preliminary study of tryptophan-related metabolites, bile acids and microbial composition. Behav Brain Res 418, 113635.

Koren, O., et al., 2012. Host remodeling of the gut microbiome and metabolic changes during pregnancy. Cell 150 (3), 470—480.

Lammert, C.R., Lukens, J.R., 2019. Modeling autism-related disorders in mice with maternal immune activation (MIA). Methods Mol Biol 1960, 227—236.

Laue, H.E., et al., 2021. Sex-specific relationships of the infant microbiome and early-childhood behavioral outcomes. Pediatr Res 92, 580—591.

Leclercq, S., et al., 2017. Low-dose penicillin in early life induces long-term changes in murine gut microbiota, brain cytokines and behavior. Nat Commun 8, 15062.

Lee, G.A., et al., 2021. Maternal immune activation causes social behavior deficits and hypomyelination in male rat offspring with an autism-like microbiota profile. Brain Sci 11 (8), 1085.

Leyrolle, Q., et al., 2021. Maternal dietary omega-3 deficiency worsens the deleterious effects of prenatal inflammation on the gut-brain axis in the offspring across lifetime. Neuropsychopharmacology 46 (3), 579—602.

Li, S., et al., 2020. Altered gut microbiota associated with symptom severity in schizophrenia. PeerJ 8, e9574.

Lord, C., et al., 2020. Autism spectrum disorder. Nat Rev Dis Prim 6 (1), 5.

Loughman, A., et al., 2020. Gut microbiota composition during infancy and subsequent behavioural outcomes. EBioMedicine 52, 102640.

Lynch, C.M.K., et al., 2021. Wrapping things up: recent developments in understanding the role of the microbiome in regulating myelination. Curr Opin Physiol 23, 100468.

Maier, L., et al., 2018. Extensive impact of non-antibiotic drugs on human gut bacteria. Nature 555 (7698), 623—628.

Martinez 2nd, K.A., et al., 2017. Increased weight gain by C-section: functional significance of the primordial microbiome. Sci Adv 3 (10), eaao1874.

McGuinness, A.J., et al., 2022. A systematic review of gut microbiota composition in observational studies of major depressive disorder, bipolar disorder and schizophrenia. Mol Psychiatr 27 (4), 1920—1935.

Minakova, E., Warner, B.B., 2018. Maternal immune activation, central nervous system development and behavioral phenotypes. Birth Defects Res 110 (20), 1539—1550.

Morais, L.H., et al., 2018. Strain differences in the susceptibility to the gut-brain axis and neurobehavioural alterations induced by maternal immune activation in mice. Behav Pharmacol 29 (2 and 3-Spec Issue), 181—198.

Morais, L.H., et al., 2020. Enduring behavioral effects induced by birth by caesarean section in the mouse. Curr Biol 30 (19), 3761—37674 e6.

Murray, N., et al., 2021. Compositional and functional alterations in the oral and gut microbiota in patients with psychosis or schizophrenia: a systematic review. HRB Open Res 4, 108.

Needham, B.D., et al., 2022. A gut-derived metabolite alters brain activity and anxiety behaviour in mice. Nature 602 (7898), 647—653.

Nikolova, V.L., et al., 2021. Perturbations in gut microbiota composition in psychiatric disorders: a review and meta-analysis. JAMA Psychiatr 78 (12), 1343—1354.

O'Mahony, S.M., et al., 2011. Maternal separation as a model of brain-gut axis dysfunction. Psychopharmacology (Berl) 214 (1), 71–88.

O'Mahony, S.M., et al., 2009. Early life stress alters behavior, immunity, and microbiota in rats: implications for irritable bowel syndrome and psychiatric illnesses. Biol Psychiatr 65 (3), 263–267.

O'Riordan, K.J., et al., 2022. Short chain fatty acids: microbial metabolites for gut-brain axis signalling. Mol Cell Endocrinol 546, 111572.

Oliphant, K., et al., 2021. Bacteroidota and Lachnospiraceae integration into the gut microbiome at key time points in early life are linked to infant neurodevelopment. Gut Microb 13 (1), 1997560.

Ou, Y., et al., 2022. Development of the gut microbiota in healthy children in the first ten years of life: associations with internalizing and externalizing behavior. Gut Microb 14 (1), 2038853.

Penders, J., et al., 2006. Factors influencing the composition of the intestinal microbiota in early infancy. Pediatrics 118 (2), 511–521.

Rao, C., et al., 2021. Multi-kingdom ecological drivers of microbiota assembly in preterm infants. Nature 591 (7851), 633–638.

Reid, B.M., et al., 2021. Microbiota-immune alterations in adolescents following early life adversity: a proof of concept study. Dev Psychobiol 63 (5), 851–863.

Renz, H., Skevaki, C., 2021. Early life microbial exposures and allergy risks: opportunities for prevention. Nat Rev Immunol 21 (3), 177–191.

Russell, J.T., et al., 2021. Antibiotics and the developing intestinal microbiome, metabolome and inflammatory environment in a randomized trial of preterm infants. Sci Rep 11 (1), 1943.

Schwarz, E., et al., 2018. Analysis of microbiota in first episode psychosis identifies preliminary associations with symptom severity and treatment response. Schizophr Res 192, 398–403.

Seki, D., et al., 2021. Aberrant gut-microbiota-immune-brain axis development in premature neonates with brain damage. Cell Host Microbe 29 (10), 1558–15572 e6.

Sgritta, M., et al., 2019. Mechanisms underlying microbial-mediated changes in social behavior in mouse models of autism spectrum disorder. Neuron 101 (2), 246–259 e6.

Shao, Y., et al., 2019. Stunted microbiota and opportunistic pathogen colonization in caesarean-section birth. Nature 574 (7776), 117–121.

Sharon, G., et al., 2019. Human gut microbiota from autism spectrum disorder promote behavioral symptoms in mice. Cell 177 (6), 1600–1618 e17.

Stokholm, J., et al., 2018. Maturation of the gut microbiome and risk of asthma in childhood. Nat Commun 9 (1), 141.

Stokholm, J., et al., 2020. Delivery mode and gut microbial changes correlate with an increased risk of childhood asthma. Sci Transl Med 12 (569), eaax9929.

Swann, J.R., Spitzer, S.O., Diaz Heijtz, R., 2020. Developmental signatures of microbiota-derived metabolites in the mouse brain. Metabolites 10 (5), 172.

Tamana, S.K., et al., 2021. Bacteroides-dominant gut microbiome of late infancy is associated with enhanced neurodevelopment. Gut Microb 13 (1), 1–17.

Teskey, G., et al., 2021. Intestinal permeability correlates with behavioural severity in very young children with ASD: a preliminary study. J Neuroimmunol 357, 577607.

Upadhyay, R.P., et al., 2020. Effect of prebiotic and probiotic supplementation on neurodevelopment in preterm very low birth weight infants: findings from a meta-analysis. Pediatr Res 87 (5), 811–822.

van de Wouw, M., et al., 2022. Associations between the gut microbiota and internalizing behaviors in preschool children. Psychosom Med 84 (2), 159–169.

Vellingiri, B., et al., 2022. An anxious relationship between autism spectrum disorder and gut microbiota: a tangled chemistry? J Clin Neurosci 99, 169–189.

Volkova, A., et al., 2021. Effects of early-life penicillin exposure on the gut microbiome and frontal cortex and amygdala gene expression. iScience 24 (7), 102797.

Vuong, H.E., et al., 2020. The maternal microbiome modulates fetal neurodevelopment in mice. Nature 586 (7828), 281–286.

Yap, C.X., et al., 2021. Autism-related dietary preferences mediate autism-gut microbiome associations. Cell 184 (24), 5916–5931 e17.

Yassour, M., et al., 2016. Natural history of the infant gut microbiome and impact of antibiotic treatment on bacterial strain diversity and stability. Sci Transl Med 8 (343), 343ra81.

Yuan, X., et al., 2018. Changes in metabolism and microbiota after 24-week risperidone treatment in drug naive, normal weight patients with first episode schizophrenia. Schizophr Res 201, 299–306.

Zhang, T., et al., 2019. Association of cesarean delivery with risk of neurodevelopmental and psychiatric disorders in the offspring: a systematic review and meta-analysis. JAMA Netw Open 2 (8), e1910236.

Zhang, T., et al., 2021. Assessment of cesarean delivery and neurodevelopmental and psychiatric disorders in the children of a population-based Swedish birth cohort. JAMA Netw Open 4 (3), e210837.

Zheng, P., et al., 2019. The gut microbiome from patients with schizophrenia modulates the glutamate-glutamine-GABA cycle and schizophrenia-relevant behaviors in mice. Sci Adv 5 (2), eaau8317.

Zhu, F., et al., 2020. Transplantation of microbiota from drug-free patients with schizophrenia causes schizophrenia-like abnormal behaviors and dysregulated kynurenine metabolism in mice. Mol Psychiatr 25 (11), 2905–2918.

Zijlmans, M.A., et al., 2015. Maternal prenatal stress is associated with the infant intestinal microbiota. Psychoneuroendocrinology 53, 233–245.

# CHAPTER 8

# Microbiota in psychiatry

## Introduction

Our understanding of how the microbiome has an influence on health and disease continues to grow and is gaining momentum in biomedical science research. The microbiome has been implicated in the etiology and pathogenesis of a number of diseases and disorders including: inflammatory bowel disease, irritable bowel syndrome, celiac disease, asthma, obesity, cancer, neurodegenerative diseases, autoimmune diseases, as well as mental health disorders, including autism spectrum disorders, anxiety, depression, and schizophrenia (Abdel-Aziz et al., 2019; Armour et al., 2019; Audet, 2019; Caminero and Verdu, 2019; Halfvarson et al., 2017; Nguyen et al., 2019; Nikolova et al., 2021; Pittayanon et al., 2019; Tremlett et al., 2017; Xu et al., 2019a). This chapter will provide an overview of the recent literature, primarily the past 5 years, that has examined the association between the microbiome and mental health.

General evidence that microbes influence mental health resulted from an examination of the relationship between antibiotics and mental health. The diversity as well as the composition of gut microbiota has been shown to be influenced by a variety of different factors including antibiotics (Iizumi et al., 2017; Jernberg et al., 2010). Antibiotics have been used to treat infections in both humans and animals for decades and the exposure of antibiotics as a treatment for infections has shown to reduce the presence of bacteria within an individual's gut microbiome. The degree to which an individual's gut microbiome is altered depends on the antibiotic's chemical structure as well as dose and the period of use and baseline composition (Jernberg et al., 2010; Rashid et al., 2015; Rashidi et al., 2021). A few key studies demonstrate a link between antibiotic use and mental health. In a nested case-control study of 10 million adults in the United Kingdom, recurrent antibiotic treatment over an extended period was associated with increased risk of both anxiety and depression (Lurie et al., 2015).

*Microbiota Brain Axis*
ISBN 978-0-12-814800-6
https://doi.org/10.1016/B978-0-12-814800-6.00001-7

A population-based study in Finland showed that exposure to antibiotics in utero and in the first 2 years of life was associated with an increased risk of psychiatric disorders including those with a childhood onset (Lavebratt et al., 2019). Similarly, an observational study using data from the United Kingdom Biobank showed that long-term antibiotic use in early life was associated with anxiety and depression (Liang et al., 2020). Further, an observational study of 14,500 children in the United States showed that the number of courses of antibiotics in children was associated with childhood-onset allergy, obesity, and attention deficit hyperactivity disorder (ADHD) (Aversa et al., 2021). Together these studies suggest that the collective influence of the microbiota on host physiology is important to consider if researchers want to advance their understanding of the biological basis of mental health.

## Microbes and mood in healthy people

In the past decade, evidence from preclinical work in rodents supports a connection between stress, microbiota, and stress-related behaviors (Bailey et al., 2011; Bharwani et al., 2016; Borre et al., 2014; Clarke et al., 2013; Cryan and Dinan, 2012; Foster, 2016; Foster et al., 2016; Foster and McVey Neufeld, 2013; Heijtz et al., 2011; Neufeld et al., 2011; Stilling et al., 2015). Translating these stress-related data to generate evidence that the microbiome influences mood is ongoing and utilizes several approaches (Table 8.1). Using these approaches, the emerging data suggest that observations in animal studies are applicable to people. One popular approach, that links microbes to mood, is to administer probiotics or prebiotics to healthy individuals and to measure the impact on stress and mood-related outcomes. The use of probiotics and prebiotics to improve mental health is attractive to the public, both in general for healthy individuals and as a treatment option for depression or other psychiatric conditions (Burokas et al., 2017; Foster, 2017). Probiotics are live microorganisms that when administered in adequate amounts confer a health benefit on the host (Hill et al., 2014). Prebiotics are dietary products that when consumed can be fermented by commensal gut bacteria and alter microbiota composition or function (Gibson et al., 2010; Sarkar et al., 2016). Studies in healthy individuals provide evidence that probiotic, as well as prebiotic, consumption leads to improved mood, reduced stress hormone levels, reduced stress reactivity, and influenced brain activity in emotional centers (Allen et al., 2016; Benton et al., 2007; Kato-Kataoka et al., 2016; Ma et al., 2021;

**Table 8.1** Evidence that microbiota influence mental health in people.

| Approach | Key findings | References |
|---|---|---|
| Administration of probiotics to healthy individuals | • Improved mood<br>• Reduced stress hormone levels<br>• Reduced stress reactivity<br>• Influenced brain activity in stress centers | (Allen et al., 2016; Benton et al., 2007; Kato–Kataoka et al., 2016; Lew et al., 2019; Ma et al., 2021; Messaoudi et al., 2011; Schmidt et al., 2015; Steenbergen et al., 2015; Tillisch et al., 2013) |
| Association of bacterial composition with brain function using imaging | • Association between enterotype with emotional response, white matter connectivity, and brain volume | (Tillisch et al., 2017) |
| Comparison of compositional analysis between healthy and depressed individuals | • Differentially abundant taxa identified in MDD, BD, ADHD, GAD, SZ | (Evans et al., 2017; Hu et al., 2019; Lai et al., 2021; Lu et al., 2019; Painold et al., 2019; Prehn–Kristensen et al., 2018; Yolken et al., 2015; Zhang et al., 2021; Zheng et al., 2019, 2020) |
| Association of bacteria composition with clinical symptoms | • Taxa identified that are associated with disease severity<br>• Taxa identified that are associated with clinical symptoms | (Chen et al., 2019, 2021; Chung et al., 2019; Jiang et al., 2018; Lai et al., 2021; Liu et al., 2020; Madan et al., 2020; Minichino et al., 2021; Stevens et al., 2020; Valles–Colomer et al., 2019; Zhang et al., 2021) |
| Microbiome functional readouts | • Alterations in tryptophan metabolism and metabolic pathways<br>• Alterations in host and gut metabolites | (Bengesser et al., 2019; Bhattacharyya et al., 2019a,b; Caspani et al., 2021; Lai et al., 2021; Philippe et al., 2021; Ren et al., 2021) |
| Fecal microbiota transfer to rodents | • Transfer of fecal samples from MDD leads to anhedonic and depressive-like phenotype<br>• Transfer of fecal samples from SZ | (Kelly et al., 2016; Li et al., 2018; Liang et al., 2019; Zheng et al., 2016, 2019; Zhu et al., 2019) |

Messaoudi et al., 2011; Schmidt et al., 2015; Steenbergen et al., 2015; Tillisch et al., 2013; Wang et al., 2019). Notably, a randomized control trial in healthy individuals showed that consumption of *Lactobacillus plantarum* for 12 weeks reduced stress and anxiety symptoms in moderately stressed adults (Lew et al., 2019). Using metagenomic sequencing, a follow-up report showed changes in microbiota composition in response to probiotic administration including an increased abundance of *Bifidobacterium adolescentis, B. longum, Faecalibacterium prausnitzii,* and *Subdoligranulum sp 60_17,* and a decreased abundance of *Roseburia faecis* and *Fusicatenibacter saccharivorans.* A key neuroimaging study showed that probiotic consumption was associated with reduced engagement of an extensive brain network in response to an emotion recognition task (Tillisch et al., 2013). Using multiple imaging modalities, a separate neuroimaging study showed an association between bacterial composition and emotional response, white matter connectivity, and brain volume in healthy individuals (Tillisch et al., 2017). In addition, an association of salivary bacterial taxa with stress was reported in healthy adults (Kohn et al., 2020). The approach to measure microbiota in saliva samples is novel, but also allowed the researchers to consider the importance of time of day by taking four different samples from individuals at different times, demonstrating a diurnal pattern in salivary microbiota composition (Kohn et al., 2020). Together, these studies demonstrate that the microbiota influences mood in healthy adults. Moving into the clinic, several studies to date demonstrate a role for microbiota—brain signaling in psychiatric disorders.

## Microbes in mood and anxiety disorders

Across biomedical research, investigators have extended these preclinical findings to clinical populations. A common approach is to compare microbiota diversity and composition between healthy individuals and clinical populations. With respect to mood and anxiety disorders, studies have demonstrated differential composition of gut microbiota in major depressive disorder (MDD) and bipolar disorder (BP), and in general anxiety disorder (GAD). A reverse translation approach to demonstrate that the microbiome has a role in a particular clinical condition is fecal microbiota transplantation (FMT). This approach transfers fecal matter collected from healthy individuals or individuals with depression or other clinical conditions to recipient animals (mice or rats) and examines the impact on behavior or host systems (see recent review, recommendations, and guidelines (Gheorghe et al., 2021; Secombe et al., 2021; Settanni et al., 2021)). There are several options for the delivery of the sample to the recipient including transfer of fecal matter to germ-free (GF) animals, to microbiota-depleted animals by administering a course of antibiotics before FMT, or to animals with a normal microbiota composition. Transfer of fecal samples from individuals with depression to rodents resulted in an anhedonic and depressive-like phenotype demonstrating an active role for the microbiome in depression and depressive symptoms (Kelly et al., 2016; Zheng et al., 2016). A follow-up study that compared MDD-FMT mice to HC-FMT mice showed a significant change in liver metabolites in MDD-FMT mice and a few alterations in hippocampus or serum, with the authors suggesting liver metabolism as a possible factor in gut—brain signaling in MDD (Li et al., 2018).

## Major depressive disorder

In the past 5 years, several studies have examined compositional differences in microbiota, specifically gut bacteria, between healthy individuals (HC) and those with major depressive disorder (MDD). While reduced alpha diversity of the microbiome has been observed in MDD subjects (Liu et al., 2020), several studies have observed no differences in alpha diversity (Chen et al., 2018, 2021; Chung et al., 2019). Alpha diversity metrics are within-sample diversity measures including total observed species count, Shannon and Simpson diversity which examines richness (number) and evenness of the community, and Chao1 which estimates species richness. In general, higher alpha diversity has been associated with a better health status. More consistently across MDD studies, differences in beta diversity and bacterial composition have been reported between HC and MDD, although the specific bacterial taxa altered in MDD were variable across studies. This variability is best explained by the small sample size in published clinical studies in MDD and the large clinical heterogeneity that exists in MDD cohorts. A recent large population study reported an association between the microbiome and quality of life and depression (Valles-Colomer et al., 2019). The relative abundances of *Faecalibacterium* and *Coprococcus* bacteria were associated with higher quality of life, and a reduction of *Coprococcus* and *Dialister* spp were linked to depression, an observation that was validated in a second cohort (Valles-Colomer et al., 2019). These population-level findings have been reproduced in a recent clinical study of female MDD individuals compared to healthy volunteers; this study used random forest models to identify bacterial genera that are enriched in healthy individuals compared to MDD that included *Faecalibacterium* and *Coprococcus* (Chen et al., 2021). This study also identified bacterial genera enriched in MDD including *Escherichia-Shigella* and *Alistipes,* taxa that are suggested to be associated with increased inflammation (Chen et al., 2021), implicating microbiota—immune—brain signaling in MDD. Notably, gut microbiota are important to the intestinal barrier and increased permeability of the intestinal barrier can contribute to inflammation. Biomarkers of intestinal permeability, including zonulin and intestinal fatty acid binding protein (I-FABP), have been associated with depression and anxiety disorders (Stevens et al., 2018). Of interest, baseline abundance of *Coprococcus* was associated with improved mood in response to prebiotic administration in obese individuals (Leyrolle et al., 2021) suggesting that an individual's microbiome may influence treatment response to microbiota-targeted (or other) treatments. Across studies, several bacterial genera have been reported to be enriched in healthy individuals or enriched in MDD subjects (Chen et al., 2021; Chung et al., 2019; Lai et al., 2021; Liu et al., 2020; Madan et al., 2020; Stevens et al., 2020; Zhang et al., 2021). These studies have relied primarily on 16S rRNA sequencing and have included a small number of individuals, and in several cases, the analyses have not effectively considered the compositional nature of microbiota or controlled for false discovery. Nonetheless, some health-related taxa and depression-related taxa are emerging. In addition to the study noted above, a number of studies, comparing healthy volunteers to MDD, have reported a reduced abundance of *Faecalibacterium* in MDD individuals and also a reduced abundance of other members of the *Ruminococcaceae* family (Chen et al., 2018, 2021; Li et al., 2020; Liu et al., 2020; Zhang et al., 2021). *Faecalibacterium prausnitzii* is an abundant commensal in the healthy human gut, plays an important role in gut physiology (Lopez-Siles et al., 2017; Miquel et al., 2014), and is a key butyrate producer that can utilize carbohydrate sources that are both host- and diet-derived (Barcenilla et al., 2000; Duncan et al., 2002; Heinken et al., 2014). Short-chain fatty acids (SCFAs) including acetate, propionate, and butyrate are metabolites that are products of commensal fermentation (Rios-Covian et al., 2016). SCFAs produced by gut bacteria influence other commensals and are important to gut physiology but are also part of microbiota—host signaling systems that extend beyond the gut (Dalile et al., 2019; Rios-Covian et al., 2016).

While it is useful to identify microbial taxa that are depression-related, it is also important to consider the clinical impact of these alterations. Some studies examine the association of specific taxa with depression severity and symptoms (Chung et al., 2019; Liu et al., 2020; Zhang et al., 2021). While the analytical approaches as well as the taxonomical level considered in the analysis vary across studies, several interesting associations have emerged (Table 8.2). There are different strategies employed to consider associations between microbiota and clinical features. Associations are examined across the entire study population, which could include HC and MDD, or within subgroups which may include HC separate from MDD, males separate from females, or subgroups based on clinical features. Across studies, several genera have been associated with depression severity including *Sutterella, Pseudobutyrivibrio, Peptococcus, Parabacteroides, Haemophilus, Flavonifractor, Eggerthella, Dorea, Clostridium XI, Butyricicoccus,* and *Blautia* (Chung et al., 2019; Liu et al., 2020; Zhang et al., 2021). Parallel analysis of female and male individuals revealed 22 bacterial clades that differed in female MDD compared to female HC and six bacterial clades that differed in male MDD compared to male HC (Chen et al., 2018). Notably, there was no overlap in these taxa suggesting that it is important to consider sex differences in microbiota in clinical studies (Chen et al., 2018). Further, *Clostridium* XIVa, Erysipelotrichaceae incertae sedis, and *Streptococcus* were associated with depression severity in females, whereas *Veillonella* and *Collinsella* were associated with depression severity in males (Chen et al., 2018). In addition, studies have identified sleep-related taxa including *Blautia, Coprococcus, Doria, Haemophilus, Intestinibacter, Streptococcus*; stress-related taxa including *Blautia, Eggerthella, Holdemania, Prevotella, Ruminococcus and Sutterella* (Chung et al., 2019; Zhang et al., 2021); age-related taxa including *Asaccharobacter, Clostridium* XIVa, Erysipelotrichaceae incertae sedis, *Streptococcus, Faecalibacterium,* and *Lachnospira* incertae sedis in MDD females and Erysipelotrichaceae incertae sedis in MDD males (Chen et al., 2018). In healthy young adults, sleep quality was positively associated with alpha diversity as well as increased abundance of *Blautia* and *Ruminococcus* and reduced abundance of *Prevotella* (Grosicki et al., 2020). Alpha diversity and microbiota species richness were shown to be associated with anxiety and depression severity in an inpatient psychiatric population (73% MDD), and notably diversity at admission was predictive of remission at the time of release (Madan et al., 2020). Other host features including sex, age, BMI, and gastrointestinal disease were not associated with microbial diversity in this study (Madan et al., 2020). In a separate study, increased alpha diversity at baseline in unmedicated MDD patients was associated with treatment response at 6 months (Bharwani et al., 2020). Moreover, alpha diversity was linked to depression severity in young adult MDD individuals (Liu et al., 2020).

**Table 8.2** Genera that are associated with clinical symptoms in psychiatry.

| Taxa | Approach | Clinical symptom | Scale | Group | References | Analytical details |
|---|---|---|---|---|---|---|
| *Acetanaerobacterium* | 16S | Depression severity | MADRS | BD | (Hu et al., 2019) | Spearman Cor |
| *Anaerotruncus* | 16S | Depression severity | MADRS | BD | (Hu et al., 2019) | Spearman Cor |
| *Actinobacter* | 16S | Depression severity | MADRS | BD | (Hu et al., 2019) | Spearman Cor |
| *Bacteroides* | 16S | Axniety severity | HAMA | GAD | (Chen et al., 2019) | Spearman Cor |
| | 16S | Depression severity | HAMD | GAD | (Chen et al., 2019) | Spearman Cor |
| *Blautia* | 16S | Sleep | PSQI | MDD | (Zhang et al., 2021) | FDR>0.05 |
| | 16S | Depression severity | BDI | MDD | (Chung et al., 2019) | Spearman Cor |
| | 16S | Stress | PSS | MDD | (Chung et al., 2019) | Spearman Cor |
| *Blautia_obeum* | 16S | Sleep | PSQI | MDD | (Zhang et al., 2021) | FDR>0.05 |
| *butyrate-producing_bacterium_L2-50* | 16S | Sleep | PSQI | MDD | (Zhang et al., 2021) | FDR>0.05 |
| *Butyricicoccus* | 16S | Depression severity | HAMD | MDD | (Zhang et al., 2021) | FDR>0.05 |
| *Coprococcus_comes* | 16S | Sleep | PSQI | MDD | (Zhang et al., 2021) | FDR>0.05 |

**Table 8.2** Genera that are associated with clinical symptoms in psychiatry.—cont'd

| Taxa | Approach | Clinical symptom | Scale | Group | References | Analytical details |
|---|---|---|---|---|---|---|
| *Clostridiales bacterium 1 7 47FAA* | metag | Anxiety severity | GAD-7 | psych | (Madan et al., 2020) | KW |
| *Clostridium XI* | 16S | Depression severity | PSS | MDD | (Chung et al., 2019) | Spearman Cor |
| *Clostridium_paraputrificum* | 16S | Insomnia | ISI | HC | (Zhang et al., 2021) | FDR corrected |
| *Clostridium symbiosum* | metag | Depression severity | PHQ-9 | MDD | (Madan et al., 2020) | KW |
| *Coprococcus* | 16S | Sleep | PSQI | MDD | (Zhang et al., 2021) | FDR>0.05 |
| *uncultured_Coprococcus_sp.* | 16S | Sleep | PSQI | MDD | (Zhang et al., 2021) | FDR>0.05 |
| *Coprococcus catus* | metag | Depression severity | PHQ-9 | MDD | (Madan et al., 2020) | KW |
| *Cronobacter* | 16S | Depression severity | MADRS | BD | (Hu et al., 2019) | Spearman Cor |
| *Dorea* | 16S | Depression severity | HAMD | MDD | (Zhang et al., 2021) | FDR>0.05 |
| | 16S | Sleep | PSQI | MDD | (Zhang et al., 2021) | FDR>0.05 |
| *Dorea_formicigenerans* | 16S | Sleep | PSQI | MDD | (Zhang et al., 2021) | FDR>0.05 |
| *Eggerthella* | 16S | Depression severity | BDI | HC, MDD | (Chung et al., 2019) | Spearman Cor |
| | 16S | Anxiety | BAI | HC, MDD | (Chung et al., 2019) | Spearman Cor |

*Continued*

**Table 8.2** Genera that are associated with clinical symptoms in psychiatry.—cont'd

| Taxa | Approach | Clinical symptom | Scale | Group | References | Analytical details |
|---|---|---|---|---|---|---|
| | 16S | Stress | PSS | HC, MDD | (Chung et al., 2019) | Spearman Cor |
| | metag | Anxiety severity | GAD-7 | psych | (Madan et al., 2020) | KW |
| *Eggerthella lenta* | metag | Anxiety severity | GAD-7 | psych | (Madan et al., 2020) | KW |
| *Escherichia shigella* | 16S | Axniety severity | HAMA | GAD | (Chen et al., 2019) | Spearman Cor |
| | 16S | Depression severity | HAMD | GAD | (Chen et al., 2019) | Spearman Cor |
| *Eubacterium coprostanoligenes group* | 16S | Axniety severity | HAMA | GAD | (Chen et al., 2019) | Spearman Cor |
| | 16S | Depression severity | HAMD | GAD | (Chen et al., 2019) | Spearman Cor |
| *Faecalibacterium* | 16S | Depression severity | PHQ-9 | BD | (Evans et al., 2017) | Regression |
| | 16S | Sleep | PSQI | BD | (Evans et al., 2017) | Regression |
| *Flavonifracter* | 16S | Depression severity | PROMIS | HC, MDD | (Liu et al., 2020) | Univariate |
| | metag | Depression severity | PHQ-9 | psych | (Madan et al., 2020) | KW |
| *Flavonifracter plauti* | metag | Depression severity | PHQ-9 | psych | (Madan et al., 2020) | KW |

**Table 8.2** Genera that are associated with clinical symptoms in psychiatry.—cont'd

| Taxa | Approach | Clinical symptom | Scale | Group | References | Analytical details |
|---|---|---|---|---|---|---|
| *Haemophilus* | 16S | Depression severity | HAMD | HC | (Zhang et al., 2021) | FDR corrected |
| | 16S | Insomnia | ISI | HC | (Zhang et al., 2021) | FDR corrected |
| *Haemophilus_parainfluenzae* | 16S | Depression severity | HAMD | HC | (Zhang et al., 2021) | FDR corrected |
| | 16S | Insomnia | ISI | HC | (Zhang et al., 2021) | FDR corrected |
| *Holdemania* | 16S | Stress | PSS | HC, MDD | (Chung et al., 2019) | Spearman Cor |
| *Intestinibacter* | 16S | Sleep | PSQI | MDD | (Zhang et al., 2021) | FDR corrected |
| | 16S | Insomnia | ISI | MDD | (Zhang et al., 2021) | FDR >0.05 |
| *Lachnospiraceae_bacterium_1_4_56FAA* | metag | Depression severity | PHQ-9 | psych | (Madan et al., 2020) | KW |
| | metag | Anxiety severity | GAD-7 | psych | (Madan et al., 2020) | KW |
| *Mitsuokella* | 16S | Axniety severity | HAMA | GAD | (Chen et al., 2019) | Spearman Cor |
| | 16S | Depression severity | HAMD | GAD | (Chen et al., 2019) | Spearman Cor |

*Continued*

**Table 8.2** Genera that are associated with clinical symptoms in psychiatry.—cont'd

| Taxa | Approach | Clinical symptom | Scale | Group | References | Analytical details |
|---|---|---|---|---|---|---|
| *Mollicutes RF39* | 16S | Axniety severity | HAMA | GAD | (Chen et al., 2019) | Spearman Cor |
| | 16S | Depression severity | HAMD | GAD | (Chen et al., 2019) | Spearman Cor |
| *Parabacteroides* | 16S | Depression severity | BDI | HC, MDD | (Chung et al., 2019) | Spearman Cor |
| | 16S | Anxiety | BAI | HC, MDD | (Chung et al., 2019) | Spearman Cor |
| *Paraprevotella* | 16S | Depression severity | HAMD | MDD | (Liskiewicz et al., 2021) | Spearman Cor |
| *Peptococcus* | 16S | Depression severity | HAMD | MDD | (Zhang et al., 2021) | FDR>0.05 |
| *Peptostreptococcaceae OTU* | 16S | Depression severity | HAMD | MDD | (Zheng et al., 2020) | Random forest |
| *Prevotella* | 16S | Depression severity | BDI | HC, MDD | (Chung et al., 2019) | Spearman Cor |
| | 16S | Stress | PSS | HC, MDD | (Chung et al., 2019) | Spearman Cor |
| | 16S | Axniety severity | HAMA | GAD | (Chen et al., 2019) | Spearman Cor |
| | 16S | Depression severity | HAMD | GAD | (Chen et al., 2019) | Spearman Cor |
| *Pseudobutyrivibrio* | 16S | Depression severity | HAMD | HC | (Zhang et al., 2021) | FDR>0.05 |

**Table 8.2** Genera that are associated with clinical symptoms in psychiatry.—cont'd

| Taxa | Approach | Clinical symptom | Scale | Group | References | Analytical details |
|------|----------|------------------|-------|-------|------------|--------------------|
| *Raoultella* | 16S | Depression severity | MADRS | BD | (Hu et al., 2019) | Spearman Cor |
| *Ruminococcus* | 16S | Stress | PSS | HC, MDD | (Chung et al., 2019) | Spearman Cor |
| | 16S | Axniety severity | HAMA | GAD | (Chen et al., 2019) | Spearman Cor |
| | 16S | Depression severity | HAMD | GAD | (Chen et al., 2019) | Spearman Cor |
| *Stenotrophomonas* | 16S | Depression severity | MADRS | BD | (Hu et al., 2019) | Spearman Cor |
| *Streptococcus_anginosus* | metag | Depression severity | PHQ-9 | psych | (Madan et al., 2020) | KW |
| | metag | Anxiety severity | PHQ-9 | psych | (Madan et al., 2020) | KW |
| *Streptococcus_salivarius_subsp._salivarius* | 16S | Sleep | PSQI | MDD | (Zhang et al., 2021) | FDR>0.05 |
| *Subdoligranulum* | 16S | Axniety severity | HAMA | GAD | (Chen et al., 2019) | Spearman Cor |
| | 16S | Depression severity | HAMD | GAD | (Chen et al., 2019) | Spearman Cor |

*Continued*

**Table 8.2** Genera that are associated with clinical symptoms in psychiatry.—cont'd

| Taxa | Approach | Clinical symptom | Scale | Group | References | Analytical details |
|------|----------|------------------|-------|-------|------------|--------------------|
| *Sutterella* | 16S | Depression severity | BDI | HC, MDD | (Chung et al., 2019) | Spearman Cor |
| | 16S | Stress | PSS | HC, MDD | (Chung et al., 2019) | Spearman Cor |
| *Veillonella* | 16S | Axniety severity | HAMA | GAD | (Chen et al., 2019) | Spearman Cor |
| | 16S | Depression severity | HAMD | GAD | (Chen et al., 2019) | Spearman Cor |

*16S*, 16S rRNA sequencing; *BAI*, Beck Anxiety Inventory; *BDI*, Beck Depression Inventory; *FDR*, false discovery rate; *HAMD*, Hamilton Depression Score; *HC*, healthy control; *ISI*, Insomnia Severity Index; *MDD*, major depressive disorder; *PROMIS*, Patient-Reported Outcomes Measurement Information System; *PSQI*, Pittsburgh Sleep Quality Index; *PSS*, Perceived Stress Scale.

Overall, work to date has provided evidence of significant associations between microbiota and MDD, with some key depression-related taxa emerging. It is possible that reduced alpha diversity or specific taxa are associated with symptoms in a subset of MDD individuals within a given study. Notably, a recent study demonstrated a link between alpha diversity and anhedonia (an established symptom of depression) in a general population cohort (Minichino et al., 2021). Moving forward, mapping microbial and clinical features to subgroups of patients has the potential to be clinically important but requires larger clinical studies as well as multivariate and integrative analytical approaches.

The above-noted studies establish a link between alterations in microbiota and MDD; however, compositional analysis alone is not sufficient. It is essential to move beyond compositional differences to consider functional readouts of the microbiome. The use of metagenomic sequencing provides higher taxonomic resolution and functional information. A recent study constructed a random forest classifier to distinguish between HC and MDD using metagenomic sequencing at the genus (AUC of 0.89) and species level (AUC of 0.997) (Lai et al., 2021). Key genera that distinguished HC from MDD included *Eggerthella, Enterococcus, Lactobacillus, Acidaminococcus, Streptococcus, Oscillibacter, Lachnoclostridium, Megasphaera,* and *Bifidobacterium* (Lai et al., 2021). In addition, they examined the gut and host tryptophan biosynthesis and metabolism pathways and identified key KEGG orthologs that were associated with depression severity measured using HAM-D and with anxiety symptoms measured using HAM-A (Lai et al., 2021). Tryptophan, an essential amino acid that comes from the food we eat is the precursor for serotonin. Within the body, majority of tryptophan is metabolized through the kynurenine pathway where only a small percentage of tryptophan is metabolized to serotonin (Badawy, 2017; Le Floc'h et al., 2011; Richard et al., 2009; Waclawikova and El Aidy, 2018). Specific to the serotonin pathway, tryptophan is able to cross the blood–brain barrier where it participates in the synthesis of serotonin within the raphe nucleus, within the brain stem (Ruddick et al., 2006; Waclawikova and El Aidy, 2018). Interestingly, about 95% of serotonin is produced and stored outside of the central nervous system, by enterochromaffin cells within the intestinal mucosa. These cells produce serotonin in response to host signals and microbial metabolites (Kennedy et al., 2017; Waclawikova and El Aidy, 2018). Regulation of serotonin production within the intestinal mucosa can have an influence on both brain chemistry and behaviors (Clarke et al., 2022; Rieder et al., 2017). The importance of microbiota in tryptophan metabolism and stress-related behaviors has been demonstrated in animal studies. In the above-noted study, the importance of microbiota regulation of tryptophan metabolism was demonstrated in MDD individuals, and several key KEGG orthologs were associated with key depression-related bacterial taxa at the genus and species level (Lai et al., 2021). Another study examined the role of tryptophan metabolism by gut microbiota by measuring the levels of 3-indoxylsulfate (an indole metabolite) in urine and demonstrated a positive correlation of indole production by gut microbiota and recurrent depressive symptoms in women in a 2-year observational case-control study (Philippe et al., 2021). Increased abundance of bacterial taxa, such as *Alistipes*, could contribute to higher indole metabolites in urine in depressed patients (Jiang et al., 2015; Philippe et al., 2021). More studies addressing host−microbe metabolism and gut−brain signaling pathways are needed to better understand the mechanisms by which microbiota contribute to mood and anxiety disorders.

Metabolomics examines the metabolites of the bacteria and/or the host. SCFAs are important bacterial metabolites produced by gut bacteria that influence other commensals and are important to gut physiology as well as part of microbiota−host signaling systems that extend beyond the gut (Dalile et al., 2019; Rios-Covian et al., 2016). Beyond SCFAs, microbially-derived molecules include neurotransmitters, indoles, bile acids, choline metabolites, lactate, and vitamins and evidence is accumulating that the microbial metabolites may contribute to the pathophysiology of psychiatric illness (e.g., in depression (Averina et al., 2020; Caspani et al., 2019)). Recent studies using a targeted approach focused on a limited panel of metabolites have demonstrated an association between metabolites and treatment response (Bhattacharyya et al., 2019a,b; Caspani et al., 2021). Notable sex differences in the ability of baseline levels of lipid metabolites to distinguish healthy controls from individuals with MDD were reported, as well as sex differences in the ability of baseline levels of lipid metabolites to predict treatment to antidepressant response (Caspani et al., 2021).

## Bipolar disorder

Bipolar disorder (BD) presents as changes in a person's mood, energy, and ability to function, including mood episodes that can be either manic/hypomanic or depressive (American Psychiatric Association, 2013). While much of the attention related to the microbiome to date in mood disorders has focused on MDD, there are a number of recent papers that have considered the role of the microbiome in BD (Bengesser et al., 2019; Cheng et al., 2020; Coello et al., 2019; Evans et al., 2017; Hu et al., 2019; Lu et al., 2019; McIntyre et al., 2021; Painold et al., 2019; Ren et al., 2021; Zheng et al., 2020). Consistently, reduced alpha diversity is reported in BD compared to HC (Hu et al., 2019; McIntyre et al., 2021). Compositional differences between healthy controls and BD have been reported using 16S rRNA sequencing (Evans et al., 2017; Hu et al., 2019; Painold et al., 2019), metagenomics (Lai et al., 2021), and targeted qPCR (Lu et al., 2019). Notably, a reduced abundance of *Faecalibacterium,* reported frequently in MDD, was also reported in BD (Evans et al., 2017; Hu et al., 2019; Painold et al., 2019), and a link between bacterial taxa and clinical symptoms in BD included *Faecalibacterium* with depression symptoms and sleep (Evans et al., 2017). *Parabacteroides, Bacteriodes,* and *Halomonas* genera were reported higher in BD; whereas, *Roseburia, Faecalibacterium, and Coprococcus* genera were reported higher in HC (Hu et al., 2019). Similar to MDD, microbiota regulation of tryptophan metabolism in BD has been suggested (Lai et al., 2021). In this study, the investigators used metagenomic data to conduct a KEGG-based analysis to demonstrate impairments in microbial tryptophan biosynthesis and metabolism pathways in BD compared to healthy controls (Lai et al., 2021). Metabolic analysis showed increased levels of urine metabolites in BD compared to HC, including betaine, glycerol, hippuric acid, indole sulfate, trimethylamine oxide, and urea (Ren et al., 2021). A lower level of inositol was also observed in BD compared to HC (Ren et al., 2021). Notably, gut microbiota were directly implicated in these results as hippuric acid, indole sulfate, and trimethylamine oxide are microbial metabolites (Ren et al., 2021). A link between microbiota diversity and methylation of the clock gene, aryl hydrocarbon receptor nuclear translocator-like (ARNTL), is of interest in BD as ARNTL methylation has previously been shown to be modified in BD, as well as regulate monoamine oxidase (Bengesser et al., 2018, 2019). Overall, the preliminary findings in BD suggest that microbiota—host interactions may contribute to dysfunctional gut—brain signaling; however, larger cohort studies are needed to validate these early findings.

## Anxiety disorder

Early work in neuroscience linked microbiota to behavior utilized GF mice (mice lacking all microbes) (Clarke et al., 2013; Gustafsson et al., 1957;

Gustafsson, 1959; Heijtz et al., 2011; Neufeld et al., 2011). These landmark studies using GF mice provided the spark needed for neuroscientists to consider how microbes may influence brain function and demonstrated that microbiota influence stress reactivity and stress-related (e.g., anxiety-like) behaviors (Clarke et al., 2013; Heijtz et al., 2011; Neufeld et al., 2011; Sudo et al., 2004). In clinical populations, anxiety symptoms are often comorbid in both physical and mental illness, and several studies have associated the microbiome with anxiety-related symptoms. Recently, reports have identified alterations in microbiome composition for individuals with general anxiety disorder (GAD) (Chen et al., 2019; Jiang et al., 2018). Several taxa were associated with clinical symptoms. Specifically, *Prevotellaceae UCG-001, Mollicutes RF39_norank, Succinivibrio, Mitsuokella, Prevotella 9, Subdoligranulum, Ruminococcaceae_NK4A214_group, Ruminococcaceae_UCG-014*, and *Eubacterium_coprostanoligenes group* were negatively associated with depressive and anxiety symptoms, whereas *Bacteroides* and *Escherichia-Shigella* were positively correlated with these symptoms (Chen et al., 2019). In addition, one study has also revealed reduced microbiota diversity and alterations in bacterial taxa in individuals with attention deficit hyperactivity disorder (ADHD) (Prehn-Kristensen et al., 2018). These are small studies and have relied on 16S rRNA gene sequencing and analytical tools that only identify taxa to the genus level but provide no direct insight into the functional changes of microbiota that may be driving effects on host physiology. Importantly, sex differences in prevalence and symptom presentation are well established in anxiety and other stress-related disorders. Evidence linking microbiome changes and inflammatory activation that are sex-specific is emerging (Audet, 2019; Chen et al., 2018) and serves as an important consideration for future studies in this area.

## Schizophrenia

In addition to MDD, the transfer of fecal matter from individuals with schizophrenia (SZ) compared to healthy controls (HC) has been shown to alter behavior and brain metabolites (Liang et al., 2019; Zheng et al., 2019). Using this reverse translation approach, SZ-FMT mice showed increased activity in the open field, reduced anxiety- and depressive-like behavior, and an increased startle response compared to CON-FMT mice; however, they did not display the expected reduced prepulse inhibition that is observed in animal models of SZ and in patient populations (Zheng et al., 2019). This highlights a limitation to this approach as the donors were not assessed on these behavioral phenotypes but instead were selected based on their PANSS scores used to diagnose SZ (Zheng et al., 2019). Transfer of microbiome fecal transplants to GF mice also resulted in altered neurochemistry in the hippocampus of recipient mice (Zheng et al., 2019). Interestingly, the transfer of fecal microbiota from drug-free individuals with schizophrenia to mice also resulted in altered behavior including in the domains of psychomotor and learning/memory function (Zhu et al., 2019).

A few reports have examined the microbiome in schizophrenia in both gut and oral sites (Castro-Nallar et al., 2015; Schwarz et al., 2018; Shen et al., 2018; Xu et al., 2019b; Yolken et al., 2015). Through the use of shotgun metagenomic sequencing, the bacteriophage virome is also reported to differ in individuals with schizophrenia compared to healthy controls, although due to the small sample size only a single phage was significant after correction for multiple comparisons (Yolken et al., 2015). Reduced diversity in schizophrenia and a schizophrenia-specific group of taxa have been suggested (Zheng et al., 2019). The impact of medication on the microbiome and on drug metabolism is an active area of research and is important in clinical psychiatry. In the above-noted population study that examined the impact of antibiotic use on psychiatric illness, an association between antibiotics and psychosis was not found (Lurie et al., 2015). However, in a screen of greater than 1000 drugs, antipsychotic drugs were overrepresented in the group of drugs that inhibited the growth of gut bacterial strains (Maier et al., 2018), emphasizing the importance of considering medication in clinical studies. In a small study, differences in microbiota composition in schizophrenia patients receiving (vs. not receiving) atypical antipsychotics were noted (Flowers et al., 2019). More work is needed to determine the mechanistic nature of microbiome—drug—brain interactions, particularly in schizophrenia.

## Future directions

Overall, evidence in healthy and clinical populations shows that microbes influence brain function and behavior. Much of the work to date has examined gut microbiome composition, and more studies are needed that utilize functional readouts such as metagenomics and metabolomics. A recent review and meta-analysis of 59 case-control studies across psychiatric disorders suggested a transdiagnostic composition across disorders characterized by reduced butyrate-producing bacteria and increased proinflammatory taxa is a subset of disorders (Nikolova et al., 2021).

Understanding the association between key taxa, functional readouts, and clinical symptoms within and across disorders in larger clinical cohorts is needed. Active studies continue to add to our knowledge of how genes and environment influence microbiota—brain interactions in mental health as well as the capacity of psychotropic drugs to remodel the gut microbiome. There are several reasons to consider the microbiome in mental health: (1) to identify biomarkers related to biological differences that allow us to identify subgroups of clinical populations and improve the ability to match individuals to the best treatment in the shortest time; (2) to identify individuals at risk for early intervention; (3) to provide novel targets for drug development; and (4) to facilitate the expansion of microbiome-targeted therapies including, but not limited to, diet, prebiotics, and probiotics. Advances in our understanding of the microbiome in mental health and disease are promising and ongoing work will advance our understanding of this dynamic area of research. Moving forward translational and multidisciplinary approaches will allow researchers and health care providers to utilize the microbiome to advance drug discovery and precision medicine approaches.

# References

Abdel-Aziz, M.I., Vijverberg, S.J.H., Neerincx, A.H., Kraneveld, A.D., Maitland-van der Zee, A.H., 2019. The crosstalk between microbiome and asthma: exploring associations and challenges. Clin Exp Allergy 49, 1067—1086.

Allen, A.P., Hutch, W., Borre, Y.E., Kennedy, P.J., Temko, A., Boylan, G., Murphy, E., Cryan, J.F., Dinan, T.G., Clarke, G., 2016. Bifidobacterium longum 1714 as a translational psychobiotic: modulation of stress, electrophysiology and neurocognition in healthy volunteers. Transl Psychiatry 6, e939.

American Psychiatric Association, A, 2013. APA Guidelines for the Undergraduate Psychology Major: Version 2.0. Retrieved from. http://www.apa.org/ed/precollege/undergrad/index.aspx.

Armour, C.R., Nayfach, S., Pollard, K.S., Sharpton, T.J., 2019. A metagenomic meta-analysis reveals functional signatures of health and disease in the human gut microbiome. MSystems 4.

Audet, M.C., 2019. Stress-induced disturbances along the gut microbiota-immune-brain axis and implications for mental health: does sex matter? Front Neuroendocrinol 54, 100772.

Averina, O.V., Zorkina, Y.A., Yunes, R.A., Kovtun, A.S., Ushakova, V.M., Morozova, A.Y., Kostyuk, G.P., Danilenko, V.N., Chekhonin, V.P., 2020. Bacterial metabolites of human gut microbiota correlating with depression. Int J Mol Sci 21.

Aversa, Z., Atkinson, E.J., Schafer, M.J., Theiler, R.N., Rocca, W.A., Blaser, M.J., LeBrasseur, N.K., 2021. Association of infant antibiotic exposure with childhood health outcomes. Mayo Clin Proc 96, 66—77.

Badawy, A.A., 2017. Tryptophan availability for kynurenine pathway metabolism across the life span: control mechanisms and focus on aging, exercise, diet and nutritional supplements. Neuropharmacology 112, 248—263.

Bailey, M.T., Dowd, S.E., Galley, J.D., Hufnagle, A.R., Allen, R.G., Lyte, M., 2011. Exposure to a social stressor alters the structure of the intestinal microbiota: implications for stressor-induced immunomodulation. Brain Behav Immun 25, 397—407.

Barcenilla, A., Pryde, S.E., Martin, J.C., Duncan, S.H., Stewart, C.S., Henderson, C., Flint, H.J., 2000. Phylogenetic relationships of butyrate-producing bacteria from the human gut. Appl Environ Microbiol 66, 1654—1661.

Bengesser, S.A., Morkl, S., Painold, A., Dalkner, N., Birner, A., Fellendorf, F.T., Platzer, M., Queissner, R., Hamm, C., Maget, A., Pilz, R., Rieger, A., Wagner-Skacel, J., Reininghaus, B., Kapfhammer, H.P., Petek, E., Kashofer, K., Halwachs, B., Holzer, P., Waha, A., Reininghaus, E.Z., 2019. Epigenetics of the molecular clock and bacterial diversity in bipolar disorder. Psychoneuroendocrinology 101, 160—166.

Bengesser, S.A., Reininghaus, E.Z., Lackner, N., Birner, A., Fellendorf, F.T., Platzer, M., Kainzbauer, N., Tropper, B., Hormanseder, C., Queissner, R., Kapfhammer, H.P., Wallner-Liebmann, S.J., Fuchs, R., Petek, E., Windpassinger, C., Schnalzenberger, M., Reininghaus, B., Evert, B., Waha, A., 2018. Is the molecular clock ticking differently in bipolar disorder? methylation analysis of the clock gene ARNTL. World J Biol Psychiatry 19, S21—S29.

Benton, D., Williams, C., Brown, A., 2007. Impact of consuming a milk drink containing a probiotic on mood and cognition. Eur J Clin Nutr 61, 355—361.

Bharwani, A., Bala, A., Surette, M., Bienenstock, J., Vigod, S.N., Taylor, V.H., 2020. Gut microbiome patterns associated with treatment response in patients with major depressive disorder. Can J Psychiatry 65, 278—280.

Bharwani, A., Mian, M.F., Foster, J.A., Surette, M.G., Bienenstock, J., Forsythe, P., 2016. Structural & functional consequences of chronic psychosocial stress on the microbiome & host. Psychoneuroendocrinology 63, 217—227.

Bhattacharyya, S., Ahmed, A.T., Arnold, M., Liu, D., Luo, C., Zhu, H., Mahmoudiandehkordi, S., Neavin, D., Louie, G., Dunlop, B.W., Frye, M.A., Wang, L., Weinshilboum, R.M., Krishnan, R.R., Rush, A.J., Kaddurah-Daouk, R., 2019a. Metabolomic signature of exposure and response to citalopram/escitalopram in depressed outpatients. Transl Psychiatry 9, 173.

Bhattacharyya, S., Dunlop, B.W., Mahmoudiandehkordi, S., Ahmed, A.T., Louie, G., Frye, M.A., Weinshilboum, R.M., Krishnan, R.R., Rush, A.J., Mayberg, H.S., Craighead, W.E., Kaddurah-Daouk, R., 2019b. Pilot study of metabolomic clusters as state markers of major depression and outcomes to CBT treatment. Front Neurosci 13, 926.

Borre, Y.E., Moloney, R.D., Clarke, G., Dinan, T.G., Cryan, J.F., 2014. The impact of microbiota on brain and behavior: mechanisms & therapeutic potential. Adv Exp Med Biol 817, 373—403.

Burokas, A., Arboleya, S., Moloney, R.D., Peterson, V.L., Murphy, K., Clarke, G., Stanton, C., Dinan, T.G., Cryan, J.F., 2017. Targeting the microbiota-gut-brain axis: prebiotics have anxiolytic and antidepressant-like effects and reverse the impact of chronic stress in mice. Biol Psychiatry 82 (7), 472—487.

Caminero, A., Verdu, E.F., 2019. Celiac disease: should we care about microbes? Am J Physiol Gastrointest Liver Physiol 317, G161—g170.

Caspani, G., Kennedy, S., Foster, J.A., Swann, J., 2019. Gut microbial metabolites in depression: understanding the biochemical mechanisms. Microb Cell 6, 454—481.

Caspani, G., Turecki, G., Lam, R.W., Milev, R.V., Frey, B.N., MacQueen, G.M., Muller, D.J., Rotzinger, S., Kennedy, S.H., Foster, J.A., Swann, J.R., 2021.

Metabolomic signatures associated with depression and predictors of antidepressant response in humans: a CAN-BIND-1 report. Commun Biol 4, 903.

Castro-Nallar, E., Bendall, M.L., Perez-Losada, M., Sabuncyan, S., Severance, E.G., Dickerson, F.B., Schroeder, J.R., Yolken, R.H., Crandall, K.A., 2015. Composition, taxonomy and functional diversity of the oropharynx microbiome in individuals with schizophrenia and controls. PeerJ 3, e1140.

Chen, J.J., Zheng, P., Liu, Y.Y., Zhong, X.G., Wang, H.Y., Guo, Y.J., Xie, P., 2018. Sex differences in gut microbiota in patients with major depressive disorder. Neuropsychiatr Dis Treat 14, 647−655.

Chen, Y.H., Bai, J., Wu, D., Yu, S.F., Qiang, X.L., Bai, H., Wang, H.N., Peng, Z.W., 2019. Association between fecal microbiota and generalized anxiety disorder: severity and early treatment response. J Affect Disord 259, 56−66.

Chen, Y.H., Xue, F., Yu, S.F., Li, X.S., Liu, L., Jia, Y.Y., Yan, W.J., Tan, Q.R., Wang, H.N., Peng, Z.W., 2021. Gut microbiota dysbiosis in depressed women: the association of symptom severity and microbiota function. J Affect Disord 282, 391−400.

Cheng, S., Han, B., Ding, M., Wen, Y., Ma, M., Zhang, L., Qi, X., Cheng, B., Li, P., Kafle, O.P., Liang, X., Liu, L., Du, Y., Zhao, Y., Zhang, F., 2020. Identifying psychiatric disorder-associated gut microbiota using microbiota-related gene set enrichment analysis. Brief Bioinform 21, 1016−1022.

Chung, Y.E., Chen, H.C., Chou, H.L., Chen, I.M., Lee, M.S., Chuang, L.C., Liu, Y.W., Lu, M.L., Chen, C.H., Wu, C.S., Huang, M.C., Liao, S.C., Ni, Y.H., Lai, M.S., Shih, W.L., Kuo, P.H., 2019. Exploration of microbiota targets for major depressive disorder and mood related traits. J Psychiatr Res 111, 74−82.

Clarke, G., Grenham, S., Scully, P., Fitzgerald, P., Moloney, R.D., Shanahan, F., Dinan, T.G., Cryan, J.F., 2013. The microbiome-gut-brain axis during early life regulates the hippocampal serotonergic system in a sex-dependent manner. Mol Psychiatry 18, 666−673.

Clarke, G., Villalobos-Manriquez, F., Marin, D.C., 2022. Tryptophan metabolism and the microbiome-gut-brain axis. In: Burnet, P.W.J. (Ed.), The Oxford Handbook of the Microbiome-Gut-Brain Axis. Oxford University Press, pp. 1−77.

Coello, K., Hansen, T.H., Sorensen, N., Munkholm, K., Kessing, L.V., Pedersen, O., Vinberg, M., 2019. Gut microbiota composition in patients with newly diagnosed bipolar disorder and their unaffected first-degree relatives. Brain Behav Immun 75, 112−118.

Cryan, J.F., Dinan, T.G., 2012. Mind-altering microorganisms: the impact of the gut microbiota on brain and behaviour. Nat Rev Neurosci 13, 701−712.

Dalile, B., Van Oudenhove, L., Vervliet, B., Verbeke, K., 2019. The role of short-chain fatty acids in microbiota-gut-brain communication. Nat Rev Gastroenterol Hepatol 16, 461−478.

Duncan, S.H., Hold, G.L., Harmsen, H.J., Stewart, C.S., Flint, H.J., 2002. Growth requirements and fermentation products of Fusobacterium prausnitzii, and a proposal to reclassify it as Faecalibacterium prausnitzii gen. nov., comb. nov. Int J Syst Evol Microbiol 52, 2141−2146.

Evans, S.J., Bassis, C.M., Hein, R., Assari, S., Flowers, S.A., Kelly, M.B., Young, V.B., Ellingrod, V.E., McInnis, M.G., 2017. The gut microbiome composition associates with bipolar disorder and illness severity. J Psychiatr Res 87, 23−29.

Flowers, S.A., Baxter, N.T., Ward, K.M., Kraal, A.Z., McInnis, M.G., Schmidt, T.M., Ellingrod, V.L., 2019. Effects of atypical antipsychotic treatment and resistant starch supplementation on gut microbiome composition in a cohort of patients with bipolar disorder or schizophrenia. Pharmacotherapy 39, 161−170.

Foster, J.A., 2016. Gut microbiome and behavior: focus on neuroimmune interactions. Int Rev Neurobiol 131, 49−65.

Foster, J.A., 2017. Targeting the microbiome for mental health: hype or hope? Biol Psychiatry 82, 456–457.

Foster, J.A., Lyte, M., Meyer, E., Cryan, J.F., 2016. Gut microbiota and brain function: an evolving field in neuroscience. Int J Neuropsychopharmacol 19.

Foster, J.A., McVey Neufeld, K.A., 2013. Gut-brain axis: how the microbiome influences anxiety and depression. Trends Neurosci 36, 305–312.

Gheorghe, C.E., Ritz, N.L., Martin, J.A., Wardill, H.R., Cryan, J.F., Clarke, G., 2021. Investigating causality with fecal microbiota transplantation in rodents: applications, recommendations and pitfalls. Gut Microb 13, 1941711.

Gibson, G.R., Scott, K.P., Rastall, R.A., Tuohy, K.M., Hotchkiss, A., Dubert-Ferrandon, A., Gareau, M.G., Murphy, E.F., Saulnier, D., Loh, G., Macfarlane, S., Delzenne, N.M., Ringel, Y., Kozianowski, G., Dickmann, R., Lenoir-Wijnkoop, I., Walker, C., Buddington, R., 2010. Dietary prebiotics: current status and new definition. Food Sci Technol Bull Funct Foods 7, 1–19.

Grosicki, G.J., Riemann, B.L., Flatt, A.A., Valentino, T., Lustgarten, M.S., 2020. Self-reported sleep quality is associated with gut microbiome composition in young, healthy individuals: a pilot study. Sleep Med 73, 76–81.

Gustafsson, B., Kahlson, G., Rosengren, E., 1957. Biogenesis of histamine studied by its distribution and urinary excretion in germ free reared and not germ free rats fed a histamine free diet. Acta Physiol Scand 41, 217–228.

Gustafsson, B.E., 1959. Lightweight stainless steel systems for rearing germfree animals. Ann N Y Acad Sci 78, 17–28.

Halfvarson, J., Brislawn, C.J., Lamendella, R., Vazquez-Baeza, Y., Walters, W.A., Bramer, L.M., D'Amato, M., Bonfiglio, F., McDonald, D., Gonzalez, A., McClure, E.E., Dunklebarger, M.F., Knight, R., Jansson, J.K., 2017. Dynamics of the human gut microbiome in inflammatory bowel disease. Nat Microbiol 2, 17004.

Heijtz, R.D., Wang, S., Anuar, F., Qian, Y., Bjorkholm, B., Samuelsson, A., Hibberd, M.L., Forssberg, H., Pettersson, S., 2011. Normal gut microbiota modulates brain development and behavior. Proc Natl Acad Sci U S A 108, 3047–3052.

Heinken, A., Khan, M.T., Paglia, G., Rodionov, D.A., Harmsen, H.J., Thiele, I., 2014. Functional metabolic map of Faecalibacterium prausnitzii, a beneficial human gut microbe. J Bacteriol 196, 3289–3302.

Hill, C., Guarner, F., Reid, G., Gibson, G.R., Merenstein, D.J., Pot, B., Morelli, L., Canani, R.B., Flint, H.J., Salminen, S., Calder, P.C., Sanders, M.E., 2014. Expert consensus document. The International Scientific Association for Probiotics and Prebiotics consensus statement on the scope and appropriate use of the term probiotic. Nat Rev Gastroenterol Hepatol 11, 506–514.

Hu, S., Li, A., Huang, T., Lai, J., Li, J., Sublette, M.E., Lu, H., Lu, Q., Du, Y., Hu, Z., Ng, C.H., Zhang, H., Lu, J., Mou, T., Lu, S., Wang, D., Duan, J., Hu, J., Huang, M., Wei, N., Zhou, W., Ruan, L., Li, M.D., Xu, Y., 2019. Gut microbiota changes in patients with bipolar depression. Adv Sci 6, 1900752.

Iizumi, T., Battaglia, T., Ruiz, V., Perez Perez, G.I., 2017. Gut microbiome and antibiotics. Arch Med Res 48, 727–734.

Jernberg, C., Lofmark, S., Edlund, C., Jansson, J.K., 2010. Long-term impacts of antibiotic exposure on the human intestinal microbiota. Microbiology 156, 3216–3223.

Jiang, H., Ling, Z., Zhang, Y., Mao, H., Ma, Z., Yin, Y., Wang, W., Tang, W., Tan, Z., Shi, J., Li, L., Ruan, B., 2015. Altered fecal microbiota composition in patients with major depressive disorder. Brain Behav Immun 48, 186–194.

Jiang, H.Y., Zhang, X., Yu, Z.H., Zhang, Z., Deng, M., Zhao, J.H., Ruan, B., 2018. Altered gut microbiota profile in patients with generalized anxiety disorder. J Psychiatr Res 104, 130–136.

Kato-Kataoka, A., Nishida, K., Takada, M., Suda, K., Kawai, M., Shimizu, K., Kushiro, A., Hoshi, R., Watanabe, O., Igarashi, T., Miyazaki, K., Kuwano, Y., Rokutan, K., 2016. Fermented milk containing Lactobacillus casei strain Shirota prevents the onset of physical symptoms in medical students under academic examination stress. Benef Microbes 7, 153—156.

Kelly, J.R., Borre, Y., O'Brien, C., Patterson, E., El Aidy, S., Deane, J., Kennedy, P.J., Beers, S., Scott, K., Moloney, G., Hoban, A.E., Scott, L., Fitzgerald, P., Ross, P., Stanton, C., Clarke, G., Cryan, J.F., Dinan, T.G., 2016. Transferring the blues: depression-associated gut microbiota induces neurobehavioural changes in the rat. J Psychiatr Res 82, 109—118.

Kennedy, P.J., Cryan, J.F., Dinan, T.G., Clarke, G., 2017. Kynurenine pathway metabolism and the microbiota-gut-brain axis. Neuropharmacology 112, 399—412.

Kohn, J.N., Kosciolek, T., Marotz, C., Aleti, G., Guay-Ross, R.N., Hong, S.H., Hansen, S., Swafford, A., Knight, R., Hong, S., 2020. Differing salivary microbiome diversity, community and diurnal rhythmicity in association with affective state and peripheral inflammation in adults. Brain Behav Immun 87, 591—602.

Lai, W.T., Deng, W.F., Xu, S.X., Zhao, J., Xu, D., Liu, Y.H., Guo, Y.Y., Wang, M.B., He, F.S., Ye, S.W., Yang, Q.F., Liu, T.B., Zhang, Y.L., Wang, S., Li, M.Z., Yang, Y.J., Xie, X.H., Rong, H., 2021. Shotgun metagenomics reveals both taxonomic and tryptophan pathway differences of gut microbiota in major depressive disorder patients. Psychol Med 51, 90—101.

Lavebratt, C., Yang, L.L., Giacobini, M., Forsell, Y., Schalling, M., Partonen, T., Gissler, M., 2019. Early exposure to antibiotic drugs and risk for psychiatric disorders: a population-based study. Transl Psychiatry 9, 317.

Le Floc'h, N., Otten, W., Merlot, E., 2011. Tryptophan metabolism, from nutrition to potential therapeutic applications. Amino Acids 41, 1195—1205.

Lew, L.C., Hor, Y.Y., Yusoff, N.A.A., Choi, S.B., Yusoff, M.S.B., Roslan, N.S., Ahmad, A., Mohammad, J.A.M., Abdullah, M., Zakaria, N., Wahid, N., Sun, Z., Kwok, L.Y., Zhang, H., Liong, M.T., 2019. Probiotic Lactobacillus plantarum P8 alleviated stress and anxiety while enhancing memory and cognition in stressed adults: a randomised, double-blind, placebo-controlled study. Clin Nutr 38, 2053—2064.

Leyrolle, Q., Cserjesi, R., Mulders, D.G.H.M., Zamariola, G., Hiel, S., Gianfrancesco, M.A., Portheault, D., Amadieu, C., Bindels, L.B., Leclercq, S., Rodriguez, J., Neyrinck, A.M., Cani, P.D., Lanthier, N., Trefois, P., Bindelle, J., Paquot, N., Cnop, M., Thissen, J.P., Klein, O., Luminet, O., Delzenne, N.M., 2021. Prebiotic effect on mood in obese patients is determined by the initial gut microbiota composition: a randomized, controlled trial. Brain Behav Immun 94, 289—298.

Li, B., Guo, K., Zeng, L., Zeng, B., Huo, R., Luo, Y., Wang, H., Dong, M., Zheng, P., Zhou, C., Chen, J., Liu, Y., Liu, Z., Fang, L., Wei, H., Xie, P., 2018. Metabolite identification in fecal microbiota transplantation mouse livers and combined proteomics with chronic unpredictive mild stress mouse livers. Transl Psychiatry 8, 34.

Li, J., Ma, Y., Bao, Z., Gui, X., Li, A.N., Yang, Z., Li, M.D., 2020. Clostridiales are predominant microbes that mediate psychiatric disorders. J Psychiatr Res 130, 48—56.

Liang, W., Huang, Y., Tan, X., Wu, J., Duan, J., Zhang, H., Yin, B., Li, Y., Zheng, P., Wei, H., Xie, P., 2019. Alterations of glycerophospholipid and fatty acyl metabolism in multiple brain regions of schizophrenia microbiota recipient mice. Neuropsychiatr Dis Treat 15, 3219—3229.

Liang, X., Ye, J., Wen, Y., Li, P., Cheng, B., Cheng, S., Liu, L., Zhang, L., Ma, M., Qi, X., Liang, C., Chu, X., Kafle, O.P., Jia, Y., Zhang, F., 2020. Long-term antibiotic use during early life and risks to mental traits: an observational study and gene-environment-wide interaction study in UK Biobank cohort. Neuropsychopharmacology 46 (6), 1086—1092.

Liskiewicz, P., Kaczmarczyk, M., Misiak, B., Wronski, M., Baba-Kubis, A., Skonieczna-Zydecka, K., Marlicz, W., Bienkowski, P., Misera, A., Pelka-Wysiecka, J., et al., 2021. Analysis of gut microbiota and intestinal integrity markers of inpatients with major depressive disorder. Prog Neuropsychopharmacol Biol Psychiatry 106, 110076. https://doi.org/10.1016/j.pnpbp.2020.110076.

Liu, R.T., Rowan-Nash, A.D., Sheehan, A.E., Walsh, R.F.L., Sanzari, C.M., Korry, B.J., Belenky, P., 2020. Reductions in anti-inflammatory gut bacteria are associated with depression in a sample of young adults. Brain Behav Immun 88, 308−324.

Lopez-Siles, M., Duncan, S.H., Garcia-Gil, L.J., Martinez-Medina, M., 2017. Faecalibacterium prausnitzii: from microbiology to diagnostics and prognostics. ISME J 11, 841−852.

Lu, Q., Lai, J., Lu, H., Ng, C., Huang, T., Zhang, H., Ding, K., Wang, Z., Jiang, J., Hu, J., Lu, J., Lu, S., Mou, T., Wang, D., Du, Y., Xi, C., Lyu, H., Chen, J., Xu, Y., Liu, Z., Hu, S., 2019. Gut microbiota in bipolar depression and its relationship to brain function: an advanced exploration. Front Psychiatr 10, 784.

Lurie, I., Yang, Y.X., Haynes, K., Mamtani, R., Boursi, B., 2015. Antibiotic exposure and the risk for depression, anxiety, or psychosis: a nested case-control study. J Clin Psychiatry 76, 1522−1528.

Ma, T., Jin, H., Kwok, L.Y., Sun, Z., Liong, M.T., Zhang, H., 2021. Probiotic consumption relieved human stress and anxiety symptoms possibly via modulating the neuroactive potential of the gut microbiota. Neurobiol Stress 14, 100294.

Madan, A., Thompson, D., Fowler, J.C., Ajami, N.J., Salas, R., Frueh, B.C., Bradshaw, M.R., Weinstein, B.L., Oldham, J.M., Petrosino, J.F., 2020. The gut microbiota is associated with psychiatric symptom severity and treatment outcome among individuals with serious mental illness. J Affect Disord 264, 98−106.

Maier, L., Pruteanu, M., Kuhn, M., Zeller, G., Telzerow, A., Anderson, E.E., Brochado, A.R., Fernandez, K.C., Dose, H., Mori, H., Patil, K.R., Bork, P., Typas, A., 2018. Extensive impact of non-antibiotic drugs on human gut bacteria. Nature 555, 623−628.

McIntyre, R.S., Subramaniapillai, M., Shekotikhina, M., Carmona, N.E., Lee, Y., Mansur, R.B., Brietzke, E., Fus, D., Coles, A.S., Iacobucci, M., Park, C., Potts, R., Amer, M., Gillard, J., James, C., Anglin, R., Surette, M.G., 2021. Characterizing the gut microbiota in adults with bipolar disorder: a pilot study. Nutr Neurosci 24, 173−180.

Messaoudi, M., Violle, N., Bisson, J.F., Desor, D., Javelot, H., Rougeot, C., 2011. Beneficial psychological effects of a probiotic formulation (Lactobacillus helveticus R0052 and Bifidobacterium longum R0175) in healthy human volunteers. Gut Microb 2, 256−261.

Minichino, A., Jackson, M.A., Francesconi, M., Steves, C.J., Menni, C., Burnet, P.W.J., Lennox, B.R., 2021. Endocannabinoid system mediates the association between gut-microbial diversity and anhedonia/amotivation in a general population cohort. Mol Psychiatry 26 (11), 6269−6276.

Miquel, S., Martin, R., Bridonneau, C., Robert, V., Sokol, H., Bermudez-Humaran, L.G., Thomas, M., Langella, P., 2014. Ecology and metabolism of the beneficial intestinal commensal bacterium Faecalibacterium prausnitzii. Gut Microb 5, 146−151.

Neufeld, K.M., Kang, N., Bienenstock, J., Foster, J.A., 2011. Reduced anxiety-like behavior and central neurochemical change in germ-free mice. Neuro Gastroenterol Motil 23, 255−264 e119.

Nguyen, T.T., Hathaway, H., Kosciolek, T., Knight, R., Jeste, D.V., 2019. Gut microbiome in serious mental illnesses: a systematic review and critical evaluation. Schizophr Res 234, 24−40.

Nikolova, V.L., Hall, M.R.B., Hall, L.J., Cleare, A.J., Stone, J.M., Young, A.H., 2021. Perturbations in gut microbiota composition in psychiatric disorders: a review and meta-analysis. JAMA Psychiatr 78, 1343—1354.

Painold, A., Morkl, S., Kashofer, K., Halwachs, B., Dalkner, N., Bengesser, S., Birner, A., Fellendorf, F., Platzer, M., Queissner, R., Schutze, G., Schwarz, M.J., Moll, N., Holzer, P., Holl, A.K., Kapfhammer, H.P., Gorkiewicz, G., Reininghaus, E.Z., 2019. A step ahead: exploring the gut microbiota in inpatients with bipolar disorder during a depressive episode. Bipolar Disord 21, 40—49.

Philippe, C., Szabo de Edelenyi, F., Naudon, L., Druesne-Pecollo, N., Hercberg, S., Kesse-Guyot, E., Latino-Martel, P., Galan, P., Rabot, S., 2021. Relation between mood and the host-microbiome co-metabolite 3-indoxylsulfate: results from the observational prospective NutriNet-sante study. Microorganisms 9.

Pittayanon, R., Lau, J.T., Yuan, Y., Leontiadis, G.I., Tse, F., Surette, M., Moayyedi, P., 2019. Gut microbiota in patients with irritable bowel syndrome-A systematic review. Gastroenterology 157, 97—108.

Prehn-Kristensen, A., Zimmermann, A., Tittmann, L., Lieb, W., Schreiber, S., Baving, L., Fischer, A., 2018. Reduced microbiome alpha diversity in young patients with ADHD. PLoS One 13, e0200728.

Rashid, M.U., Zaura, E., Buijs, M.J., Keijser, B.J., Crielaard, W., Nord, C.E., Weintraub, A., 2015. Determining the long-term effect of antibiotic administration on the human normal intestinal microbiota using culture and pyrosequencing methods. Clin Infect Dis 60 Suppl. 2, S77—S84.

Rashidi, A., Ebadi, M., Rehman, T.U., Elhusseini, H., Nalluri, H., Kaiser, T., Holtan, S.G., Khoruts, A., Weisdorf, D.J., Staley, C., 2021. Gut microbiota response to antibiotics is personalized and depends on baseline microbiota. Microbiome 9, 211.

Ren, Y., Chen, Z.Z., Sun, X.L., Duan, H.J., Tian, J.S., Wang, J.Y., Yang, H., 2021. Metabolomic analysis to detect urinary molecular changes associated with bipolar depression. Neurosci Lett 742, 135515.

Richard, D.M., Dawes, M.A., Mathias, C.W., Acheson, A., Hill-Kapturczak, N., Dougherty, D.M., 2009. L-tryptophan: basic metabolic functions, behavioral research and therapeutic indications. Int J Tryptophan Res 2, 45—60.

Rieder, R., Wisniewski, P.J., Alderman, B.L., Campbell, S.C., 2017. Microbes and mental health: a review. Brain Behav Immun 66, 9—17.

Rios-Covian, D., Ruas-Madiedo, P., Margolles, A., Gueimonde, M., de Los Reyes-Gavilan, C.G., Salazar, N., 2016. Intestinal short chain fatty acids and their link with diet and human health. Front Microbiol 7, 185.

Ruddick, J.P., Evans, A.K., Nutt, D.J., Lightman, S.L., Rook, G.A., Lowry, C.A., 2006. Tryptophan metabolism in the central nervous system: medical implications. Expert Rev Mol Med 8, 1—27.

Sarkar, A., Lehto, S.M., Harty, S., Dinan, T.G., Cryan, J.F., Burnet, P.W., 2016. Psychobiotics and the manipulation of bacteria-gut-brain signals. Trends Neurosci 39, 763—781.

Schmidt, K., Cowen, P.J., Harmer, C.J., Tzortzis, G., Errington, S., Burnet, P.W., 2015. Prebiotic intake reduces the waking cortisol response and alters emotional bias in healthy volunteers. Psychopharmacology 232, 1793—1801.

Schwarz, E., Maukonen, J., Hyytiainen, T., Kieseppa, T., Oresic, M., Sabunciyan, S., Mantere, O., Saarela, M., Yolken, R., Suvisaari, J., 2018. Analysis of microbiota in first episode psychosis identifies preliminary associations with symptom severity and treatment response. Schizophr Res 192, 398—403.

Secombe, K.R., Al-Qadami, G.H., Subramaniam, C.B., Bowen, J.M., Scott, J., Van Sebille, Y.Z.A., Snelson, M., Cowan, C., Clarke, G., Gheorghe, C.E., Cryan, J.F., Wardill, H.R., 2021. Guidelines for reporting on animal fecal transplantation (GRAFT) studies: recommendations from a systematic review of murine transplantation protocols. Gut Microb 13, 1979878.

Settanni, C.R., Ianiro, G., Bibbo, S., Cammarota, G., Gasbarrini, A., 2021. Gut microbiota alteration and modulation in psychiatric disorders: current evidence on fecal microbiota transplantation. Prog Neuro-Psychopharmacol Biol Psychiatry 109, 110258.

Shen, Y., Xu, J., Li, Z., Huang, Y., Yuan, Y., Wang, J., Zhang, M., Hu, S., Liang, Y., 2018. Analysis of gut microbiota diversity and auxiliary diagnosis as a biomarker in patients with schizophrenia: a cross-sectional study. Schizophr Res 197, 470—477.

Steenbergen, L., Sellaro, R., van Hemert, S., Bosch, J.A., Colzato, L.S., 2015. A randomized controlled trial to test the effect of multispecies probiotics on cognitive reactivity to sad mood. Brain Behav Immun 48, 258—264.

Stevens, B.R., Goel, R., Seungbum, K., Richards, E.M., Holbert, R.C., Pepine, C.J., Raizada, M.K., 2018. Increased human intestinal barrier permeability plasma biomarkers zonulin and FABP2 correlated with plasma LPS and altered gut microbiome in anxiety or depression. Gut 67, 1555—1557.

Stevens, B.R., Roesch, L., Thiago, P., Russell, J.T., Pepine, C.J., Holbert, R.C., Raizada, M.K., Triplett, E.W., 2020. Depression phenotype identified by using single nucleotide exact amplicon sequence variants of the human gut microbiome. Mol Psychiatry 26 (8), 4277—4287.

Stilling, R.M., Ryan, F.J., Hoban, A.E., Shanahan, F., Clarke, G., Claesson, M.J., Dinan, T.G., Cryan, J.F., 2015. Microbes & neurodevelopment—absence of microbiota during early life increases activity-related transcriptional pathways in the amygdala. Brain Behav Immun 50, 209—220.

Sudo, N., Chida, Y., Aiba, Y., Sonoda, J., Oyama, N., Yu, X.N., Kubo, C., Koga, Y., 2004. Postnatal microbial colonization programs the hypothalamic-pituitary-adrenal system for stress response in mice. J Physiol 558, 263—275.

Tillisch, K., Labus, J., Kilpatrick, L., Jiang, Z., Stains, J., Ebrat, B., Guyonnet, D., Legrain-Raspaud, S., Trotin, B., Naliboff, B., Mayer, E.A., 2013. Consumption of fermented milk product with probiotic modulates brain activity. Gastroenterology 144, 1394—1401, 1401 e1391—1394.

Tillisch, K., Mayer, E.A., Gupta, A., Gill, Z., Brazeilles, R., Le Neve, B., van Hylckama Vlieg, J.E.T., Guyonnet, D., Derrien, M., Labus, J.S., 2017. Brain structure and response to emotional stimuli as related to gut microbial profiles in healthy women. Psychosom Med 79, 905—913.

Tremlett, H., Bauer, K.C., Appel-Cresswell, S., Finlay, B.B., Waubant, E., 2017. The gut microbiome in human neurological disease: a review. Ann Neurol 81, 369—382.

Valles-Colomer, M., Falony, G., Darzi, Y., Tigchelaar, E.F., Wang, J., Tito, R.Y., Schiweck, C., Kurilshikov, A., Joossens, M., Wijmenga, C., Claes, S., Van Oudenhove, L., Zhernakova, A., Vieira-Silva, S., Raes, J., 2019. The neuroactive potential of the human gut microbiota in quality of life and depression. Nat Microbiol 4, 623—632.

Waclawikova, B., El Aidy, S., 2018. Role of microbiota and tryptophan metabolites in the remote effect of intestinal inflammation on brain and depression. Pharmaceuticals 11.

Wang, H., Braun, C., Murphy, E.F., Enck, P., 2019. Bifidobacterium longum 1714 strain modulates brain activity of healthy volunteers during social stress. Am J Gastroenterol 114, 1152—1162.

Xu, M., Xu, X., Li, J., Li, F., 2019a. Association between gut microbiota and autism spectrum disorder: a systematic review and meta-analysis. Front Psychiatr 10, 473.

Xu, R., Wu, B., Liang, J., He, F., Gu, W., Li, K., Luo, Y., Chen, J., Gao, Y., Wu, Z., Wang, Y., Zhou, W., Wang, M., 2019b. Altered gut microbiota and mucosal immunity in patients with schizophrenia. Brain Behav Immun 85, 120—127.

Yolken, R.H., Severance, E.G., Sabunciyan, S., Gressitt, K.L., Chen, O., Stallings, C., Origoni, A., Katsafanas, E., Schweinfurth, L.A., Savage, C.L., Banis, M., Khushalani, S., Dickerson, F.B., 2015. Metagenomic sequencing indicates that the oropharyngeal phageome of individuals with schizophrenia differs from that of controls. Schizophr Bull 41, 1153–1161.

Zhang, Q., Yun, Y., An, H., Zhao, W., Ma, T., Wang, Z., Yang, F., 2021. Gut microbiome composition associated with major depressive disorder and sleep quality. Front Psychiatr 12, 645045.

Zheng, P., Yang, J., Li, Y., Wu, J., Liang, W., Yin, B., Tan, X., Huang, Y., Chai, T., Zhang, H., Duan, J., Zhou, J., Sun, Z., Chen, X., Marwari, S., Lai, J., Huang, T., Du, Y., Zhang, P., Perry, S.W., Wong, M.L., Licinio, J., Hu, S., Xie, P., Wang, G., 2020. Gut microbial signatures can discriminate unipolar from bipolar depression. Adv Sci 7, 1902862.

Zheng, P., Zeng, B., Liu, M., Chen, J., Pan, J., Han, Y., Liu, Y., Cheng, K., Zhou, C., Wang, H., Zhou, X., Gui, S., Perry, S.W., Wong, M.L., Licinio, J., Wei, H., Xie, P., 2019. The gut microbiome from patients with schizophrenia modulates the glutamate-glutamine-GABA cycle and schizophrenia-relevant behaviors in mice. Sci Adv 5, eaau8317.

Zheng, P., Zeng, B., Zhou, C., Liu, M., Fang, Z., Xu, X., Zeng, L., Chen, J., Fan, S., Du, X., Zhang, X., Yang, D., Yang, Y., Meng, H., Li, W., Melgiri, N.D., Licinio, J., Wei, H., Xie, P., 2016. Gut microbiome remodeling induces depressive-like behaviors through a pathway mediated by the host's metabolism. Mol Psychiatry 21, 786–796.

Zhu, F., Guo, R., Wang, W., Ju, Y., Wang, Q., Ma, Q., Sun, Q., Fan, Y., Xie, Y., Yang, Z., Jie, Z., Zhao, B., Xiao, L., Yang, L., Zhang, T., Liu, B., Guo, L., He, X., Chen, Y., Chen, C., Gao, C., Xu, X., Yang, H., Wang, J., Dang, Y., Madsen, L., Brix, S., Kristiansen, K., Jia, H., Ma, X., 2019. Transplantation of microbiota from drug-free patients with schizophrenia causes schizophrenia-like abnormal behaviors and dysregulated kynurenine metabolism in mice. Mol Psychiatry 25 (11), 2905–2918.

# CHAPTER 9

# Microbiota—brain interactions in aging and neurodegeneration

## Introduction

The characteristics of healthy aging are determined by a number of features, including genetic, environmental, and lifestyle factors (Ghosh et al., 2022). Age-related deterioration of brain function manifests across a number of interrelated molecular and cellular pathways which together represent risk factors for cognitive decline and neurodegeneration (Lopez-Otin et al. 2013, 2023; Lopez-Otin and Kroemer 2021). The increased risk of developing brain disorders as we age is reflected in behavioral changes and fueled by altered brain plasticity, dysregulation of the immune system, and dysfunction of the hypothalamic—pituitary—adrenal (HPA) axis (Prenderville et al., 2015). Disruption of these core signaling pathways can thus be viewed in terms of how the microbiota—gut—brain axis facilitates the maintenance of brain homeostasis and regulates adequate responses to stress during the aging process. In this chapter, we focus on the changes in the gut microbiome as the host ages and in age-related disease, accompanied by reflections on how the gut microbiome can, in turn, sustain or modify the hallmarks of healthy aging in the host via the gut—brain axis (see Fig. 9.1).

## The aging gut microbiome: focus on health

The physiological hallmarks of healthy biological aging are accompanied by alterations in the gut microbiome. The majority of studies investigating gut microbiome alterations in older people are based on 16s sequencing approaches and often include a focus on centenarians. For example, analysis of fecal samples from 367 healthy Japanese subjects between the ages of 0 and 104 revealed an increase in gut microbiota $\alpha$-diversity (PD whole tree, Chao1, number of observed species, Shannon index) with aging until the centenarian stage (Odamaki et al., 2016). Greater community richness and diversity were also reported in a long-living group ($\geq$90 years including 67

*Microbiota Brain Axis*
ISBN 978-0-12-814800-6
https://doi.org/10.1016/B978-0-12-814800-6.00009-1

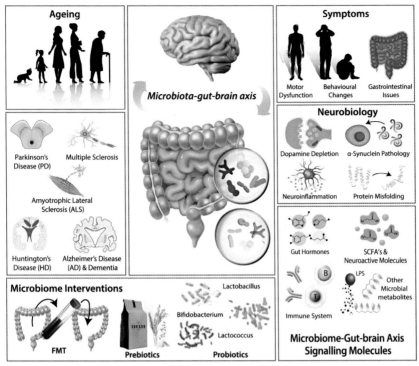

**Figure 9.1** *Microbiota—brain interactions in aging and neurodegeneration.* Both health and unhealthy aging are associated with compositional and functional reconfiguration of the gut microbiota, alterations which can be exacerbated and pathological in age-related neurodegenerative disorders. These disease-associated microbiomes have been associated with symptom expression and neurobiological features of age-related disorders. The mechanisms underpinning these observations may relate to microbial metabolites such as SCFAs or other microbial metabolites and proinflammatory states. This is the basis for consideration of various interventions based on targeted restoration of microbial functions to prevent, delay, or reverse the consequences of the impact of increasing age on microbiota-to-brain communication.

healthy centenarians and nonagenarians) compared to a younger age group (54 elderly and 47 young adults) in individuals from the Sichuan province in China, a feature that was also found in the reanalysis of an Italian dataset of centenarians and semisupercentenarians (Kong et al., 2016). In a study of participants recruited from three cities from Jiangsu Province in China, Shannon diversity was increased in the healthy elderly relative to healthy 20-year-olds but decreased relative to those aged 13—14 (Bian et al., 2017). Greater gut microbiome diversity may thus be a hallmark of *healthy* long-living individuals, in contrast to the reduced microbial diversity that has been noted for those with decreased overall health status in many aging studies (Kong et al., 2019).

Although these studies do not all report the same compositional alterations, some common relative abundance alterations in some taxa have been noted. Examples of consistent compositional patterns of microbiome taxa alterations with increased aging include a decreased relative abundance of *Prevotella* (Ghosh et al., 2022; Kim et al., 2019), *Faecalibacterium* (Badal et al., 2020; Ghosh et al., 2022), and *Bifidobacterium* (Odamaki et al., 2016; Xu et al., 2019; Zhang et al., 2021). Increased relative abundances of *Akkermansia* (Biagi et al., 2016; Iwauchi et al., 2019; Kim et al., 2019; Kong et al., 2016; Rampelli et al., 2020) and *Christensenellaceae* (Biagi et al., 2016; Iwauchi et al., 2019; Kim et al., 2019; Kong et al., 2016; Odamaki et al., 2016; Xu et al., 2019) have also been consistently reported in a number of studies. Many of these taxa are associated with the regulation of host immune function although further work is required to identify the profile of activity associated with this role.

Increasing compositional uniqueness arising from the decreased abundance of core taxa (bacterial taxa commonly detected in all healthy individuals, see Chapter 2), a signal first visible in mid-to-late adulthood, may be a feature of healthy aging that is predictive of survival in older adults with implications for microbial regulation of immune function (Wilmanski et al., 2021). The decline in core taxa appears to be initiated at 40 or 50 years of age, a time of life also marked by the early signs of immune system aging, and occurs gradually over time (Marquez et al., 2020; Wilmanski et al., 2022). From a functional perspective, many studies noted a decrease in core short-chain fatty acid (SCFA) producers (Badal et al., 2020; Ghosh et al., 2022), microbial metabolites of importance for microbiota-to-brain communication (see Chapter 4). Aging has also been associated with decreased synthesis of bacterial proteins involved in tryptophan and indole production, which was reflected in a progressive decrease with increasing age in fecal concentrations of these metabolites (Ruiz-Ruiz et al., 2020). Other metabolites of note include trimethylamine-N-oxide (TMAO), which was increased with age in both humans and mice (Ke et al., 2018).

## Unhealthy aging and the gut microbiome

Frailty is an aspect of unhealthy aging that has been frequently investigated in relation to the gut microbiome, and it is in this context that reduced microbial diversity has most frequently been reported. For example, the microbial α-diversity of elderly people in long-term residential care was significantly lower than that of the community-dwelling group in an Irish population (Claesson et al., 2012). This decreased α-diversity was associated with increased inflammation, decreased healthy food intake, and reductions in overall health status. These long-term residential care individuals displayed a gradual shift away from their initial microbiota that was not recorded in the community-dwelling group (Jeffery et al., 2016). Interestingly, both groups showed evidence of microbiota temporal instability, and this was particularly evident in those with low initial microbiota diversity (Jeffery et al., 2016). Biological, but not chronological, age was correlated with a decrease in microbiota α-diversity in 85 community-dwelling adults ranging in age from 43 to 79 (Maffei et al., 2017). Biological aging in this study was based on a frailty index, although changes in the gut microbiota might also have the potential to differentiate biological from chronological aging.

Frailty as a feature of unhealthy aging is also associated with reductions in the relative abundance of SCFA producers such as *Faecalibacterium prausnitzii* (Jackson et al., 2016) and an increased relative abundance of *Ruminococcus* (Jackson et al., 2016), which is also a butyrate-producing bacteria. A decreased relative abundance of *Faecalibacterium* has also been associated with reduced physical activity, another aspect of unhealthy aging (Fart et al., 2020; Langsetmo et al., 2019). Cognitive decline has been associated with reduced α-diversity (Verdi et al., 2018) as well as decreased relative abundance of *Akkermansia* in otherwise healthy older adults (Anderson et al., 2017; Manderino et al., 2017). *Akkermansia muciniphila* is a very abundant species in the human intestinal microbiota and has been extensively studied for its antiinflammatory potential (Cani and de Vos 2017).

## Aging and the gut—brain axis

The microbiota alterations noted above coincide with physiological aging processes across the pillars of the gut—brain axis. The hallmarks of health have been considered in terms of spatial organizational features of compartmentalization, maintenance of homeostasis over time, and adequate responses to stress (Lopez-Otin and Kroemer 2021). Interconnected with these organizational features are more specific hallmarks of relevance to aging including genomic instability, telomere attrition, epigenetic alterations, loss of proteostasis, disabled macroautophagy, deregulated nutrient-sensing, mitochondrial dysfunction, cellular senescence, stem cell exhaustion, altered intercellular communication, chronic inflammation, and compositional and functional alterations in the gut microbiome (Lopez-Otin et al., 2013, 2023). The composition and function of the microbiome in older people also reflect host behavior, lifestyle factors like diet and exercise, prior xenobiotic exposure, and social networks (Clarke et al., 2019; Ghosh et al., 2022; Matsuzaki et al., 2023; Sherwin et al., 2019).

Many of these features include provision for reciprocal communication with the gut microbiota. Compartmentalization, for example, requires integrity of blood—brain barrier function which is known to be disrupted with aging and in aging-related disorders and is influenced by the gut microbiota and microbial metabolites (Hoyles et al., 2021; Knox et al. 2022a, 2022b; Stachulski et al., 2023). This concept has also now been extended to the blood—cerebrospinal fluid barrier (Gelb and Lehtinen 2023; Knox et al., 2023; Xie et al., 2023). Meanwhile, the interface between the gut microbiota, the intestinal mucosal barrier, and the gastrointestinal neuroimmune system has been viewed as a common pathway to neurodegenerative diseases (Pellegrini et al., 2018).

Increased aging is also associated with progressive disruption and deterioration of gastrointestinal physiology which will have implications for gut microbiome composition (Ghosh et al., 2022). Alterations in gastrointestinal motility are common in the elderly (Firth and Prather 2002). Age-associated physiological changes in the gastrointestinal tract include cellular and molecular alterations in the enteric nervous system (ENS) and the microbiota in turn affect many aspects of the ENS and intestinal motility (Nguyen et al., 2023; Sharkey and Mawe 2023). Some of these age-related alterations can also be region dependent, with changes in neuromuscular structure and function detected in the ascending but not descending colon (Broad et al., 2019). As for other demographics, we still await corresponding information about biogeographical alterations in gut microbiota composition and function across the gastrointestinal tract due to the current reliance on information gathered from the sequencing of fecal samples.

Other pillars of the gut—brain axis can also be implicated in the reciprocal dynamics of age-related microbiome alterations. This includes the immune system (Bosco and Noti 2021; Hooper et al., 2012) and may be particularly important in the context of immunosenescence and inflammaging (Liu et al., 2023). These immune alterations are important for age-related changes in behavior and brain function and occur in addition to dysfunction across other relevant hallmarks of health and aging such as hypothalamic—pituitary—adrenal (HPA) axis, dysregulation of neurotransmission and neurotrophic factor signaling, genetic and epigenetic changes, oxidative stress and metabolic changes (Prenderville et al., 2015). For example, there is evidence for both stress-related changes in microbiota composition and function as well as microbial priming of HPA axis responses (Clarke et al., 2013; Foster et al., 2017; Lyte et al., 2020; Sudo et al., 2004). Impaired gut—brain communication may thus be viewed as an age-related change that occurs in the context of compositional and functional gut microbiome alterations, with complex and dynamic implications for reciprocal host—microbe dialog that are linked to several aspects of physical, physiological, and cognitive decline (Ghosh et al., 2022). Parsing the different contributions of hosts and microbes to these hallmarks of healthy and unhealthy aging remains a challenge. In this regard, studies of the microbiome in aged-related disorders have provided another important piece of the puzzle.

## Microbiome and Parkinson's disease

Parkinson's disease (PD) is the second-most common neurodegenerative disorder and is characterized by impairments in motor function including tremors, rigidity, and slowness of movement (Poewe et al., 2017) as well as nonmotor symptoms (Felice et al., 2016). Neuronal loss in the substantia nigra in the brains of PD patients leads to dopamine depletion and the emergence of the cardinal motor features (O'Neill 2019). Interestingly, the histological hallmark of PD in the CNS, misfolded alpha-synuclein aggregates that form Lewy bodies and neurites, can be detected in the ENS before clinical diagnosis (Travagli et al., 2020). The compositional gut microbiota alterations reported in PD were the subject of recent systematic reviews and meta-analyses, with enrichment of the genera *Lactobacillus*, *Akkermansia*, and *Bifidobacterium* and depletion of bacteria belonging to the Lachnospiraceae family and the *Faecalibacterium* genus being the most consistent signals (Li et al., 2023; Nuzum et al., 2020; Romano et al., 2021). Differentially abundant taxa between control subjects and PD patients, such as the increases in the genera *Bifidobacterium* and decreases in *Roseburia* and *Prevotella*, were noted to persist over 2 years in a study of 64 PD patients and 64 healthy controls (Aho et al., 2019).

The reduction in SCFA producers is consistent with observations reported for healthy and unhealthy aging and could be linked to intestinal inflammation, and has also been noted in two independent cohorts of treatment-naive de novo PD subjects (Boertien et al., 2022). A lower relative abundance of *Butyricimonas synergistica*, a butyrate-producer, was also associated with worse nonmotor symptoms in the PD group of 18 individuals (Nuzum et al., 2023). Studies that have measured fecal SCFA concentrations have also reported reductions in PD patients compared to controls (Aho et al., 2021; Unger et al., 2016). In addition to the enrichment with proinflammatory species, a metagenomic analysis of 90 PD and 234 control individuals also indicated dysregulation in microbial genes and pathways involved in the synthesis and metabolism of neuroactive molecules such as dopamine, glutamate, gamma-aminobutyric acid (GABA), and serotonin (Wallen et al., 2022). Plasma levels of the microbial metabolite TMAO have also been associated with PD severity and progression (Chen et al., 2020).

The evaluation of microbiota-to-brain communication as an important mechanism in PD pathology draws on data from preclinical studies. Studying the mechanisms underpinning neurodegeneration in PD is often based on neurotoxin-based models. For example, in the 1-methyl-4-phenyl-1,2,3,6-tetrahydropyridine (MPTP) mouse model, motor impairment and striatal neurotransmitter decrease are associated with a remodeled gut microbiota (Sun et al., 2018). The motor deficits and neuroinflammation in a rodent model of PD based on α-synuclein pathology in the CNS have also been associated with the gut microbiota (Sampson et al., 2016). Consistent with the emergence of ENS pathology before CNS signs and symptoms, there is also accumulating evidence that the gut microbiota could be a potential trigger for the gastrointestinal and CNS pathology in PD (Challis et al., 2020; O'Donovan et al., 2020; Stockdale et al., 2021). These observations make important contributions to the debate about whether PD originates in the brain or begins in the gut, drawing the gut microbiota into that conversation (Nuzum et al., 2022).

Fecal microbiota transplantation (FMT) in rodents is an experimental strategy often used to investigate the potential causal role of disease-associated microbiota configurations (Gheorghe et al., 2021; Secombe et al., 2021). This approach was used with the remodeled gut microbiota from the MPTP model of PD mice to verify a role for microbial-induced motor impairment and striatal neurotransmitter decrease in recipient mice. Moreover, transplantation of the gut microbiota from normal mice into the recipient mice administered MPTP was neuroprotective (Sun et al., 2018). It has also been shown that colonization of α-synuclein overexpressing mice with a microbiota harvested from PD patients enhances physical impairments in the recipient animals compared to microbiota transplants donated from healthy human donors (Sampson et al., 2016). Encouraging findings from a study targeting the gut microbiota of α-synuclein overexpressing mice with a prebiotic high-fiber diet reported attenuated motor deficits and reduced α-synuclein aggregation in the substantia nigra (Abdel-Haq et al., 2022).

Clinical and preclinical observations indicate a role for the gut microbiota in the PD symptom profile of motor dysfunction, pain, behavioral changes, and gastrointestinal issues. From a neurobiological perspective, there is also evidence for microbial regulation of dopamine depletion, α-synuclein pathology, and neuroinflammation. This latter finding in particular is consistent with previous evidence for microbial regulation of microglial maturation, activity, and function (Erny et al. 2015, 2021; Lynch et al., 2021). The compositional and functional features of the gut microbiota in PD support a proinflammatory phenotype. Of further note is that the microbial metabolite p-Cresol, which is increased in PD, is associated with dopaminergic dysfunction (Cirstea et al., 2020). It is also interesting to note that gut bacterial enzymes, tyrosine decarboxylases, can restrict the bioavailability of levodopa, which is the key current treatment option for PD (Maini Rekdal et al., 2019; O'Neill 2019; van Kessel et al., 2019). Taken together, these observations open up the possibility of therapeutic targeting of the gut microbiota to turn the tables in the battle against neuropathology in PD (Barros-Santos and Clarke 2023).

## Microbiome and Alzheimer's disease

Alzheimer's disease (AD), the most common neurodegenerative disorder, is a common cause of cognitive impairment and dementia in the elderly and is associated with a neurobiological profile of β-amyloid-containing plaques and tau-containing neurofibrillary tangles (Knopman et al., 2021). In terms of microbiota summary statistics, it has been reported that there is a decrease in Phylogenetic Diversity in AD participants compared to controls, a feature which occurred in conjunction with a decreased relative abundance of *Firmicutes* and *Bifidobacterium* and an increased relative abundance of *Bacteroidetes* (Vogt et al., 2017). Studies in Chinese populations found a decreased relative abundance of *Bacteroidetes* and an increased relative abundance of *Actinobacteria* (Zhuang et al., 2018), or a reduced relative abundance of *Firmicutes* and an increased relative abundance of *Proteobacteria* in AD patients compared with health controls (Liu et al., 2019).

The microbiome composition of those with AD clustered away from those without dementia with increased relative abundances of *Bacteroides* spp, *Alistipes* spp, *Odoribacter* spp, and *Barnesiella* spp, and a decreased relative abundance of *Lachnoclostridium* spp (Haran et al., 2019). Interestingly, the AD microbiome was associated with dysregulation of the antiinflammatory p-glycoprotein pathway when assessed in in vitro T84 intestinal epithelial cell functional assays. The proinflammatory genus *Collinsella* was positively correlated with the APOE rs429358 risk allele in both discovery and replication cohorts (Cammann et al., 2023). Recently, it has also been suggested that gut microbiomes of 49 people with biomarker evidence of preclinical AD for a sample of 164 cognitively normal individuals had a different composition from that of healthy individuals (Ferreiro et al., 2023). While α-diversity was similar between groups, there were specific gut bacterial taxa associated with preclinical AD including *Dorea formicigenerans*, *Oscillibacter*, *Faecalibacterium prausnitzii*, *Coprococcus*, and *Anaerostipes hadrus*. Microbial pathways most associated with preclinical AD included those for L-arginine, L-ornithine, and 4-aminobutanoate degradation (Ferreiro et al., 2023).

As for other neurodegenerative disorders, both systemic inflammation and neuroinflammation have been identified as potential drivers of AD, factors that are often evaluated in preclinical models of AD. The Aβ-precursor protein (APP) transgenic mouse model of AD displayed a distinct gut microbiota composition compared to nontransgenic wild-type mice, and germ-free APP transgenic mice exhibited a marked reduction in cerebral Aβ-amyloid pathology compared to control mice with an intact intestinal microbiota (Harach et al., 2017). Of further note in the study from Harach and colleagues is that germ-free APP transgenic mice in receipt of an FMT from conventionally raised APP transgenic mice exhibited increased cerebral Aβ-pathology, and FMT of microbiota from wild-type donor mice was less effective in increasing cerebral Aβ-level in recipient animals. FMT of the gut microbiota from AD patients into APP/PS1 double transgenic mice induced more severe cognitive impairment in the recipient animals accompanied by activation of microglia and increased expression of neuroinflammatory factors in the hippocampus (Shen et al., 2020). A genetically engineered mouse model of tauopathy expressing human ApoE isoforms exhibited reduced gliosis, tau pathology, and neurodegeneration in a sex- and ApoE isoform-dependent manner when the mice were reared under germ-free conditions or had their gut microbiota depleted with antibiotics (Seo et al., 2023). This study also implicated SCFAs in the capacity of the gut microbiota to influence neuroinflammation and tau-mediated neurodegeneration, common suspects across clinical and pre-clinical studies. It has also been shown that the gut microbiota plays an important role in astrocyte phenotypes and microglial-astrocyte communication in male APPPS1-21 mice (Chandra et al., 2023).

## Other neurodegenerative disorders: ALS/MS/ Huntington's

A decrease in butyrate-producing bacteria and the *Firmicutes/Bacteroidetes* ratio has also been reported in a small study of nine healthy controls cohabiting with 19 patients with amyotrophic lateral sclerosis (ALS) (Rowin et al., 2017), a disorder characterized by muscle weakness and eventual paralysis due to degeneration of both upper and lower motor neurons (Hardiman et al., 2017). In the transgenic SOD1G93A mouse model of ALS, compositional alterations in the gut microbiota preceded motor dysfunction, muscle atrophy, and peripheral and central inflammation (Figueroa-Romero et al., 2019). In C9orf72-mutated mice, a genetic variant that contributes to ALS and frontotemporal dementia, a number of experimental approaches (microbiota depletion using broad-spectrum antibiotics, FMT) to reduce the abundance of immune-stimulating bacteria was protective against premature mortality and reduced both systemic inflammation and autoimmunity (Burberry et al., 2020).

Huntington's disease (HD) arises due to a mutant form of the multi-functional protein huntingtin, which leads to a combination of motor, cognitive, and behavioral features (Bates et al., 2015). In a study of 33 HD patients and 33 sex- and age-matched healthy controls, $\alpha$-diversity (richness) was increased and accompanied by relative abundance alterations in several taxa including at the genus level where *Intestinimonas* was positively correlated with total functional capacity scores and IL-4 levels, and *Bilophila* were negatively correlated with proinflammatory IL-6 levels (Du et al., 2020). HD was also associated with reduced $\alpha$-diversity which was reflected in reduced functional pathways with increased relative abundance of microbial enzymes glutathione transferase and homoserine O-succinyltransferase (Wasser et al., 2020). This study also reported correlations between a lower abundance of *E. hallii* with more severe motor signs and a significant negative relationship between the relative abundance of *E. hallii* and estimated proximity to disease onset. Microbiome profiling of the R6/1 transgenic mouse model of HD reported increased microbial diversity only in male HD mice, and an increase in the relative abundance of *Bacteriodetes* and a decrease in the relative abundance of *Firmicutes* in the HD gut microbiome of both sexes (Kong et al., 2020). These compositional alterations were noted at week 12, which coincided with the expression of motor deficits. When R6/1 mice were exposed to environmental enrichment and exercise before the onset of the compositional gut microbiota alterations and motor deficits, Bacteroidales, Lachnospirales, and Oscillospirales bacteria were affected by the intervention (Gubert et al., 2022). The authors proposed the gut microbiome as a possible mediator of the benefits of these interventions, although these measurements were conducted at a stage when the early signs of the disease were not yet impacted (Gubert et al., 2022).

In young adults, multiple sclerosis (MS) is the most common chronic inflammatory, demyelinating, and neurodegenerative CNS disorder (Filippi et al., 2018). Epstein—Barr virus (EBV) has recently been proposed as the leading cause of MS (Bjornevik et al., 2022). The gut microbiome still plays a role, at least at the symptom level, and in a longitudinal assessment of a cohort of 111 MS patients, the *Bacteroides 2* enterotype prevalence in the gut microbial composition at baseline was related to worse long-term disability (Devolder et al., 2023). Phenotypic subcategorization of MS patients also identified subtype-specific microbiome biomarkers linked to disease activity and severity, with microbial richness lower in interferon-treated and untreated relapsing-remitting MS during relapse compared to benign MS as one example (Reynders et al., 2020). In the experimental-autoimmune-encephalomyelitis (EAE) animal model of MS, antibiotic-induced microbiota depletion from weaning delays the onset of clinical symptoms as well as altering adult immunological and neurobehavioral responses (Zeraati et al., 2019). Germ-free (GF) mice are also protected from the full severity of symptom expression in the EAE model, providing further evidence supporting the role for the gut microbiome in symptom development (Berer et al., 2011; Lee et al., 2011).

## Microbiome and aging: longevity in model organisms

Microbiota—brain axis research on simpler genetic model organisms such as zebrafish, *Drosophila melanogaster*, and *Caenorhabditis elegans* can offer many important mechanistic insights (Nagpal and Cryan 2021). *C. elegans*, in particular, has shown practical advantages and utility in the context of microbe—host interactions that promote longevity with contributions to worm fitness and lifespan extension. It is therefore a promising alternative model to study if and how host—microbiome interactions are important for lifespan. In this regard, worms grown with Chryseobacterium sp. CHNTR56 MYb120 exhibited lifespan extension linked to vitamin B6 synthesis (Nagpal and Cryan 2021). An *Escherichia coli* strain that secretes colanic acid was associated with a dose-dependent lifespan-extending effect in *C. elegans* (Hartsough et al., 2020). The bacteria-derived metabolite methylglyoxal was also shown to modulate the longevity of *C. elegans*, an effect the authors linked to TORC2/SGK-1/DAF-16 signaling and illustrated translational relevance in the assessment of cellular senescence in human dermal fibroblasts (Shin et al., 2020).

## Microbiome and aging: turning back the clock in preclinical models

The association between the gut microbiome and aging in health and disease has prompted the study of whether age-related deterioration in host physiology and brain function can be prevented, delayed or reversed by targeting the gut microbiota. FMT from young donors into aging recipient mice reversed aging-associated differences in peripheral and brain immunity, and in the hippocampal metabolome and transcriptome (Boehme et al., 2021). FMT between young and aged mice either accelerated or reversed age-associated CNS inflammation, retinal inflammation, and cytokine signaling coincident with parallel intestinal barrier permeability alterations (Parker et al., 2022). FMT from aged donor mice to young recipients led to impaired spatial learning and memory whereas anxiety, explorative behavior, and locomotor activity were not modified (D'Amato et al., 2020). Given the possibility of transitional time windows that gate the emergence of initial age-related phenotypes, it has been shown prebiotics can beneficially modulate the peripheral immune response and alter neuroinflammation in middle-aged mice (Boehme et al., 2020). The aged gut microbiota has also been associated with social behavior deficits (Cruz-Pereira et al., 2023), consistent with the observation that the prebiotic FOS-Inulin was able to alleviate age-related social interaction deficits in response to a social defeat stress exposure (Cruz-Pereira et al., 2022).

Of course, it is important to note that there are many age-related microbiome alterations that occur in the context of health, and introducing microbes from aged animals has not always proven beneficial, prompting suggestions that it is not straightforward to predict the consequences of turning the microbial clock back or forward (Wilmanski et al., 2022). For example, GF mice in receipt of a donor gut microbiota from old mice displayed increased neurogenesis in the hippocampus as well as increased intestinal growth (Kundu et al., 2019).

## Conclusions, perspectives, and future directions

It is increasingly coming into focus that one of the key determinants of healthy aging and the continuum from age-related decline to age-related disorders is the microbiome. Both clinical and preclinical research supports the hypothesis that aging-related health loss is correlated with a compositional reconfiguration of the gut microbiome. Common threads identified include a reduction in SCFA producers and compositional realignments that favor proinflammatory states which are often exacerbated in unhealthy aging and in age-related neurodegenerative disorders. There are also indications that there may be transitional windows during which the first signs of the impact of increasing age on gut microbiota form and function may be visible. Harnessing the gut microbiome as a predictive biomarker reservoir thus holds the potential to allow for earlier intervention for individuals identified to be at risk of common neurogenerative disorders.

It is important to realize that age-related alterations in the gut microbiome are heavily personalized and influenced by host physiology interfacing with a lifetime of personal lifestyle-linked factors including assembly trajectory in early life, diet, medication, physical activity, and social networks. The focus on summary microbiome statistics has not provided clear answers about the best targets, and it is not at all clear, for example, if generically increasing $\alpha$-diversity will be beneficial since that is a metric that has been reported to be either increased, decreased, or unchanged in age-related neurodegenerative disorders. This has seen the focus shift from diversity toward the identification of disease-associated and health-associated taxa. Detailed participant phenotyping is critical because specific associations have been noted depending on symptom severity and disease subtype. The accrual of additional functional readouts about microbiome metabolic pathways will also be important.

These approaches may ultimately yield the information necessary to guide a more targeted restoration of the key microbial functions associated with healthier brain aging. There are a number of open questions currently delaying diagnostics and the successful therapeutic targeting of the gut microbiome to produce clinical benefits for older people. Some of these questions may require answers at the strain level to guide when and how to restore the aged microbiome and to understand how to maintain it in an optimal state for brain health.

# References

Abdel-Haq, R., et al., 2022. A prebiotic diet modulates microglial states and motor deficits in alpha-synuclein overexpressing mice. Elife 11.

Aho, V.T.E., et al., 2019. Gut microbiota in Parkinson's disease: temporal stability and relations to disease progression. EBioMedicine 44, 691–707.

Aho, V.T.E., et al., 2021. Relationships of gut microbiota, short-chain fatty acids, inflammation, and the gut barrier in Parkinson's disease. Mol Neurodegener 16 (1), 6.

Anderson, J.R., et al., 2017. A preliminary examination of gut microbiota, sleep, and cognitive flexibility in healthy older adults. Sleep Med 38, 104–107.

Badal, V.D., et al., 2020. The gut microbiome, aging, and longevity: a systematic review. Nutrients 12 (12).

Barros-Santos, T., Clarke, G., 2023. Gut-initiated neuroprotection in Parkinson's disease: when microbes turn the tables in the battle against neuroinflammation. Brain Behav Immun 108, 350–352.

Bates, G.P., et al., 2015. Huntington disease. Nat Rev Dis Prim 1, 15005.

Berer, K., et al., 2011. Commensal microbiota and myelin autoantigen cooperate to trigger autoimmune demyelination. Nature 479 (7374), 538–541.

Biagi, E., et al., 2016. Gut microbiota and extreme longevity. Curr Biol 26 (11), 1480–1485.

Bian, G., et al., 2017. The gut microbiota of healthy aged Chinese is similar to that of the healthy young. mSphere 2 (5).

Bjornevik, K., et al., 2022. Longitudinal analysis reveals high prevalence of Epstein-Barr virus associated with multiple sclerosis. Science 375 (6578), 296–301.

Boehme, M., et al., 2020. Mid-life microbiota crises: middle age is associated with pervasive neuroimmune alterations that are reversed by targeting the gut microbiome. Mol Psychiatr 25 (10), 2567–2583.

Boehme, M., et al., 2021. Microbiota from young mice counteracts selective age-associated behavioral deficits. Nat Aging 1 (8), 666–676.

Boertien, J.M., et al., 2022. Fecal microbiome alterations in treatment-naive de novo Parkinson's disease. NPJ Parkinsons Dis 8 (1), 129.

Bosco, N., Noti, M., 2021. The aging gut microbiome and its impact on host immunity. Gene Immun 22 (5–6), 289–303.

Broad, J., et al., 2019. Changes in neuromuscular structure and functions of human colon during ageing are region-dependent. Gut 68 (7), 1210–1223.

Burberry, A., et al., 2020. C9orf72 suppresses systemic and neural inflammation induced by gut bacteria. Nature 582 (7810), 89–94.

Cammann, D., et al., 2023. Genetic correlations between Alzheimer's disease and gut microbiome genera. Sci Rep 13 (1), 5258.

Cani, P.D., de Vos, W.M., 2017. Next-generation beneficial microbes: the case of Akkermansia muciniphila. Front Microbiol 8, 1765.

Challis, C., et al., 2020. Gut-seeded alpha-synuclein fibrils promote gut dysfunction and brain pathology specifically in aged mice. Nat Neurosci 23 (3), 327—336.

Chandra, S., et al., 2023. The gut microbiome regulates astrocyte reaction to Abeta amyloidosis through microglial dependent and independent mechanisms. Mol Neurodegener 18 (1), 45.

Chen, S.J., et al., 2020. The gut metabolite trimethylamine N-oxide is associated with Parkinson's disease severity and progression. Mov Disord 35 (11), 2115—2116.

Cirstea, M.S., et al., 2020. Microbiota composition and metabolism are associated with gut function in Parkinson's disease. Mov Disord 35 (7), 1208—1217.

Claesson, M.J., et al., 2012. Gut microbiota composition correlates with diet and health in the elderly. Nature 488 (7410), 178—184.

Clarke, G., et al., 2019. Gut reactions: breaking down xenobiotic-microbiome interactions. Pharmacol Rev 71 (2), 198—224.

Clarke, G., et al., 2013. The microbiome-gut-brain axis during early life regulates the hippocampal serotonergic system in a sex-dependent manner. Mol Psychiatr 18 (6), 666—673.

Cruz-Pereira, J.S., et al., 2023. Age-associated deficits in social behaviour are microbiota-dependent. Brain Behav Immun 110, 119—124.

Cruz-Pereira, J.S., et al., 2022. Prebiotic supplementation modulates selective effects of stress on behavior and brain metabolome in aged mice. Neurobiol Stress 21, 100501.

D'Amato, A., et al., 2020. Faecal microbiota transplant from aged donor mice affects spatial learning and memory via modulating hippocampal synaptic plasticity- and neurotransmission-related proteins in young recipients. Microbiome 8 (1), 140.

Devolder, L., et al., 2023. Gut microbiome composition is associated with long-term disability worsening in multiple sclerosis. Gut Microb 15 (1), 2180316.

Du, G., et al., 2020. Altered gut microbiota related to inflammatory responses in patients with Huntington's disease. Front Immunol 11, 603594.

Erny, D., et al., 2015. Host microbiota constantly control maturation and function of microglia in the CNS. Nat Neurosci 18 (7), 965—977.

Erny, D., et al., 2021. Microbiota-derived acetate enables the metabolic fitness of the brain innate immune system during health and disease. Cell Metabol 33 (11), 2260—22676 e7.

Fart, F., et al., 2020. Differences in gut microbiome composition between senior orienteering athletes and community-dwelling older adults. Nutrients 12 (9).

Felice, V.D., et al., 2016. Microbiota-gut-brain signalling in Parkinson's disease: implications for non-motor symptoms. Parkinsonism Relat Disorders 27, 1—8.

Ferreiro, A.L., et al., 2023. Gut microbiome composition may be an indicator of preclinical Alzheimer's disease. Sci Transl Med 15 (700), eabo2984.

Figueroa-Romero, C., et al., 2019. Temporal evolution of the microbiome, immune system and epigenome with disease progression in ALS mice. Dis Model Mech 13 (2).

Filippi, M., et al., 2018. Multiple sclerosis. Nat Rev Dis Prim 4 (1), 43.

Firth, M., Prather, C.M., 2002. Gastrointestinal motility problems in the elderly patient. Gastroenterology 122 (6), 1688—1700.

Foster, J.A., Rinaman, L., Cryan, J.F., 2017. Stress and the gut-brain axis: regulation by the microbiome. Neurobiol Stress 7, 124—136.

Gelb, S., Lehtinen, M.K., 2023. Snapshot: choroid plexus brain barrier. Cell 186 (16), 3522—35222 e1.

Gheorghe, C.E., et al., 2021. Investigating causality with fecal microbiota transplantation in rodents: applications, recommendations and pitfalls. Gut Microb 13 (1), 1941711.

Ghosh, T.S., Shanahan, F., O'Toole, P.W., 2022. The gut microbiome as a modulator of healthy ageing. Nat Rev Gastroenterol Hepatol 19 (9), 565—584.

Gubert, C., et al., 2022. Gene-environment-gut interactions in Huntington's disease mice are associated with environmental modulation of the gut microbiome. iScience 25 (1), 103687.

Harach, T., et al., 2017. Reduction of Abeta amyloid pathology in APPPS1 transgenic mice in the absence of gut microbiota. Sci Rep 7, 41802.

Haran, J.P., et al., 2019. Alzheimer's disease microbiome is associated with dysregulation of the anti-inflammatory P-glycoprotein pathway. mBio 10 (3).

Hardiman, O., et al., 2017. Amyotrophic lateral sclerosis. Nat Rev Dis Prim 3, 17071.

Hartsough, L.A., et al., 2020. Optogenetic control of gut bacterial metabolism to promote longevity. Elife 9.

Hooper, L.V., Littman, D.R., Macpherson, A.J., 2012. Interactions between the microbiota and the immune system. Science 336 (6086), 1268–1273.

Hoyles, L., et al., 2021. Regulation of blood-brain barrier integrity by microbiome-associated methylamines and cognition by trimethylamine N-oxide. Microbiome 9 (1), 235.

Iwauchi, M., et al., 2019. Relationship between oral and gut microbiota in elderly people. Immun Inflamm Dis 7 (3), 229–236.

Jackson, M.A., et al., 2016. Signatures of early frailty in the gut microbiota. Genome Med 8 (1), 8.

Jeffery, I.B., Lynch, D.B., O'Toole, P.W., 2016. Composition and temporal stability of the gut microbiota in older persons. ISME J 10 (1), 170–182.

Ke, Y., et al., 2018. Gut flora-dependent metabolite Trimethylamine-N-oxide accelerates endothelial cell senescence and vascular aging through oxidative stress. Free Radic Biol Med 116, 88–100.

Kim, B.S., et al., 2019. Comparison of the gut microbiota of centenarians in longevity villages of South Korea with those of other age groups. J Microbiol Biotechnol 29 (3), 429–440.

Knopman, D.S., et al., 2021. Alzheimer disease. Nat Rev Dis Prim 7 (1), 33.

Knox, E.G., et al., 2022a. The blood-brain barrier in aging and neurodegeneration. Mol Psychiatr 27 (6), 2659–2673.

Knox, E.G., et al., 2022b. Microbial-derived metabolites induce actin cytoskeletal rearrangement and protect blood-brain barrier function. iScience 25 (12), 105648.

Knox, E.G., et al., 2023. The gut microbiota is important for the maintenance of blood-cerebrospinal fluid barrier integrity. Eur J Neurosci 57 (2), 233–241.

Kong, F., et al., 2019. Identification of gut microbiome signatures associated with longevity provides a promising modulation target for healthy aging. Gut Microb 10 (2), 210–215.

Kong, F., et al., 2016. Gut microbiota signatures of longevity. Curr Biol 26 (18), R832–R833.

Kong, G., et al., 2020. Microbiome profiling reveals gut dysbiosis in a transgenic mouse model of Huntington's disease. Neurobiol Dis 135, 104268.

Kundu, P., et al., 2019. Neurogenesis and prolongevity signaling in young germ-free mice transplanted with the gut microbiota of old mice. Sci Transl Med 11 (518).

Langsetmo, L., et al., 2019. The association between objectively measured physical activity and the gut microbiome among older community dwelling men. J Nutr Health Aging 23 (6), 538–546.

Lee, Y.K., et al., 2011. Proinflammatory T-cell responses to gut microbiota promote experimental autoimmune encephalomyelitis. Proc Natl Acad Sci U S A 108 Suppl 1 (Suppl. 1), 4615–4622.

Li, Z., et al., 2023. Gut bacterial profiles in Parkinson's disease: a systematic review. CNS Neurosci Ther 29 (1), 140–157.

Liu, P., et al., 2019. Altered microbiomes distinguish Alzheimer's disease from amnestic mild cognitive impairment and health in a Chinese cohort. Brain Behav Immun 80, 633–643.

Liu, Z., et al., 2023. Immunosenescence: molecular mechanisms and diseases. Signal Transduct Targeted Ther 8 (1), 200.

Lopez-Otin, C., Kroemer, G., 2021. Hallmarks of health. Cell 184 (1), 33–63.

Lopez-Otin, C., et al., 2013. The hallmarks of aging. Cell 153 (6), 1194–1217.

Lopez-Otin, C., et al., 2023. Hallmarks of aging: an expanding universe. Cell 186 (2), 243–278.

Lynch, C.M.K., Clarke, G., Cryan, J.F., 2021. Powering up microbiome-microglia interactions. Cell Metabol 33 (11), 2097–2099.

Lyte, J.M., et al., 2020. Gut-brain axis serotonergic responses to acute stress exposure are microbiome-dependent. Neuro Gastroenterol Motil 32 (11), e13881.

Maffei, V.J., et al., 2017. Biological aging and the human gut microbiota. J Gerontol A Biol Sci Med Sci 72 (11), 1474–1482.

Maini Rekdal, V., et al., 2019. Discovery and inhibition of an interspecies gut bacterial pathway for Levodopa metabolism. Science 364 (6445).

Manderino, L., et al., 2017. Preliminary evidence for an association between the composition of the gut microbiome and cognitive function in neurologically healthy older adults. J Int Neuropsychol Soc 23 (8), 700–705.

Marquez, E.J., et al., 2020. Sexual-dimorphism in human immune system aging. Nat Commun 11 (1), 751.

Matsuzaki, R., et al., 2023. Pesticide exposure and the microbiota-gut-brain axis. ISME J 17 (8), 1153–1166.

Nagpal, J., Cryan, J.F., 2021. Microbiota-brain interactions: moving toward mechanisms in model organisms. Neuron 109 (24), 3930–3953.

Nguyen, T.T., et al., 2023. The aging enteric nervous system. Int J Mol Sci 24 (11).

Nuzum, N.D., et al., 2020. Gut microbiota differences between healthy older adults and individuals with Parkinson's disease: a systematic review. Neurosci Biobehav Rev 112, 227–241.

Nuzum, N.D., et al., 2022. To the gut microbiome and beyond: the brain-first or body-first hypothesis in Parkinson's disease. Front Microbiol 13, 791213.

Nuzum, N.D., et al., 2023. Differences in the gut microbiome across typical ageing and in Parkinson's disease. Neuropharmacology 235, 109566.

O'Donovan, S.M., et al., 2020. Nigral overexpression of alpha-synuclein in a rat Parkinson's disease model indicates alterations in the enteric nervous system and the gut microbiome. Neuro Gastroenterol Motil 32 (1), e13726.

O'Neill, C., 2019. Gut microbes metabolize Parkinson's disease drug. Science 364 (6445), 1030–1031.

Odamaki, T., et al., 2016. Age-related changes in gut microbiota composition from newborn to centenarian: a cross-sectional study. BMC Microbiol 16, 90.

Parker, A., et al., 2022. Fecal microbiota transfer between young and aged mice reverses hallmarks of the aging gut, eye, and brain. Microbiome 10 (1), 68.

Pellegrini, C., et al., 2018. Interplay among gut microbiota, intestinal mucosal barrier and enteric neuro-immune system: a common path to neurodegenerative diseases? Acta Neuropathol 136 (3), 345–361.

Poewe, W., et al., 2017. Parkinson disease. Nat Rev Dis Prim 3, 17013.

Prenderville, J.A., et al., 2015. Adding fuel to the fire: the impact of stress on the ageing brain. Trends Neurosci 38 (1), 13–25.

Rampelli, S., et al., 2020. Shotgun metagenomics of gut microbiota in humans with up to extreme longevity and the increasing role of xenobiotic degradation. mSystems 5 (2).

Reynders, T., et al., 2020. Gut microbiome variation is associated to Multiple Sclerosis phenotypic subtypes. Ann Clin Transl Neurol 7 (4), 406−419.

Romano, S., et al., 2021. Meta-analysis of the Parkinson's disease gut microbiome suggests alterations linked to intestinal inflammation. NPJ Parkinsons Dis 7 (1), 27.

Rowin, J., et al., 2017. Gut inflammation and dysbiosis in human motor neuron disease. Phys Rep 5 (18).

Ruiz-Ruiz, S., et al., 2020. Functional microbiome deficits associated with ageing: chronological age threshold. Aging Cell 19 (1), e13063.

Sampson, T.R., et al., 2016. Gut microbiota regulate motor deficits and neuroinflammation in a model of Parkinson's disease. Cell 167 (6), 1469−1480 e12.

Secombe, K.R., et al., 2021. Guidelines for reporting on animal fecal transplantation (GRAFT) studies: recommendations from a systematic review of murine transplantation protocols. Gut Microb 13 (1), 1979878.

Seo, D.O., et al., 2023. ApoE isoform- and microbiota-dependent progression of neurodegeneration in a mouse model of tauopathy. Science 379 (6628), eadd1236.

Sharkey, K.A., Mawe, G.M., 2023. The enteric nervous system. Physiol Rev 103 (2), 1487−1564.

Shen, H., et al., 2020. New mechanism of neuroinflammation in Alzheimer's disease: the activation of NLRP3 inflammasome mediated by gut microbiota. Prog Neuro-Psychopharmacol Biol Psychiatry 100, 109884.

Sherwin, E., et al., 2019. Microbiota and the social brain. Science 366 (6465).

Shin, M.G., et al., 2020. Bacteria-derived metabolite, methylglyoxal, modulates the longevity of C. elegans through TORC2/SGK-1/DAF-16 signaling. Proc Natl Acad Sci U S A 117 (29), 17142−17150.

Stachulski, A.V., et al., 2023. A host-gut microbial amino acid co-metabolite, p-cresol glucuronide, promotes blood-brain barrier integrity in vivo. Tissue Barriers 11 (1), 2073175.

Stockdale, S.R., et al., 2021. Alpha-synuclein alters the faecal viromes of rats in a gut-initiated model of Parkinson's disease. Commun Biol 4 (1), 1140.

Sudo, N., et al., 2004. Postnatal microbial colonization programs the hypothalamic-pituitary-adrenal system for stress response in mice. J Physiol 558 (Pt 1), 263−275.

Sun, M.F., et al., 2018. Neuroprotective effects of fecal microbiota transplantation on MPTP-induced Parkinson's disease mice: gut microbiota, glial reaction and TLR4/TNF-alpha signaling pathway. Brain Behav Immun 70, 48−60.

Travagli, R.A., Browning, K.N., Camilleri, M., 2020. Parkinson disease and the gut: new insights into pathogenesis and clinical relevance. Nat Rev Gastroenterol Hepatol 17 (11), 673−685.

Unger, M.M., et al., 2016. Short chain fatty acids and gut microbiota differ between patients with Parkinson's disease and age-matched controls. Parkinsonism Relat Disorders 32, 66−72.

van Kessel, S.P., et al., 2019. Gut bacterial tyrosine decarboxylases restrict levels of levodopa in the treatment of Parkinson's disease. Nat Commun 10 (1), 310.

Verdi, S., et al., 2018. An investigation into physical frailty as a link between the gut microbiome and cognitive health. Front Aging Neurosci 10, 398.

Vogt, N.M., et al., 2017. Gut microbiome alterations in Alzheimer's disease. Sci Rep 7 (1), 13537.

Wallen, Z.D., et al., 2022. Metagenomics of Parkinson's disease implicates the gut microbiome in multiple disease mechanisms. Nat Commun 13 (1), 6958.

Wasser, C.I., et al., 2020. Gut dysbiosis in Huntington's disease: associations among gut microbiota, cognitive performance and clinical outcomes. Brain Commun 2 (2), fcaa110.

Wilmanski, T., Gibbons, S.M., Price, N.D., 2022. Healthy aging and the human gut microbiome: why we cannot just turn back the clock. Nat Aging 2 (10), 869—871.

Wilmanski, T., et al., 2021. Gut microbiome pattern reflects healthy ageing and predicts survival in humans. Nat Metab 3 (2), 274—286.

Xie, J., et al., 2023. Gut microbiota regulates blood-cerebrospinal fluid barrier function and Abeta pathology. EMBO J 42 (17), e111515.

Xu, C., Zhu, H., Qiu, P., 2019. Aging progression of human gut microbiota. BMC Microbiol 19 (1), 236.

Zeraati, M., et al., 2019. Gut microbiota depletion from early adolescence alters adult immunological and neurobehavioral responses in a mouse model of multiple sclerosis. Neuropharmacology 157, 107685.

Zhang, X., et al., 2021. Sex- and age-related trajectories of the adult human gut microbiota shared across populations of different ethnicities. Nat Aging 1 (1), 87—100.

Zhuang, Z.Q., et al., 2018. Gut microbiota is altered in patients with Alzheimer's disease. J Alzheimers Dis 63 (4), 1337—1346.

# CHAPTER 10

# Pharmacological treatments and the microbiome—Antibiotics and nonantibiotic drugs

## Introduction

Trillions of microbes in the human gastrointestinal tract have an enormous metabolic potential that contributes to nutrient metabolism but also have a significant impact on drug metabolism. Recent advances demonstrate that our microbes influence drug action and may contribute to individual differences in drug response. The ability of microbiota to metabolize drugs has been known for more than 40 years (Scheline, 1968, 1973), and yet, the clinical importance of how our microbes and their metabolic potential influence drug action has been underrecognized in many areas of biomedical research. Understanding the importance of how host and microbial factors contribute to individual differences in drug response is critical to advancing precision medicine approaches. The focus in this chapter is on pharmaceuticals, including antibiotic and nonantibiotic drugs, their relationship with gut microbiota, as well as implications for microbiota-brain communication in health and disease.

Drug—microbe interactions can influence the metabolism of orally administered drugs through both direct and indirect mechanisms (Fig. 10.1) (Haiser and Turnbaugh, 2012, 2013). Orally administered drugs interact with microbiota in the small and/or large intestine; microbiota can influence activity, bioavailability, and toxicity of drugs and drugs influence microbiota composition and function (Clarke et al., 2019; Haiser and Turnbaugh, 2013; Jourova et al., 2016; Koppel et al., 2017). The chemical modifications and detailed mechanisms by which gut microbes metabolize xenobiotics (ingested foreign substances that include dietary compounds, environmental chemicals, and pharmaceuticals) is well-described in recent review papers (Clarke et al., 2019; Flowers et al., 2020; Koppel et al., 2017; Yip and Chan, 2015). Importantly, microbiota influence pharmaceutical drug metabolism within the context of host metabolic processes and in

*Microbiota Brain Axis*
ISBN 978-0-12-814800-6
https://doi.org/10.1016/B978-0-12-814800-6.00003-0

195

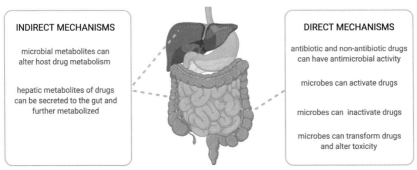

**INDIRECT MECHANISMS**

microbial metabolites can alter host drug metabolism

hepatic metabolites of drugs can be secreted to the gut and further metabolized

**DIRECT MECHANISMS**

antibiotic and non-antibiotic drugs can have antimicrobial activity

microbes can activate drugs

microbes can inactivate drugs

microbes can transform drugs and alter toxicity

**Figure 10.1** Gut microbiota metabolize oral drugs by direct and indirect mechanisms. Microbiota in the small and large intestine encounter orally administered drugs and influence host metabolism by direct mechanisms (*right panel*). Microbe-drug—host interactions in the gut and liver impact drug metabolism through indirect mechanisms (*left panel*). (Based on (Haiser, H.J., Turnbaugh, P.J., 2012. Is it time for a metagenomic basis of therapeutics? Science 336, 1253—1255; Haiser, H.J., Turnbaugh, P.J., 2013. Developing a metagenomic view of xenobiotic metabolism. Pharmacol Res 69, 21—31), created with Biorender.com.)

order to understand how host and microbes systems influence drug metabolism and host response, a basic understanding of the main concepts and processes involved in host drug metabolism is needed. Oral drugs are often absorbed in the small intestine and acted on by host metabolic enzymes. The liver plays a central role in drug metabolism by the host. Bidirectional pathways between the gut and the liver are important for metabolism, as nutrients and metabolites are transported from the GI tract to the liver via the portal vein and bile salts and other molecules are transported to the GI tract via the biliary tract (Bjorkholm et al., 2009; Tripathi et al., 2018). Microbiota indirectly influence drug metabolism by influencing host systems. For example, studies in germ-free (GF) mice show that microbiota influence expression of liver genes including target genes that are involved in drug metabolism and thereby can indirectly alter host drug metabolic processes (Bjorkholm et al., 2009; Jourova et al., 2017). Microbiota can also influence drug clearance by host systems. A good example of the influence of gut microbiota on drug clearance is acetaminophen (Clayton et al., 2009). This clinical study showed that urinary levels of the gut metabolite p-cresol predicted the post-drug urinary levels of acetaminophen metabolites. Acetaminophen is absorbed in the small intestine and metabolized primarily in the liver. This study suggests that competition for sulfonation enzymes, in individuals with higher pre-drug levels of p-cresol, influenced the ability of the liver to metabolize acetaminophen. In this case and in other cases where microbiota influence drug metabolism, it is possible that manipulating the microbiota could improve drug efficacy or reduce adverse drug reactions (Clayton et al., 2009). In addition to oral drugs, intravenous drugs and/or their metabolites can reach the large intestine via the biliary duct; where they are either excreted or in some cases further metabolized by microbiota (Koppel et al., 2017). Notably, bacterial metabolic enzymes outnumber host metabolic enzymes and microbial metabolism of many clinically important drugs has been reported including antibiotic and nonantibiotic drugs (Jourova et al., 2016). The focus of this chapter is on the impact of oral drugs on microbiota composition and function and how microbe-drug interactions impact the microbiota-gut-brain axis.

## Antibiotics impact the microbiome and the microbiota-brain axis

The discovery of penicillin in 1928 was a game-changer in treating infectious disease (Markel, 2013). Since its initial clinical use in 1942, penicillin and other antibiotics have been used throughout the world to treat infectious diseases (Markel, 2013). There are more than 20 different classes of antibiotics and each antibiotic targets specific bacteria (Werth, 2020). The main action of antibiotics is to kill bacteria; however, in addition to the targeted pathogenic bacteria, commensal bacteria are directly impacted. Oral antibiotics have a significant impact on gut microbiota with a general reduction in microbial diversity, composition, and function, including microbe-host interactions. Understanding the impact of short-term and long-term exposure to antibiotics on the microbiome is important since changes in microbiota and microbe-host interactions may contribute to health status or risk of disease.

### Exposure to antibiotics early in life—Animal studies

Animal studies show that exposure to antibiotics during early life leads to long-term changes in microbiome composition and function (Champagne-Jorgensen et al., 2020; Cho et al., 2012; Cox et al., 2014; Lebovitz et al., 2019; Leclercq et al., 2017). In order to examine exposure to antibiotics early in life in animal studies, antibiotics are administered to the dame during pregnancy and/or during the preweaning period, and the impact on offspring is measured at different times during development. Collectively these studies demonstrate that antibiotic-related alterations in microbiota in the pre- and postnatal period impact offspring development including behavior, adiposity and metabolism, peripheral and brain immune signaling, neurogenesis, as well as other brain pathways (Champagne-Jorgensen et al., 2020; Cho et al., 2012; Cox et al., 2014; Lebovitz et al., 2019; Leclercq et al., 2017; Mohle et al., 2016). The timing of the antibiotic exposure, the dose, and the choice of antibiotic treatment influence the short-term and long-term outcomes.

Exposure to subthreshold therapeutic doses of oral antibiotic penicillin via drinking water in the last gestational week and through the pre-weaning period, in Balb/C mice, altered gut microbiota composition in dams and in male and female offspring (Leclercq et al., 2017). Behavioral testing during adolescence (6 weeks of age) showed significant changes in antibiotic-exposed offspring including reduced anxietylike behavior in male offspring, reduced sociability (not separated by sex), and reduced social novelty (not separated by sex). In addition, antibiotic-exposed male offspring showed increased resilience to social defeat stress (Leclercq et al., 2017). Although the specific gut-brain signaling pathway linking antibiotic-exposure to behavior was not examined in this study, the impact of antibiotics on brain systems included increased cytokine expression in frontal cortex in antibiotic-exposed mice, a change that occurred in the absence of similar changes in peripheral cytokine levels and in the absence of gut inflammation. The brain-specific cytokine changes were associated with increased expression of arginine vasopressin receptor 1b (AVPR1b), known to influence aggressive behavior, suggesting that antibiotic-related brain-specific gene expression changes may contribute to the alterations in response to social defeat stress observed. In addition, increased expression of tight junction proteins in the hippocampus were observed suggesting a long-term change in blood—brain barrier permeability in response to oral exposure to antibiotics (Leclercq et al., 2017). In a subsequent study, exposure to low dose penicillin in the last gestational week (only) showed sex-specific behavioral effects in adolescent Balb/C mice but did not reproduce the changes in brain cytokine expression or AVPR1b mRNA expression observed following antibiotic exposure in the last gestational week through the preweaning period (Champagne-Jorgensen et al., 2020; Leclercq et al., 2017) showing that timing of exposure as well as the duration influences the long-term consequences on the offspring. The role of microbiota specifically in these effects is suggested since coadministration of the probiotic, *Lactobacillus rhamnosus* JB-1, at the same time as low-dose penicillin to dams reduced the impact of antibiotic exposure on the offspring (Leclercq et al., 2017).

Interestingly, exposure to low-dose penicillin in the last gestational week alone resulted in an altered splenic immune phenotype in male mice that included a decreased relative proportion of FOXP3+ regulatory T cells and increased relative proportions of Ly6C$^{lo}$+CD11b+ monocytes and IL12+F4/80+CD11b+ monocytes (Champagne-Jorgensen et al., 2020). Immune-brain cross-talk is well known to influence CNS systems and behavior, and it is likely that microbiota-immune-brain signaling is a key pathway that mediates antibiotic-related changes in brain function. In the perinatal period, the microbiome is essential to the development and maturation of the peripheral immune system and the maturation of neuroimmune signaling systems (Pronovost and Hsiao, 2019). Perturbations of the microbiome, including administration of antibiotics, during embryonic and postnatal development are shown to alter both the mucosal immune system and the peripheral immune system (for review see Pronovost and Hsiao, 2019). A few studies have also reported alterations in neuroimmune systems, such as microglia, in GF mice (Castillo-Ruiz et al., 2018; Erny et al., 2015; Thion et al., 2018). Recently, antibiotic treatment of dams, 1 week prior to breeding, through gestation, and until weaning in C57Bl/6 mice resulted in an altered/dystrophic microglial morphology as well as increased microglial activation at postnatal day 22 in antibiotic-exposed offspring (Lebovitz et al., 2019). Probiotic treatment in combination with antibiotic treatment prevented the observed changes in microglial development; however, antibiotic-related reductions in fecal short chain fatty acids (SCFA) were still present following cotreatment with probiotics and antibiotics (Lebovitz et al., 2019). Together these studies show that antibiotic treatment during pregnancy and early postnatal life in rodents alters the microbiome composition and diversity of the pups and impacts neurodevelopment and behavior.

## Exposure to antibiotics early in life—Clinical studies

In the animal studies, the impact of antibiotic administration during pregnancy and postnatal development is most often examined in healthy mice. The early life subtherapeutic antibiotic treatment mouse model, used in several of the studies noted above, was developed based on the broad administration of low doses of antibiotics to promote the growth of farm animals in the United States (Cho et al., 2012). This approach is an excellent choice to examine how perturbations to the microbiome influence the development of host systems, and to examine the host—microbe mechanisms that mediate the outcomes in offspring following early life exposure to antibiotics. The translational value of these studies, related to clinical use of antibiotics in pregnancy and in early life, is limited, primarily due to the fact that antibiotic use in pregnancy and early postnatal life in people is targeted at infection or risk of infection, which can also have a direct or indirect impact on development of the offspring. In fact, an extended body of work looking at the impact of maternal infection on neurodevelopment demonstrates that the activation of the maternal and/or the fetal immune system has enduring consequences on neurodevelopment (Bergdolt and Dunaevsky, 2019; Filiano et al., 2015; Gumusoglu and Stevens, 2019; Harvey and Boksa, 2012; Jiang et al., 2018b; Meyer, 2014; Pronovost and Hsiao, 2019). Related to maternal infection and risk of mental health disorders, a large national study in Denmark showed that maternal infection and antiinfective treatment was associated with an increased risk of mental disorders in offspring (Lydholm et al., 2019). The discussion below considers key findings to help understand how antibiotic-mediated changes in microbiota impact neurodevelopmental outcomes through review of clinical studies that examined the impact of antibiotic administration directly to infants in the postnatal period or indirectly, during pregnancy or via breast milk postnatally.

Microbiome colonization and maturation during early life influence human development (Dominguez-Bello et al., 2019). Evidence of an important mechanistic link between microbiota and brain development is slowly emerging (Cowan et al., 2020; Diaz Heijtz, 2016). Several factors influence the infant microbiome including mode of delivery, dietary factors, as well as antibiotic use (Senn et al., 2020). Notably, studies have shown that exposure to antibiotics during the first few months following birth influences the colonization and maturation of an infant's microbiome. Preterm infants are often exposed to antibiotics and studies have considered the impact of antibiotics on microbiome colonization and maturation in preterm infants in order to understand how changes may contribute to risk of disease. To note, the progression of microbiota colonization and maturation is influenced by gestational age at birth in preterm infants (La Rosa et al., 2014). A study that examined preterm infants that received repeated administrations of antibiotics compared to preterm infants that received antibiotics only in the first week of life showed that exposure to some antibiotics (meropenem, cefotaxime, ticarcillin-clavulanate) reduced microbiome species richness that was accompanied by increased abundance of key genera including *Escherichia, Klebsiella, Enterobacter, Enterococcus,* and *Staphylococcus,* that is, bacteria taxa capable of causing human infections

(Gibson et al., 2016). In a recent comparison of preterm infants to full-term infants, several taxa discriminated preterm and full-term individuals for up to 4 years after birth, although a comprehensive analysis of antibiotic use was not conducted (Fouhy et al., 2019). Preterm birth is associated with increased risk of neurological complications including increased risk of attention deficit hyperactivity disorder and reduced cognitive performance (reviewed in Lu and Claud, 2019). Evidence that altered microbiota influences brain development was demonstrated, indirectly, by transferring preterm human infant microbiota to pregnant GF mice. In comparison to mice colonized with microbiota from infants with good growth, mice colonized with microbiota from infants with low growth showed reduced levels of neuronal and myelination markers, as well as increased levels of neuroinflammation (Lu et al., 2018). It is important to note that several factors, including antibiotics, contribute to developmental outcomes; more research is needed to determine the contribution of antibiotic treatment to neurodevelopment and later risk of disease. A more comprehensive understanding of antibiotic-microbe-neurodevelopment interactions will help develop clinical guidelines and intervention strategies to reduce the long-term impact of antibiotic exposure early in life.

Understanding how alterations in microbiome colonization and maturation influence risk of later disease is a key topic in pediatric research. The impact of antibiotic treatment on the later development of allergy and asthma, as well as metabolic disorders such as obesity has been examined (Ahmadizar et al., 2018; Renz and Skevaki, 2021). Less attention has considered the impact of antibiotic treatment on brain development and risk of psychiatric or neurological disorders. To date a few studies have considered the link between antibiotic treatment and brain development in general. A recent report examined the composition of gut microbiota at 1 year of age and behavioral outcomes at 2 years of age in a subset of individuals enrolled in the longitudinal Australian birth cohort study, The Barwon Infant Study. This study demonstrated an association between reduced abundance of *Prevotella* and behavioral outcomes related to internalizing behavior at 2 years of age (Loughman et al., 2020). To note, the best predictor of reduced abundance of *Prevotella* at 1 year of age was exposure to antibiotics (Loughman et al., 2020). The observed association between exposure to antibiotics in the first year of life with cognitive scores and emotional behaviors assessed at 3.5, 7, and 11 years of age provides additional evidence that disruption of the microbiome can influence brain development (Slykerman et al., 2017). In this study, it was exposure in the first year of life impacted cognitive scores but antibiotic exposure during pregnancy and later in childhood was not found to be associated with outcomes measured (Slykerman et al., 2017). An additional study showed that antibiotic treatment during the perinatal period was associated with higher electroencephalography (EEG) power in newborns and a later increase in attention problems at 4–5 years of age (Firestein et al., 2019). The relationship of altered behavioral outcomes following antibiotics and the impact on later mental health has been considered in a few studies (Lavebratt et al., 2019; Liang et al., 2021). A recent population study that considered one million births in Finland over a 16-year period, demonstrated that antibiotic exposure in utero and during the first 2 years of life was associated with a modest increased risk of a psychiatric disorder in offspring including development of sleep disorders, ADHD, conduct disorder, mood and anxiety disorders as well as other childhood onset emotional disorders (Lavebratt et al., 2019). Similarly, an observational study in the UK Biobank cohort showed an association between long-term antibiotic use during early life and both anxiety and depression (Liang et al., 2021). In this study, genetic variation was considered and several gene-antibiotic use associations were identified, including some genes that had been previously associated with anxiety and depression (Liang et al., 2021). Overall, the work to date suggests a connection between antibiotic use, brain development, and risk of psychiatric disorders; however, clearly separating the role of antibiotics from infection, as well as other factors that influence microbiota-brain development, is needed in future studies.

## Exposure in adolescence and adulthood

Experimental studies that examine exposure to antibiotics post-weaning provide insight into antibiotic use during adolescence and also provide a more direct assessment of the impact of oral antibiotics on host systems. Administration of several different antibiotics at a subthreshold treatment dose in female C57Bl/6 mice for 55 days following weaning via the drinking water resulted in alterations in gut microbiota composition, SCFA, and host systems (Cho et al., 2012). Notably, low-dose antibiotics increased adiposity and increased levels of key liver genes involved in fatty acid and lipid metabolism (Cho et al., 2012). Changes in gut microbiota composition in cecal samples were accompanied by alterations in SCFA including increased levels of acetate and butyrate. The results showed increased bacterial gene counts of butyrl coA transferase and formyltetrahydrofolate synthetase, two prokaryotic genes involved in SCFA synthesis (Cho et al., 2012).

This suggests that low-dose antibiotic exposure can lead to adiposity and altered host metabolism that is mediated in part by a microbiota-related change in SCFA synthesis (Cho et al., 2012). Additional studies link long-term exposure to low-dose antibiotics to metabolic-related disorders later in life. For example, low-dose penicillin exposure from birth through adulthood increased the effect of a high-fat diet in adulthood (Cox et al., 2014). Fecal transplant experiments in this model supported the identification of protective taxa, including *Lactobacillus, Allobaculum, Candidatus arthromitus,* and *Rikenellaceae,* that were reduced following long-term exposure to low-dose penicillin and associated with later metabolic outcomes (Cox et al., 2014). Exposure to ciprofloxacin for 14 days in female adult C57BL/6 mice showed significant changes in gut microbiota composition and parallel changes in the fecal metabolome (Choo et al., 2017). This study also included treatment with the clinically relevant antibiotic vancomycin-imipenem and showed antibiotic-specific shifts in microbiota diversity and composition, changes that did not normalize when examined 9 days after cessation of antibiotic treatment. Alterations in several fecal metabolites were observed following both ciprofloxacin and vancomycin-imipenem treatment. In comparison to taxa composition, few metabolite differences were observed 9 days after treatment, however, antibiotic-related microbiota-metabolite associations were observed after 14 days of antibiotic treatment that were consistent with predicted metabolic pathways generated using PICRUSt (Choo et al., 2017).

Several studies, including those described above, demonstrate an impact of antibiotic treatment in adolescence and adults on microbiota composition, as well as microbiome function. In parallel, with those influences, it is important to determine the impact of antibiotic exposure, during adolescence and adulthood, on behavior and brain systems. Long-term treatment of adult female C57Bl/6 mice with broad spectrum antibiotics was shown to reduce neurogenesis and impaired performance in the novel object recognition tasks that measure hippocampal-dependent behavior (Mohle et al., 2016). Through a series of additional approaches, this study also demonstrated that interventions that alleviated antibiotic-induced deficits (exercise and probiotics) relied on microbiota-monocyte interactions, highlighting the importance of microbiota-immune pathways in regulating brain systems and behavior (Mohle et al., 2016). The impact of antibiotics on cognitive behavior was also shown by administration of a cocktail of broad spectrum antibiotics to adult male C57Bl/6 mice that resulted in alterations in microbiota diversity and composition that was accompanied by impaired performance in novel object recognition (Frohlich et al., 2016). These investigators demonstrated that the altered microbiota composition was accompanied by decreased levels of microbial metabolite levels in colonic luminal contents, including SCFA, and reduced levels of circulating plasma metabolites (Frohlich et al., 2016). Antibiotic treatment altered tight junction protein and neural signaling-related RNA expression in both the amygdala and hippocampus demonstrating a brain-specific impact of antibiotic exposure (Frohlich et al., 2016).

The above studies examined the impact of antibiotics in healthy adult mice, however, microbiota-brain interactions are also an active area of research in animal models of brain-related diseases. A recent report demonstrated that exposure to antibiotics post-weaning through adulthood in female mice delayed the development of experimental autoimmune encephalomyelitis (EAE) symptoms (Zeraati et al., 2019). In comparison to non-EAE mice not treated with antibiotics, EAE mice not treated with antibiotics showed deficits in spontaneous alteration and spatial memory tests. Antibiotic treatment exaggerated these deficits in EAE mice, but did not impact cognitive behavior in non-EAE mice (Zeraati et al., 2019). Antibiotic treatment also influenced depressive-like behavior in EAE mice, whereas an impact of exploratory and anxietylike behavior was observed in both non-EAE and EAE mice exposed to antibiotics (Zeraati et al., 2019). The combination of antibiotics and other perturbations of the microbiome in animal models of disease provides a more informative view of how microbe—host interactions may present differently in the context of disease; an observation that may have great clinical significance and directly impact the advancement of precision medicine approaches.

Clinical studies examining the impact of antibiotic exposure in adolescents and adults on brain systems are limited. Certainly, antibiotic exposure impacts microbiota diversity, composition, and function throughout the lifespan, with class, dose, and length of exposure influencing the short-term and long-term effects. The impact of repeated exposure to a 5-day course of ciprofloxacin 6 months apart on microbiome composition in three adult individuals showed that the postantibiotic microbiota composition was similar but not identical to the preantibiotic microbiota composition (Dethlefsen and Relman, 2011). In this study, more than 50 samples across the three individuals were assessed over a 10-month period. Antibiotic shifts in taxa abundance normalized by 1 week following antibiotic treatment, although interindividual differences existed. Interestingly, many of the taxa that were reduced in the postantibiotic microbiome were abundant taxa that are known butyrate producers (Dethlefsen and Relman, 2011).

To date several studies have reported altered microbiota composition and diversity in both psychiatric and neurological diseases (Chen et al., 2018a,b, 2019; Jiang et al., 2018a; Mason et al., 2020; Nguyen et al., 2018; Scheperjans et al., 2015; Xu et al., 2020). Whether or not exposure to antibiotics contributes to risk of these conditions has only been considered in a few studies to date. A nested case-control study that utilized The Health Improvement Network database (more than 10 million individuals in UK) reported an increased risk of anxiety and depression in individuals that were exposed to recurrent antibiotic treatment (Lurie et al., 2015). While this study did not show an association of antibiotic use and psychosis, altered microbiota composition in schizophrenia and other psychotic disorders has been reported (Nguyen et al., 2018; Xu et al., 2020). As noted above, a population-based study in Finland showed that exposure to antibiotics in utero and in the first 2 years of life increased risks for several psychiatric disorders with childhood onset, including ADHD, conduct disorder, as well as mood and anxiety disorders (Lavebratt et al., 2019). Similarly, an observational study using data from the UK biobank cohort revealed a link between long-term antibiotic use in early life with anxiety and depression (Liang et al., 2021).

## The impact of nonantibiotic drugs on microbiota

Nonantibiotic drugs can impact the microbiome through direct and indirect mechanisms (Fig. 10.1) (Haiser and Turnbaugh, 2012, 2013). Like antibiotics, nonantibiotic drugs of several classes have a direct antimicrobial action on microbiota (Maier et al., 2018; Munoz-Bellido et al., 2000). In a large drug screen, over 1000 human administered drugs, representing a broad range of therapeutic classes, were tested for their impact on bacterial growth of 40 isolates grown in anaerobic conditions, representing 21 genera and 38 species (Maier et al., 2018). Of the 835 host-targeted drugs tested, 204 drugs had antimicrobial activity to one or more bacterial species, including 48 drugs known to act on the nervous system (Table 10.1) (Maier et al., 2018; WHO, 2019). As noted by the authors, antipsychotic drugs were overrepresented in the list; however, active antimicrobial drugs also included anticholinergic and dopaminergic agents, as well as psychostimulants and antidepressants (Table 10.1) (Maier et al., 2018; WHO, 2019).

**Table 10.1** Human drugs with known action on the nervous system that show antimicrobial activity.

| Prestwick_ID | Chemical_name | ATC codes | n_hit[a] | ATC level 2 | ATC level 3 | ATC level 4 |
|---|---|---|---|---|---|---|
| Prestw-1346 | Sumatriptan succinate | N02CC01 | 1 | Analgesics | Antimigraine preparations | Selective serotonin (5HT1) agonists |
| Prestw-992 | Pizotifen malate | N02CX01 | 1 | Analgesics | Antimigraine preparations | Selective serotonin (5HT1) agonists |
| Prestw-293 | Glafenine hydrochloride | N02BG03 | 1 | Analgesics | Other analgesics and antipyretics | Other analgesics and antipyretics |
| Prestw-32 | Epirizole | N02BB | 3 | Analgesics | Other analgesics and antipyretics | Pyrazolones |
| Prestw-703 | Famprofazone | N02BB | 1 | Analgesics | Other analgesics and antipyretics | Pyrazolones |
| Prestw-1041 | Etomidate | N01AX07 | 4 | Anesthetics | Anesthetics, general | Other general anesthetics |
| Prestw-199 | Prilocaine hydrochloride | N01BB04 | 1 | Anesthetics | Anesthetics, local | Amides |
| Prestw-491 | Metixene hydrochloride | N04AA03 | 9 | Anti-parkinson drugs | Anticholinergic agents | Tertiary amines |
| Prestw-840 | Ethopropazine hydrochloride | N04AA05 | 11 | Anti-parkinson drugs | Anticholinergic agents | Tertiary amines |

| | | | | | Anti-parkinson drugs | Dopaminergic agents | Other dopaminergic agents |
|---|---|---|---|---|---|---|---|
| Prestw-1403 | Entacapone | N04BX02 | 10 | | Anti-parkinson drugs | Dopaminergic agents | Other dopaminergic agents |
| Prestw-101 | R(−) Apomorphine hydrochloride hemihydrate | G04BE07 N04BC07 | 1 | | Anti-parkinson drugs | Dopaminergic agents | Dopamine agonists |
| Prestw-278 | Cinnarizine | N07CA02 | 3 | | Other nervous system drugs | Antivertigo preparations | Antivertigo preparations |
| Prestw-312 | Flunarizine dihydrochloride | N07CA03 | 1 | | Other nervous system drugs | Antivertigo preparations | Antivertigo preparations |
| Prestw-167 | Riluzole hydrochloride | N07XX02 | 2 | | Other nervous system drugs | Other nervous system drugs | Other nervous system drugs |
| Prestw-269 | Clomipramine hydrochloride | N06AA04 | 1 | | Psychoanaleptics | Antidepressants | Nonselective monoamine reuptake inhibitors |
| Prestw-930 | Protriptyline hydrochloride | N06AA11 | 8 | | Psychoanaleptics | Antidepressants | Nonselective monoamine reuptake inhibitors |
| Prestw-102 | Amoxapine | N06AA17 | 2 | | Psychoanaleptics | Antidepressants | Nonselective monoamine reuptake inhibitors |

*Continued*

**Table 10.1** Human drugs with known action on the nervous system that show antimicrobial activity.—cont'd

| Prestwick_ID | Chemical_name | ATC codes | n_hit[a] | ATC level 2 | ATC level 3 | ATC level 4 |
|---|---|---|---|---|---|---|
| Prestw-1014 | Sertraline | N06AB06 | 3 | Psychoanaleptics | Antidepressants | Nonselective monoamine reuptake inhibitors |
| Prestw-1152 | Nefazodone HCl | N06AX06 | 1 | Psychoanaleptics | Antidepressants | Other antidepressants |
| Prestw-292 | Trazodone hydrochloride | A06AC07 N06AX05 | 1 | Psychoanaleptics | Antidepressants | Other antidepressants |
| Prestw-386 | GBR 12909 dihydrochloride | N06BX | 3 | Psychoanaleptics | Psychostimulants, agents used for ADHD and nootropics | Other psychostimulants and nootropics |
| Prestw-1288 | Idebenone | N06BX13 | 3 | Psychoanaleptics | Psychostimulants, agents used for ADHD and nootropics | Other psychostimulants and nootropics |
| Prestw-288 | Spiperone | N05AD | 1 | Psycholeptics | Antipsychotics | Butyrophenone derivatives |
| Prestw-115 | Haloperidol | N05AD01 | 2 | Psycholeptics | Antipsychotics | Butyrophenone derivatives |
| Prestw-509 | Bromperidol | N05AD06 | 1 | Psycholeptics | Antipsychotics | Butyrophenone derivatives |

| Prestw–132 | Loxapine succinate | N05AH01 | 7 | Psycholeptics | Antipsychotics | Diazepines, oxazepines, thiazepines, and oxepines |
| Prestw–1323 | Quetiapine hemifumarate | N05AH04 | 1 | Psycholeptics | Antipsychotics | Diazepines, oxazepines, thiazepines, and oxepines |
| Prestw–906 | Fluspirilen | N05AG01 | 2 | Psycholeptics | Antipsychotics | Diphenylbutylpiperidine derivatives |
| Prestw–308 | Pimozide | N05AG02 | 7 | Psycholeptics | Antipsychotics | Diphenylbutylpiperidine derivatives |
| Prestw–1341 | Sertindole | N05AE03 | 9 | Psycholeptics | Antipsychotics | Indole derivatives |
| Prestw–1188 | Ziprasidone hydrochloride | N05AE04 | 1 | Psycholeptics | Antipsychotics | Indole derivatives |
| Prestw–1231 | Asenapine maleate | N05AH05 | 3 | Psycholeptics | Antipsychotics | Other antipsychotics |
| Prestw–375 | Methiothepin maleate | N05AX | 10 | Psycholeptics | Antipsychotics | Other antipsychotics |
| Prestw–1368 | Zotepine | N05AX11 | 7 | Psycholeptics | Antipsychotics | Other antipsychotics |
| Prestw–1229 | Aripiprazole | N05AX12 | 3 | Psycholeptics | Antipsychotics | Other antipsychotics |

*Continued*

**Table 10.1** Human drugs with known action on the nervous system that show antimicrobial activity.—cont'd

| Prestwick_ID | Chemical_name | ATC codes | n_hit[a] | ATC level 2 | ATC level 3 | ATC level 4 |
|---|---|---|---|---|---|---|
| Prestw–64 | Chlorpromazine hydrochloride | N05AA01 | 6 | Psycholeptics | Antipsychotics | Phenothiazines with aliphatic side–chain |
| Prestw–797 | Methotrime prazine maleat salt | N05AA02 | 1 | Psycholeptics | Antipsychotics | Phenothiazines with aliphatic side–chain |
| Prestw–693 | Promazine hydrochloride | N05AA03 | 11 | Psycholeptics | Antipsychotics | Phenothiazines with aliphatic side–chain |
| Prestw–576 | Acetopromazine maleate salt | N05AA04 | 2 | Psycholeptics | Antipsychotics | Phenothiazines with aliphatic side–chain |
| Prestw–53 | Triflupromazine hydrochloride | N05AA05 | 2 | Psycholeptics | Antipsychotics | Phenothiazines with aliphatic side–chain |
| Prestw–320 | Fluphenazine dihydrochloride | N05AB02 | 5 | Psycholeptics | Antipsychotics | Phenothiazines with piperazine structure |
| Prestw–125 | Perphenazine | N05AB03 | 4 | Psycholeptics | Antipsychotics | Phenothiazines with piperazine structure |
| Prestw–399 | Prochlorpera zine dimaleate | N05AB04 | 6 | Psycholeptics | Antipsychotics | Phenothiazines with piperazine structure |

| Prestw-313 | Trifluoperazine dihydrochloride | N05AB06 | 6 | Psycholeptics | Antipsychotics | Phenothiazines with piperazine structure |
|---|---|---|---|---|---|---|
| Prestw-78 | Thioridazine hydrochloride | N05AC02 | 13 | Psycholeptics | Antipsychotics | Phenothiazines with piperazine structure |
| Prestw-529 | Mesoridazine besylate | N05AC03 | 3 | Psycholeptics | Antipsychotics | Phenothiazines with piperazine structure |
| Prestw-998 | Zuclopenthixol dihydrochloride | N05AF02 N05AF05 | 7 | Psycholeptics | Antipsychotics | Thioxanthene derivatives |
| Prestw-348 | Chlorprothixene hydrochloride | N05AF03 | 8 | Psycholeptics | Antipsychotics | Thioxanthene derivatives |

[a]Number of bacterial species impacted by drug as reported in Maier et al. (2018).
Data compiled from (Maier, L., Pruteanu, M., Kuhn, M., Zeller, G., Telzerow, A., Anderson, E.E., Brochado, A.R., Fernandez, K.C., Dose, H., Mori, H., et al., 2018. Extensive impact of non-antibiotic drugs on human gut bacteria. Nature 555, 623–628) using WHO Collaborating Center for Drug Statistics Methodology, Guidelines for ATC classification and DDD assignment, 2020. Oslo, 2019; https://www.whocc.no/atc_ddd_index/

A recent study using the TwinsUK cohort, of more than 2700 individuals, examined microbiota associations with multiple diseases and also considered the association of microbiota and 51 common prescription drugs (Jackson et al., 2018). This study observed significant medication-microbiota associations in 19 drugs including significant reduced abundance (FDR <0.05) of *Turibacteriaceae*, *Mollicutes*, and *Clostridiaceae* and selective serotonin reuptake inhibitors (SSRIs) usage (Jackson et al., 2018). In addition, a weak association between reduced abundance (FDR <0.1) of *Streptococcaceae*, *Enterobacteriaceae*, and *Lactobacillaceae* and tricyclic usage was observed (Jackson et al., 2018). Interestingly, there were no significant positive associations observed between antidepressants and microbiota taxa (Jackson et al., 2018). This work included 38 common diseases and therefore provided a good overview and is an essential reference for interconnectivity of medication and microbiota in a variety of clinical conditions. Notably, it demonstrated that it is important to consider the relationship between microbiota and medication in clinical studies. In another study that examined microbiota—drug associations in three Dutch cohorts, including a general population cohort, a cohort of patients with IBD, and a cohort of patients with IBS; 19 of 41 drugs included were identified to be associated with microbiota composition (Vich Vila et al., 2020). Moreover, in addition to compositional changes observed with medication, functional changes in 411 microbial pathways were associated with 11 drugs. While proton-pump inhibitors, metformin, antibiotics, and laxatives showed the highest number of associations, it is notable that within the IBS cohort SSRI antidepressant use was associated with increased abundance of *Eubacterium ramulus* (Vich Vila et al., 2020). Benzodiazepine use in IBD was associated with increased abundance of *Haemophilus parainfluenzae,* a taxa previously shown to be increased in IBD individuals compared to healthy individuals. This highlights the point that some taxa may be disease-associated and some may be medication-associated. The authors also reported differences in drug—microbiota associations in the case of multidrug use compared to single-drug use, and suggest that consideration of the microbiome may help our understanding of drug-related side effects (Vich Vila et al., 2020). In a group of hospitalized elderly individuals, polypharmacy was associated with reduced microbiota diversity (Ticinesi et al., 2017). Several taxa were associated with specific drug classes, with most robust associations observed with protein pump inhibitors, antidepressants, and antipsychotics (Ticinesi et al., 2017).

Knowing that nonantibiotic drugs can have antimicrobial activity and thereby influence the microbiota composition is an important observation. It also provides the rationale to conduct a more in-depth examination of how bacteria chemically modify drugs and to identify functional microbial genes and pathways that contribute to microbe-host-drug interactions. An in vitro assay of 76 bacterial species and 271 drugs identified 176 drugs that were metabolized by bacteria (Zimmermann et al., 2019). Analysis of the functional groups present on these drugs suggested that specific chemical structures, including ester, amide, nitro, azo, and urea groups, as well as lactones, predisposed compounds for microbial metabolism (Zimmermann et al., 2019). To translate the drug metabolism findings to an in vivo model, the researchers examined the ability of *Clostridium scindens,* identified in their screen, to metabolize corticosteroid dexamethasone metabolism in GF mice. Oral administration of dexamethasone to GF mice or mice mono-colonized with *C. scindens* demonstrated the ability of *C. scindens* to metabolize dexamethasone to its androgen metabolite in vivo (Zimmermann et al., 2019). Translating this finding to healthy human samples revealed a more complex picture. Substantial individual differences in the ability of fecal samples from healthy donors to metabolize dexamethasone was observed, and more importantly, the capacity for these samples to metabolize dexamethasone was not related to the presence of *C. scindens* (Zimmermann et al., 2019). Since their initial screen was conducted using a specific strain of *C. scindens,* the authors highlight the importance of strain-specific gene expression in the ability of a fecal sample to metabolize dexamethasone, and also the potential for other bacterial taxa to metabolize dexamethasone in the healthy human gut. Further the specific genes and functional pathways present will influence individual differences in how microbiota metabolize drugs (Zimmermann et al., 2019). To further examine this, the researchers used gain of function approach to identify genes that confer drug metabolism capacity to specific taxa. In their screen, *Bacteroides thetaiotaomicron* metabolized 46 drugs and examined this further in their gain of function experiments. The authors identified 17 *B. thetaiotaomicron* gene products that metabolized 18 different drugs to distinct metabolites (Zimmermann et al., 2019). Using a similar approach, they established that enrichment of key enzyme genes in bacteria explained their ability to metabolize a certain drug. In addition, the community of co-occurring taxa was an important factor that influenced drug metabolism highlighting that microbe—microbe interactions are important in this domain (Zimmermann et al., 2019).

The importance of the community of bacteria and the dynamic ecology of the gastrointestinal tract is highlighted in the comprehensive study of microbiome drug metabolism noted above (Zimmermann et al., 2019). Another important consideration is the ability of both microbes and the host to metabolize many nonantibiotic drugs, such that it is important to identify the impact of cometabolism on drug metabolism, drug action, and drug toxicity (Li et al., 2016). Specific to neuroactive drugs, researchers have identified key human gut taxa that have the ability to metabolize levodopa (Maini Rekdal et al., 2019; van Kessel et al., 2019). Levadopa (L-3,4- dihydroxyphenylalanine or L-dopa) is the primary drug used to treat Parkinson's disease (PD), which requires the drug to enter the brain and be converted to dopamine to be effective. Peripheral metabolism of L-dopa by intestinal microbes or host enzymes impacts drug availability. Recent studies have identified specific bacterial taxa and enzymes that are able to metabolize dopamine and have associated the presence of these taxa/genes with drug metabolism in patient samples (Maini Rekdal et al., 2019; van Kessel et al., 2019). Further, metabolism of dopamine by microbiota can also increase drug toxicity if converted to m-tyramine (Weersma et al., 2020). Additional work to identify both bacterial and host biomarkers of drug metabolism has the potential to improve treatment approaches in PD and other conditions.

## Future directions

More than 40 years ago, microbiologists demonstrated that microbiota can metabolize xenobiotics including human-targeted pharmaceutical drugs (Scheline, 1968, 1973). This chapter provided an overview of how orally administered drugs, antibiotics and nonantibiotics, influence microbiota composition and function in the small and large intestine, how microbiota influence drug metabolism, and how microbe—drug interactions influence behavior and brain systems. Drug-microbe-host interactions are important to consider as the direct and indirect impact of these interactions includes drug bioavailability, drug toxicity, drug action, drug clearance, and drug response. Individual differences in microbiota composition are influenced by genetic and environmental factors, and an individual's microbiome is unique. These individual differences influence all aspects of drug-microbe-host interactions and must be considered in the context of treatment response and side effects. There is an extended literature related to the impact of antibiotics on microbiota, and its related impact on host systems in health and disease. In addition to the direct effect of antibiotics on bacteria, antibiotics can influence the expression of antibiotic-resistance genes in bacteria and understanding how this influences gut—brain signaling has not been considered. Research into the impact of nonantibiotic drugs on microbiota composition and function is in its infancy, particularly when considering neurological and mental health. Recent animal studies show that antidepressant drugs including fluoxetine, escitalopram, and ketamine alter microbiota composition and can influence host systems (Cussotto et al., 2019a,b; Getachew et al., 2018; Lyte et al., 2019). Translating this work into clinical trials is an important next step in advancing the potential for considering individual differences in drug-microbe-host interactions to be utilized for precision medicine approaches to brain health.

## References

Ahmadizar, F., Vijverberg, S.J.H., Arets, H.G.M., de Boer, A., Lang, J.E., Garssen, J., Kraneveld, A., Maitland-van der Zee, A.H., 2018. Early-life antibiotic exposure increases the risk of developing allergic symptoms later in life: a meta-analysis. Allergy 73, 971—986.

Bergdolt, L., Dunaevsky, A., 2019. Brain changes in a maternal immune activation model of neurodevelopmental brain disorders. Prog Neurobiol 175, 1—19.

Bjorkholm, B., Bok, C.M., Lundin, A., Rafter, J., Hibberd, M.L., Pettersson, S., 2009. Intestinal microbiota regulate xenobiotic metabolism in the liver. PLoS One 4, e6958.

Castillo-Ruiz, A., Mosley, M., George, A.J., Mussaji, L.F., Fullerton, E.F., Ruszkowski, E.M., Jacobs, A.J., Gewirtz, A.T., Chassaing, B., Forger, N.G., 2018. The microbiota influences cell death and microglial colonization in the perinatal mouse brain. Brain Behav Immun 67, 218—229.

Champagne-Jorgensen, K., Mian, M.F., Kay, S., Hanani, H., Ziv, O., McVey Neufeld, K.A., Koren, O., Bienenstock, J., 2020. Prenatal low-dose penicillin results in long-term sex-specific changes to murine behaviour, immune regulation, and gut microbiota. Brain Behav Immun 84, 154—163.

Chen, J.J., Zheng, P., Liu, Y.Y., Zhong, X.G., Wang, H.Y., Guo, Y.J., Xie, P., 2018a. Sex differences in gut microbiota in patients with major depressive disorder. Neuropsychiatric Dis Treat 14, 647—655.

Chen, Y.H., Bai, J., Wu, D., Yu, S.F., Qiang, X.L., Bai, H., Wang, H.N., Peng, Z.W., 2019. Association between fecal microbiota and generalized anxiety disorder: severity and early treatment response. J Affect Disord 259, 56—66.

Chen, Z., Li, J., Gui, S., Zhou, C., Chen, J., Yang, C., Hu, Z., Wang, H., Zhong, X., Zeng, L., et al., 2018b. Comparative metaproteomics analysis shows altered fecal microbiota signatures in patients with major depressive disorder. Neuroreport 29, 417—425.

Cho, I., Yamanishi, S., Cox, L., Methe, B.A., Zavadil, J., Li, K., Gao, Z., Mahana, D., Raju, K., Teitler, I., et al., 2012. Antibiotics in early life alter the murine colonic microbiome and adiposity. Nature 488, 621—626.

Choo, J.M., Kanno, T., Zain, N.M., Leong, L.E., Abell, G.C., Keeble, J.E., Bruce, K.D., Mason, A.J., Rogers, G.B., 2017. Divergent relationships between fecal microbiota and metabolome following distinct antibiotic-induced disruptions. mSphere 2.

Clarke, G., Sandhu, K.V., Griffin, B.T., Dinan, T.G., Cryan, J.F., Hyland, N.P., 2019. Gut reactions: breaking down xenobiotic-microbiome interactions. Pharmacol Rev 71, 198−224.

Clayton, T.A., Baker, D., Lindon, J.C., Everett, J.R., Nicholson, J.K., 2009. Pharmaco-metabonomic identification of a significant host-microbiome metabolic interaction affecting human drug metabolism. Proc Natl Acad Sci U S A 106, 14728−14733.

Cowan, C.S.M., Dinan, T.G., Cryan, J.F., 2020. Annual Research review: Critical windows - the microbiota-gut-brain axis in neurocognitive development. J Child Psychol Psychiatry Allied Discip 61, 353−371.

Cox, L.M., Yamanishi, S., Sohn, J., Alekseyenko, A.V., Leung, J.M., Cho, I., Kim, S.G., Li, H., Gao, Z., Mahana, D., et al., 2014. Altering the intestinal microbiota during a critical developmental window has lasting metabolic consequences. Cell 158, 705−721.

Cussotto, S., Clarke, G., Dinan, T.G., Cryan, J.F., 2019a. Psychotropics and the microbiome: a chamber of secrets. Psychopharmacology (Berl) 236, 1411−1432.

Cussotto, S., Strain, C.R., Fouhy, F., Strain, R.G., Peterson, V.L., Clarke, G., Stanton, C., Dinan, T.G., Cryan, J.F., 2019b. Differential effects of psychotropic drugs on microbiome composition and gastrointestinal function. Psychopharmacology (Berl) 236, 1671−1685.

Dethlefsen, L., Relman, D.A., 2011. Incomplete recovery and individualized responses of the human distal gut microbiota to repeated antibiotic perturbation. Proc Natl Acad Sci U S A 108 (Suppl. 1), 4554−4561.

Diaz Heijtz, R., 2016. Fetal, neonatal, and infant microbiome: perturbations and subsequent effects on brain development and behavior. Semin Fetal Neonatal Med 21, 410−417.

Dominguez-Bello, M.G., Godoy-Vitorino, F., Knight, R., Blaser, M.J., 2019. Role of the microbiome in human development. Gut 68, 1108−1114.

Erny, D., Hrabe de Angelis, A.L., Jaitin, D., Wieghofer, P., Staszewski, O., David, E., Keren-Shaul, H., Mahlakoiv, T., Jakobshagen, K., Buch, T., et al., 2015. Host microbiota constantly control maturation and function of microglia in the CNS. Nat Neurosci 18, 965−977.

Filiano, A.J., Gadani, S.P., Kipnis, J., 2015. Interactions of innate and adaptive immunity in brain development and function. Brain Res 1617, 18−27.

Firestein, M.R., Myers, M.M., Austin, J., Stark, R.I., Barone, J.L., Ludwig, R.J., Welch, M.G., 2019. Perinatal antibiotics alter preterm infant EEG and neurobehavior in the Family Nurture Intervention trial. Dev Psychobiol 61, 661−669.

Flowers, S.A., Bhat, S., Lee, J.C., 2020. Potential implications of gut microbiota in drug pharmacokinetics and bioavailability. Pharmacotherapy 40 (7), 704−712.

Fouhy, F., Watkins, C., Hill, C.J., O'Shea, C.A., Nagle, B., Dempsey, E.M., O'Toole, P.W., Ross, R.P., Ryan, C.A., Stanton, C., 2019. Perinatal factors affect the gut microbiota up to four years after birth. Nat Commun 10, 1517.

Frohlich, E.E., Farzi, A., Mayerhofer, R., Reichmann, F., Jacan, A., Wagner, B., Zinser, E., Bordag, N., Magnes, C., Frohlich, E., et al., 2016. Cognitive impairment by antibiotic-induced gut dysbiosis: analysis of gut microbiota-brain communication. Brain Behav Immun 56, 140−155.

Getachew, B., Aubee, J.I., Schottenfeld, R.S., Csoka, A.B., Thompson, K.M., Tizabi, Y., 2018. Ketamine interactions with gut-microbiota in rats: relevance to its antidepressant and anti-inflammatory properties. BMC Microbiol 18, 222.

Gibson, M.K., Wang, B., Ahmadi, S., Burnham, C.A., Tarr, P.I., Warner, B.B., Dantas, G., 2016. Developmental dynamics of the preterm infant gut microbiota and antibiotic resistome. Nat Microbiol 1, 16024.

Gumusoglu, S.B., Stevens, H.E., 2019. Maternal inflammation and neurodevelopmental programming: a review of preclinical outcomes and implications for translational psychiatry. Biol Psychiatr 85, 107–121.

Haiser, H.J., Turnbaugh, P.J., 2012. Is it time for a metagenomic basis of therapeutics? Science 336, 1253–1255.

Haiser, H.J., Turnbaugh, P.J., 2013. Developing a metagenomic view of xenobiotic metabolism. Pharmacol Res 69, 21–31.

Harvey, L., Boksa, P., 2012. Prenatal and postnatal animal models of immune activation: relevance to a range of neurodevelopmental disorders. Dev Neurobiol 72, 1335–1348.

Jackson, M.A., Verdi, S., Maxan, M.E., Shin, C.M., Zierer, J., Bowyer, R.C.E., Martin, T., Williams, F.M.K., Menni, C., Bell, J.T., et al., 2018. Gut microbiota associations with common diseases and prescription medications in a population-based cohort. Nat Commun 9, 2655.

Jiang, H.Y., Zhang, X., Yu, Z.H., Zhang, Z., Deng, M., Zhao, J.H., Ruan, B., 2018a. Altered gut microbiota profile in patients with generalized anxiety disorder. J Psychiatr Res 104, 130–136.

Jiang, N.M., Cowan, M., Moonah, S.N., Petri, W.A., 2018b. The impact of systemic inflammation on neurodevelopment. Trends Mol Med 24, 794–804.

Jourova, L., Anzenbacher, P., Anzenbacherova, E., 2016. Human gut microbiota plays a role in the metabolism of drugs. Biomed Pap Med Fac Univ Palacky Olomouc Czech Repub 160, 317–326.

Jourova, L., Anzenbacher, P., Liskova, B., Matuskova, Z., Hermanova, P., Hudcovic, T., Kozakova, H., Hrncirova, L., Anzenbacherova, E., 2017. Colonization by nonpathogenic bacteria alters mRNA expression of cytochromes P450 in originally germ-free mice. Folia Microbiol (Praha) 62, 463–469.

Koppel, N., Maini Rekdal, V., Balskus, E.P., 2017. Chemical transformation of xenobiotics by the human gut microbiota. Science 356.

La Rosa, P.S., Warner, B.B., Zhou, Y., Weinstock, G.M., Sodergren, E., Hall-Moore, C.M., Stevens, H.J., Bennett Jr., W.E., Shaikh, N., Linneman, L.A., et al., 2014. Patterned progression of bacterial populations in the premature infant gut. Proc Natl Acad Sci U S A 111, 12522–12527.

Lavebratt, C., Yang, L.L., Giacobini, M., Forsell, Y., Schalling, M., Partonen, T., Gissler, M., 2019. Early exposure to antibiotic drugs and risk for psychiatric disorders: a population-based study. Transl Psychiatry 9, 317.

Lebovitz, Y., Kowalski, E.A., Wang, X., Kelly, C., Lee, M., McDonald, V., Ward, R., Creasey, M., Mills, W., Gudenschwager Basso, E.K., et al., 2019. Lactobacillus rescues postnatal neurobehavioral and microglial dysfunction in a model of maternal microbiome dysbiosis. Brain Behav Immun 81, 617–629.

Leclercq, S., Mian, F.M., Stanisz, A.M., Bindels, L.B., Cambier, E., Ben-Amram, H., Koren, O., Forsythe, P., Bienenstock, J., 2017. Low-dose penicillin in early life induces long-term changes in murine gut microbiota, brain cytokines and behavior. Nat Commun 8, 15062.

Li, H., He, J., Jia, W., 2016. The influence of gut microbiota on drug metabolism and toxicity. Expet Opin Drug Metabol Toxicol 12, 31–40.

Liang, X., Ye, J., Wen, Y., Li, P., Cheng, B., Cheng, S., Liu, L., Zhang, L., Ma, M., Qi, X., et al., 2021. Long-term antibiotic use during early life and risks to mental traits: an observational study and gene-environment-wide interaction study in UK Biobank cohort. Neuropsychopharmacology 46, 1086–1092.

Loughman, A., Ponsonby, A.L., O'Hely, M., Symeonides, C., Collier, F., Tang, M.L.K., Carlin, J., Ranganathan, S., Allen, K., Pezic, A., et al., 2020. Gut microbiota composition during infancy and subsequent behavioural outcomes. EBioMedicine 52, 102640.

Lu, J., Claud, E.C., 2019. Connection between gut microbiome and brain development in preterm infants. Dev Psychobiol 61, 739—751.

Lu, J., Lu, L., Yu, Y., Cluette-Brown, J., Martin, C.R., Claud, E.C., 2018. Effects of intestinal microbiota on brain development in humanized gnotobiotic mice. Sci Rep 8, 5443.

Lurie, I., Yang, Y.X., Haynes, K., Mamtani, R., Boursi, B., 2015. Antibiotic exposure and the risk for depression, anxiety, or psychosis: a nested case-control study. J Clin Psychiatry 76, 1522—1528.

Lydholm, C.N., Kohler-Forsberg, O., Nordentoft, M., Yolken, R.H., Mortensen, P.B., Petersen, L., Benros, M.E., 2019. Parental infections before, during, and after pregnancy as risk factors for mental disorders in childhood and adolescence: a Nationwide Danish study. Biol Psychiatry 85, 317—325.

Lyte, M., Daniels, K.M., Schmitz-Esser, S., 2019. Fluoxetine-induced alteration of murine gut microbial community structure: evidence for a microbial endocrinology-based mechanism of action responsible for fluoxetine-induced side effects. PeerJ 7, e6199.

Maier, L., Pruteanu, M., Kuhn, M., Zeller, G., Telzerow, A., Anderson, E.E., Brochado, A.R., Fernandez, K.C., Dose, H., Mori, H., et al., 2018. Extensive impact of non-antibiotic drugs on human gut bacteria. Nature 555, 623—628.

Maini Rekdal, V., Bess, E.N., Bisanz, J.E., Turnbaugh, P.J., Balskus, E.P., 2019. Discovery and inhibition of an interspecies gut bacterial pathway for Levodopa metabolism. Science 364.

Markel, H., 2013. The Real Story Behind Penicillin. NewsHour Productions LLC.

Mason, B.L., Li, Q., Minhajuddin, A., Czysz, A.H., Coughlin, L.A., Hussain, S.K., Koh, A.Y., Trivedi, M.H., 2020. Reduced anti-inflammatory gut microbiota are associated with depression and anhedonia. J Affect Disord 266, 394—401.

Meyer, U., 2014. Prenatal poly(I:C) exposure and other developmental immune activation models in rodent systems. Biol Psychiatr 75, 307—315.

Mohle, L., Mattei, D., Heimesaat, M.M., Bereswill, S., Fischer, A., Alutis, M., French, T., Hambardzumyan, D., Matzinger, P., Dunay, I.R., et al., 2016. Ly6C(hi) monocytes provide a link between antibiotic-induced changes in gut microbiota and adult hippocampal neurogenesis. Cell Rep 15, 1945—1956.

Munoz-Bellido, J.L., Munoz-Criado, S., Garcia-Rodriguez, J.A., 2000. Antimicrobial activity of psychotropic drugs: selective serotonin reuptake inhibitors. Int J Antimicrob Agents 14, 177—180.

Nguyen, T.T., Kosciolek, T., Eyler, L.T., Knight, R., Jeste, D.V., 2018. Overview and systematic review of studies of microbiome in schizophrenia and bipolar disorder. J Psychiatr Res 99, 50—61.

Pronovost, G.N., Hsiao, E.Y., 2019. Perinatal interactions between the microbiome, immunity, and neurodevelopment. Immunity 50, 18—36.

Renz, H., Skevaki, C., 2021. Early life microbial exposures and allergy risks: opportunities for prevention. Nat Rev Immunol 21 (3), 177—191.

Scheline, R.R., 1968. Drug metabolism by intestinal microorganisms. J Pharm Sci 57, 2021—2037.

Scheline, R.R., 1973. Metabolism of foreign compounds by gastrointestinal microorganisms. Pharmacol Rev 25, 451—523.

Scheperjans, F., Aho, V., Pereira, P.A., Koskinen, K., Paulin, L., Pekkonen, E., Haapaniemi, E., Kaakkola, S., Eerola-Rautio, J., Pohja, M., et al., 2015. Gut microbiota are related to Parkinson's disease and clinical phenotype. Mov Disord 30, 350—358.

Senn, E., Symeonides, C., Vuillermin, P., Ponsonby, A.L., Barwon Infant Study Investigator, G., 2020. Early life microbial exposure, child neurocognition and behaviour at 2 years of age: a birth cohort study. J Paediatr Child Health 56, 590—599.

Slykerman, R.F., Thompson, J., Waldie, K.E., Murphy, R., Wall, C., Mitchell, E.A., 2017. Antibiotics in the first year of life and subsequent neurocognitive outcomes. Acta Paediatr 106, 87—94.

Thion, M.S., Low, D., Silvin, A., Chen, J., Grisel, P., Schulte-Schrepping, J., Blecher, R., Ulas, T., Squarzoni, P., Hoeffel, G., et al., 2018. Microbiome influences prenatal and adult microglia in a sex-specific manner. Cell 172, 500—516 e516.

Ticinesi, A., Milani, C., Lauretani, F., Nouvenne, A., Mancabelli, L., Lugli, G.A., Turroni, F., Duranti, S., Mangifesta, M., Viappiani, A., et al., 2017. Gut microbiota composition is associated with polypharmacy in elderly hospitalized patients. Sci Rep 7, 11102.

Tripathi, A., Debelius, J., Brenner, D.A., Karin, M., Loomba, R., Schnabl, B., Knight, R., 2018. The gut-liver axis and the intersection with the microbiome. Nat Rev Gastro-enterol Hepatol 15, 397—411.

van Kessel, S.P., Frye, A.K., El-Gendy, A.O., Castejon, M., Keshavarzian, A., van Dijk, G., El Aidy, S., 2019. Gut bacterial tyrosine decarboxylases restrict levels of levodopa in the treatment of Parkinson's disease. Nat Commun 10, 310.

Vich Vila, A., Collij, V., Sanna, S., Sinha, T., Imhann, F., Bourgonje, A.R., Mujagic, Z., Jonkers, D., Masclee, A.A.M., Fu, J., et al., 2020. Impact of commonly used drugs on the composition and metabolic function of the gut microbiota. Nat Commun 11, 362.

Weersma, R.K., Zhernakova, A., Fu, J., 2020. Interaction between drugs and the gut microbiome. Gut 69 (8), 1510—1519.

Werth, B.J., 2020. Overview of antibiotics. In: Merck Manual. Merck & Co., Inc, Kenil-worth, NJ.

WHO Collaborating Centre for Drug Statistics Methodology, 2019. Guidelines for ATC Classification and DDD Assignment, 2020 (Oslo).

Xu, R., Wu, B., Liang, J., He, F., Gu, W., Li, K., Luo, Y., Chen, J., Gao, Y., Wu, Z., et al., 2020. Altered gut microbiota and mucosal immunity in patients with schizophrenia. Brain Behav Immun 85, 120—127.

Yip, L.Y., Chan, E.C., 2015. Investigation of host-gut microbiota modulation of thera-peutic outcome. Drug Metab Dispos 43, 1619—1631.

Zeraati, M., Enayati, M., Kafami, L., Shahidi, S.H., Salari, A.A., 2019. Gut microbiota depletion from early adolescence alters adult immunological and neurobehavioral re-sponses in a mouse model of multiple sclerosis. Neuropharmacology 157, 107685.

Zimmermann, M., Zimmermann-Kogadeeva, M., Wegmann, R., Goodman, A.L., 2019. Mapping human microbiome drug metabolism by gut bacteria and their genes. Nature 570, 462—467.

# CHAPTER 11

# Microbial-related treatments

## Introduction

The emerging evidence of a role for the aberrant host—microbe dialogue in stress-related disorders of the gut—brain axis makes the gut microbiota an attractive therapeutic target to produce beneficial effects across gastrointestinal and behavioral symptoms (Cryan et al., 2019; Zhou and Foster, 2015). The concept of psychobiotics was introduced initially as a "live organism that, when ingested in adequate amounts, produces a health benefit in patients suffering from psychiatric illness" (Dinan et al., 2013). A revised definition now considers the range of options available in this regard, taking into account "any exogenous influence whose effect on the brain is bacterially-mediated" (Sarkar et al., 2016).

This chapter will review the range of interventions that can be categorized as psychobiotics including probiotics, prebiotics, synbiotics, postbiotics, fermented food, and fecal microbiota transplantation (FMT). The evidence from preclinical and clinical studies is evaluated with an emphasis on the microbiota—gut—brain axis signaling pathways that might be recruited to yield benefits for brain function and behavior (see Fig. 11.1). We also consider how these interventions can be delivered be it via supplements, functional foods, or improvements to dietary intake.

## Probiotics

The International Scientific Association for Probiotics and Prebiotics (ISAPP) defines probiotics as "live microorganisms that, when administered in adequate amounts, confer a health benefit on the host" (Hill et al., 2014). Once ingested, probiotics can be considered transient members of the gut microbiome to complement the normal activity of the resident microbiome (Derrien and van Hylckama Vlieg, 2015). The transient nature of the colonization was illustrated in a study with the *Bifidobacterium longum* subsp.

*Microbiota Brain Axis*
ISBN 978-0-12-814800-6
https://doi.org/10.1016/B978-0-12-814800-6.00006-6

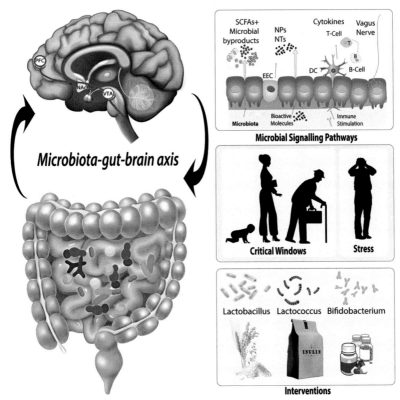

**Figure 11.1** *Psychobiotic interventions for modification of microbiota-to-brain communication.* Psychobiotic interventions include a number of options such as probiotics, prebiotics, or fermented foods. The behavioral benefits associated with these interventions may arise due to the modification of gut—brain axis signaling pathways by host—microbe interactions. These interventions have been studied during critical windows across the lifespan, in different contexts such as during the experience of stress and in stress-related disorders.

*longum* 35624 strain (formerly *B. infantis* 35624 (Allen et al., 2017; Lewis et al., 2016)), which confirmed the increase in fecal excretion levels following initiation of treatment and the subsequent decline toward baseline in the 2 weeks following dosing cessation (Charbonneau et al., 2013).

The psychobiotic concept includes a subset of probiotics with potential mental health benefits (Dinan et al., 2013; Sarkar et al., 2016). Evidence continues to accumulate from preclinical and clinical studies to support the idea that either specific single strains of probiotics or combinations of strains exert beneficial behavioral effects, the validation of which is critical to satisfy the ISAPP definition.

There are a variety of behavioral screening assays with predictive validity for more traditional drug discovery efforts that have been employed in the assessment of candidate probiotics, particularly in the context of anxiety- and depression-related behaviors (Cryan and Sweeney, 2011; Slattery and Cryan, 2012). For example, the elevated plus maze (EPM) and forced swim test (FST) were used to identify the anxiolytic and antidepressant properties of *Lactobacillus rhamnosus* (JB-1) in healthy adult male BALB/c mice (Bravo et al., 2011). *B. longum* subsp. *longum 35624(previously B. infantis 35624)* exerted antidepressant effects by reversing the behavioral deficits induced by maternal separation in male Sprague—Dawley rats (Desbonnet et al., 2010) but this probiotic had no effects on FST behaviors in naive rats on day 3 or day 14 following the commencement of treatment (Desbonnet et al., 2008). *B. longum* and *L. rhamnosus* (strains unspecified) reversed the behavioral impact of chronic unpredictable mild stress (CUMS) in the sucrose preference test (SPT) and the FST in male Wistar rats (Li et al., 2019). *B. breve* CCFM1025, but not *B. breve* FHLJDQ3M5, was shown to reduce chronic stress-induced depression- and anxiety-like behaviors in C57BL/6J male mice (Tian et al., 2020, 2021). L. reuteri (ATCC 23272) administration was able to ameliorate the escape behavior induced by chronic stress (Marin et al., 2017).

Combinations of probiotic strains have also shown potential and a formulation consisting of *L. helveticus* R0052 and *B. longum* R0175 was associated with anxiolytic-like activity in rats in the conditioned defensive burying test (Messaoudi et al., 2011b). Lifelong supplementation of Long—Evans hooded rats (male and female) with this combination of strains also reduced anxiety-like behaviors in adulthood as measured in the open field test and light—dark box (Myles et al., 2020). A multispecies probiotic formulation consisting of eight bacterial strains (*B. bifidum* W23, *B. lactis* W52, *L. acidophilus* W37, *L. brevis* W63, *L. casei* W56, *L. salivarius* W24, *Lactococcus Lactis* W19, *Lc. Lactis* W58) reduced depressive-like behavior in the FST in male Sprague—Dawley rats fed a control or high-fat diet (Abildgaard et al., 2017). There was no effect of this multispecies probiotic preparation on anxiety-like behaviors in the open field test in this study (Abildgaard et al., 2017). A combination of *L. rhamnosus* R0011 and *L. helveticus* R0052, when administered to infant rats throughout a maternal separation protocol, was able to restore normal developmental trajectories of emotion-related behaviors normally compromised by this early-life stress exposure (Cowan et al., 2016).

The effects of probiotics are strain specific, and this was neatly illustrated in a series of studies with *B. longum* 1714 and *B. breve* 1205 in adult male BALB/c mice. While both strains had positive effects on cognition in the novel object recognition, it was only the *B. longum* 1714 strain that produced clear benefits in learning and memory during fear conditioning and in the Barnes maze (Savignac et al., 2015). Meanwhile, both of these strains produced anxiolytic effects in the open field and marble burying test but only *B. breve* 1205 did so in the EPM, and it was only *B. longum* 1714 that was associated with antidepressant effects in the tail suspension test (Savignac et al., 2014). Further evidence that probiotic effects are also domain specific comes from a study with *B. longum subsp. longum* 35624 (previously *B. infantis* 35624), which produced anxiolytic but not antidepressant effects in male and female Sprague—Dawley rats after chronic corticosterone exposure (Haas et al., 2020). This strain also produced visceral antinociceptive effects in male Sprague—Dawley and Wistar—Kyoto rats (McKernan et al., 2010).

The impact of psychobiotics has also been evaluated in zebrafish, a valuable model species for neurobehavioral studies (Audira; et al., 2018; Nagpal and Cryan, 2021), where *L. rhamnosus* IMC 501 impacted on shoaling behavior, interpreted as increased exploratory behavior (Borrelli et al., 2016; Cuomo et al., 2021). *L. plantarum* also reduced anxiety-related behavior (Davis et al., 2016). Consumption of a probiotic mixture of *L. rhamnosus* CECT8361 and *B. longum* CECT7347 strains for 4 months also reduced anxiety-like behaviors of zebrafish (Valcarce et al., 2020). *L. acidophilus* NCFM, *L. rhamnosus* HN001, and *B. animalis* subsp. lactis Bi-07 increased intestinal motility zebrafish (Wang et al., 2020). *Caenorhabditis elegans* has also been employed as a model to understand host—microbe interactions (Kumar et al., 2020; Nagpal and Cryan, 2021; Shapira, 2017) with a number of strains such as *Clostridium butyricum* MIYAIRI 588, *L. fermentum* strain JDFM216 and *L. gasseri* SBT2055 (LG2055) associated with increased lifespan (Kato et al., 2018; Nakagawa et al., 2016; Park et al., 2018).

Psychobiotic administration can also be considered during different time windows and in different contexts. Treatment with *L. reuteri* MM4-1A (ATCC-PTA-6475) during adulthood was able to rescue deficits in social behaviors in several mouse models of autism spectrum disorder (ASD) (Sgritta et al., 2019). The same strain was also able to rescue maternal high-fat diet (MHFD)-induced deficits in social behavior in the offspring of C57Bl6/J mice (Buffington et al., 2016). Similarly, *L. reuteri* 6475 ingestion during adulthood was able to rescue the social deficits, but not hyperactivity, in the *Cntnap2−/−* mouse model for neurodevelopmental disorders (Buffington et al., 2021). *B. breve* M16V, given in drinking water to the nursing dams starting from birth and throughout the lactation period till weaning at P21 when male offspring were weaned onto the corresponding treatment, was able to rescue the behavioral deficits induced by cesarean section (CS) mode of birth (Morais et al., 2020). This included the maternal attachment deficits and neonatal recognition abilities in the CS pups at day 10 but not early-life communication and anxiety, or social recognition impairment in adult mice (Morais et al., 2020).

Taken together, these studies in model species reveal several important insights about the potential benefits of candidate probiotics with psychobiotic properties. Effects have been observed across species in zebrafish, *C. elegans*, rats and mice. The behavioral benefits are clearly strain specific and are evident in studies administering both single-strain and multistrain preparations. In the latter studies, it is often not obvious which strain is responsible for the beneficial effects. A number of behavioral domains are affected, including behaviors relevant to anxiety, depression, cognition, pain, and social behavior. In some studies, these behaviors are impacted in healthy animals while in others, the aberrant behaviors induced by various exposures, such as chronic stress or high-fat diet, are reversed or at least reduced. Most of these studies use male animals. Outside of the open questions surrounding sex-specific effects, the field at the minute is moving toward a more detailed understanding of the mechanisms of action and whether the observed effects will translate from bench to bedside.

A number of the microbiota-to-brain signaling pathways identified in Chapter 4 have been implicated in the beneficial impact of specific psychobiotic interventions in preclinical studies. This includes a number of studies which have documented a role for the vagus nerve (Bharwani et al., 2020; Bravo et al., 2011; Sgritta et al., 2019) as well as tryptophan metabolism (Desbonnet et al., 2008; Marin et al., 2017), the HPA axis (Ait-Belgnaoui et al., 2018), and the immune system (Abildgaard et al., 2017; Arseneault-Breard et al., 2012; Bharwani et al., 2017; Desbonnet et al., 2008; Mohammadi et al., 2019). It has also been noted that administration of *L. rhamnosus* (JB-1) to male BALB/c mice elevated both brain GABA levels (Janik et al., 2016) and altered the expression of CNS GABA receptors (Bravo et al., 2011). *L. reuteri* MM4-1A (ATCC-PTA-6475) administration was unable to impact on social behavior in mice lacking the oxytocin receptor (Sgritta et al., 2019). Genomic and metabolic features of psychobiotics involving neuromodulatory metabolites implicated in the mechanism of action include an enhanced capacity for short-chain fatty acid (SCFA) biosynthesis and tryptophan metabolism (Tian et al. 2020, 2021). Soluble mediators derived from *L. rhamnosus* GG were associated with the reduction of visceral hypersensitivity induced by that candidate psychobiotic (McVey Neufeld et al., 2020).

From a translational perspective, a number of the psychobiotic strains identified in preclinical studies have now been evaluated in human subjects, often with mixed results. *B. longum* 1714 produced an attenuation of cortisol output in response to the socially evaluated cold pressor test, reductions in daily reported perceived stress, subtle improvements in hippocampus-dependent visuospatial memory performance, and enhanced frontal midline electroencephalographic mobility consistent with the profile identified in preclinical studies (Allen et al., 2016). This strain was also shown to modulate neural responses during social stress measured using magnetoencephalography in tandem with the "Cyberball game" (Wang et al., 2019). *B. longum* 1714 also improved sleep quality and duration in male students during an exam period as a stress exposure but did not alleviate symptoms of chronic stress, depression, or any measure of cognitive performance (Moloney et al., 2021). The translational psychological benefits of a combination of *L. helveticus* R0052 and *B. longum* R0175 for anxiety scores and stress responses have also been reported (Messaoudi et al., 2011a, 2011b). Meanwhile, and despite a considerable background of efficacy in preclinical studies, *L. rhamnosus* (JB-1) did not impact stress-related measures, HPA response, inflammation, resting neural activity, or cognitive performance in healthy male participants (Kelly et al., 2017).

Interestingly, stress induction was required to unmask the benefits of the ecologic barrier (A multispecies preparation including *B. bifidum* W23, *B. lactis* W51, *B. lactis* W52, *L. acidophilus* W37, *L. brevis* W63, *L. casei* W56, *L. salivarius* W24, *L. lactis* W19, and *L. lactis* W58), which had no effects on neurocognitive measures of emotion reactivity, emotion regulation, and cognitive control but did show a significant stress-related increase in working memory performance associated neural changes in the frontal cortex (Papalini et al., 2019). Participants who consumed this multispecies probiotic intervention also showed a significantly reduced overall cognitive reactivity to sad mood (Steenbergen et al., 2015). Academic stress-induced increases in salivary cortisol levels were significantly suppressed in healthy students following supplementation with

*L. casei* strain Shirota (Takada et al., 2016) and with *L. plantarum* 299v (Andersson et al., 2016). *L. plantarum* DR7 was associated with a reduction of stress and anxiety measures and improvements in multiple aspects of memory and cognition, enhanced serotonergic signaling, and reductions in cortisol and proinflammatory cytokines in stressed but otherwise healthy adults (Chong et al., 2019).

Consumption of fermented milk products, containing *B. animalis* subsp *Lactis*, *S. thermophiles*, L. bulgaricus, and *L. lactis* subsp *Lactis*, by participants who underwent fMRI to measure brain response to an emotional faces attention task and resting brain activity was associated with the reduced task-related response of a distributed functional network and alterations in resting brain activity (Tillisch et al., 2013). Supplementation with a milk drink containing *L. casei* as part of a double-blind randomized controlled trial (RCT) improved mood scores in an elderly cohort, particularly in participants who scored the lowest in mood scores at baseline (Benton et al., 2007). A multispecies probiotic capsule (containing *L. casei*, *L. acidophilus*, *L. rhamnosus*, *L. bulgaricus*, B. breve, B. longum, and S. thermophilus) and a probiotic yogurt (containing *L. acidophilus* LA5 and *B. lactis* BB12) both improved depression, anxiety, and stress scale (DASS) scores in petrochemical workers (Mohammadi et al., 2016).

The administration of psychobiotics has also been considered in pathological situations or in states of altered physiology. *B. longum* NCC3001 reduced depression but not anxiety scores in patients with irritable bowel syndrome (IBS) and was associated with reduced responses to negative emotional stimuli in the amygdala and fronto-limbic regions in an fMRI analysis (Pinto-Sanchez et al., 2017). Women who ingested *L. rhamnosus* HN001 had significantly lower depression and anxiety scores in the postpartum period (Slykerman et al., 2017). Supplementation of an MDD cohort with a combination of three strains (*L. acidophilus*, *L. casei*, and *B. bifidum*) in an RCT conducted over 8 weeks resulted in reduced depression scores in self-report scales (Akkasheh et al., 2016). A combination of *L. helveticus* R0052 and *B. longum* R0175 also showed improvements in depression scores (Kazemi et al., 2019) although this beneficial effect was not observed in an earlier study with the same combination but in study participants with at least moderate baseline scores on self-report mood measures (Romijn et al., 2017). There was an improvement in cognitive function and a significant decrease in kynurenine concentration in a group of depressed patients receiving *L. Plantarum* 299v (Rudzki et al., 2019).

Less attention to date has focused on how this might be applied in clinical practice and in patients with a clinically diagnosed psychiatric disorder rather than assessments of anxiety and depression scores. One example relates to an open-label study with *C. butyricum* MIYAIRI 588 as adjunctive therapy in treatment-resistant major depressive disorder patients and in which an improvement in depression was reported (Miyaoka et al., 2018). Although Positive and Negative Syndrome Scale (PANSS) scores did not change following adjunctive treatment of patients diagnosed with schizophrenia or schizoaffective disorder with *L. rhamnosus* GG and *B. lactis* Bb12, patients receiving the probiotic supplement were less likely to report severe bowel difficulties and immunomodulatory properties were also reported (Tomasik et al., 2015). The use of these strains was also associated with a lower rate of rehospitalization in patients with acute mania recently discharged (Dickerson et al., 2018).

Other CNS disorders to be investigated include Alzheimer's disease and the impact of psychobiotics on cognitive performance. A probiotic milk containing *L. acidophilus*, *L. casei*, *B. bifidum*, and *L. fermentum* resulted in improvements in Mini—Mental State Exam (MMSE) scores (Akbari et al., 2016). Capsules containing either *L. fermentum*, *L. plantarum*, and *B. lactis* or *L. acidophilus*, *B. bifidum*, and *B. longum* had no impact on the performance of participants in the test your memory (TYM) cognitive assessment (Agahi et al., 2019). Cognitive performance was also assessed in the context of neurocognitive performance in HIV-1 Infected Individuals following supplementation with a probiotic cocktail (containing *L. plantarum* DSM 24730, *S. thermophilus* DSM 24731, *B. breve* DSM 24732, *L. paracasei* DSM 24733, *L. delbrueckii* subsp. *bulgaricus* DSM 24734, *L. acidophilus* DSM 24735, *B. longum* DSM 24736, and *B. infantis* DSM 24737) with improvements noted across a range of cognitive domains including memory, attention, and executive function (Ceccarelli et al., 2017).

In summary, the clinical studies to date offer some validation for the promising body of preclinical research with candidate probiotics, and it is encouraging to note some of the translational successes and proof-of-principle studies that targeting the gut microbiota with live biotherapeutics (a biological product containing live organisms) can impact on brain activity, function and a variety of behavioral domains. However, it remains a body of work that comes with some important caveats. Many of the studies are small in size and often focus on the modification of depression and anxiety symptoms in either healthy controls or in disorders with comorbid psychological features but without a formal clinical diagnosis of a psychiatric condition. There is tremendous variation in the strains assessed and no real consensus on whether single or multistrain preparations are the best option. Where multistrain preparations are evaluated, the precise rationale for the composition of the cocktail is often absent and it is not always clear that all members of the consortium are necessary or sufficient for the observed effects. Open questions remain in relation to the optimum dose or duration of treatment, and it remains difficult at this stage to make any evidence-based recommendations for the use of a particular strain, or combination of strains, in a specific pathological condition or indication.

The strain (or preparation) specific effects make it difficult to reliably assess the impact of each specific intervention, especially given the lack of standardization in this literature. Nevertheless, some authors have attempted to create a synthesis of the data via systematic review and meta-analysis using probiotics as an umbrella term. These studies have offered mixed conclusions. One such approach reported that probiotics were associated with a significant reduction in depression in both the healthy and patients with major depressive disorder, with these benefits seen (MDD) in the population aged under 60 but not in those people aged over 65 (Huang et al., 2016). The results of a systematic review that included 10 randomized controlled trials, a mixture of studies in those with a clinical diagnosis and healthy participants led the authors to conclude that there was some limited support for the use of some probiotics in reducing human anxiety and depression (Pirbaglou et al., 2016). A systematic review of 10 studies analyzed for the effects of probiotics on mood, anxiety, and cognition indicated that the majority of the studies found positive results on all measures of depressive symptoms, albeit while noting concerns in relation to variability across the strain of probiotic, the dosing, and duration of treatment (Wallace and Milev, 2017). In contrast, a meta-analysis focused specifically on depressive symptoms looked at 10 randomized controlled trials with a total of 1349 and found that probiotic supplementation had no impact but that an increased focus on those with clinically diagnosed depression was required (Ng et al., 2018). A meta-analysis focused on the cognitive benefits of probiotics found no significant effect for any intervention for global cognition or for any specific cognitive domain and again noted limitations of the research evaluated with respect to the number of small and short-term studies and heterogeneity of study population, cognitive tests, and intervention (Marx et al., 2020). From a more mechanistic perspective, a meta-analysis that included 12 probiotic interventions concluded that there was preliminary evidence to support the modulation of metabolic activity along the kynurenine pathway of tryptophan metabolism (Purton et al., 2021).

Most of the probiotic options evaluated to date have included a narrow range of those with a long history of safe administration. Future options might consider the use of next generation probiotics, based on the use of organisms with relatively unknown characteristics for investigation as novel live biotherapeutics (O'Toole et al., 2017). Some promising examples from cognate areas include the evaluation of microbes which have noted to be depleted or deficient in different disease states such as *Akkermansia muciniphila* in obesity (Depommier et al., 2019) and *Faecalibacterium prausnitzii* for gastrointestinal health in inflammatory bowel disease (IBD) and beyond (Miquel et al., 2013). The challenge with this approach lies in whether these agents should be considered more as a pharmaceutical than a dietary supplement and the regulatory framework that will be required before they reach the market (O'Toole et al., 2017). Interestingly, the European Food Safety Authority (EFSA) recently concluded that *A. muciniphila* is safe as a food ingredient.

## Prebiotics

ISAPP defines a prebiotic as "a substrate that is selectively utilized by host microorganisms conferring a health benefit," and they are found in a wide variety of fruits, vegetables, and grains, and human milk (Gibson et al., 2017). Examples of prebiotics include fibers such as inulin, fructo-oligosaccharides, galacto-oligosaccharides (GOSs), and resistant starch (Long-Smith et al., 2020). As many of these prebiotics are not absorbed in the small intestine, they can be selectively fermented by gut microbes and offer an advantage over probiotics in their potential to improve gut microbial status more globally (Berding et al., 2021a).

In the preclinical literature, several studies support the use of prebiotics as psychobiotics to manipulate microbe-to-brain communication pathways (Sarkar et al., 2016). Examples include the anxiolytic and antidepressant effects of fructo-oligosaccharides (FOSs) and GOSs, alone or in combination, and their capacity to reverse the impact of chronic stress exposures in male C57BL/6J mice (Burokas et al., 2017). Early-life supplementation of the diet a blend of two prebiotics, GOS and polydextrose (PDX), and the glycoprotein lactoferrin (LAC) attenuated stress-induced learned helplessness in male F344 rats after exposure to inescapable stress later in life (Mika et al., 2017).

Supplementation from birth with a prebiotic mixture of short-chain galacto-oligosaccharides (scGOS) and long-chain fructo-oligossaccharides (lcFOS), which induced an expansion of bifidobacteria, was sufficient to reverse some of the behavioral consequences of c-section delivery (Morais et al., 2020). Maternal B-GOS supplementation was also associated with increased exploratory behavior in the offspring of CD1 female mice (Hebert et al., 2021), an interesting observation in light of the demonstration of the possibility of increasing prebiotic intake via a perinatal educational dietary intervention (Dawson et al., 2019, 2021).

There are also a number of studies using prebiotics in human intervention studies although it is a less extensively deployed strategy compared to probiotic studies. Recently, polydextrose (a dietary fiber) supplementation resulted in a modest improvement in performance in the domains of cognitive flexibility and sustained attention (Berding et al., 2021b). B-GOS, but not FOS, was shown to reduce the salivary cortisol awakening response in healthy volunteers (Schmidt et al., 2015). This B-GOS prebiotic was also associated with improvements in antisocial behavior in children with autistic spectrum disorder (Grimaldi et al., 2018). It was reported that GOS, at 7 g/day but not 3.5 g/day, was able to reduce anxiety scores in a mixed IBS population (Silk et al., 2009). However, this needs to be considered in the context that it is often considered beneficial to restrict the intake of fermentable carbohydrates in IBS via the low FODMAP diet (Staudacher et al., 2021). It was also reported that GOS produced no improvements in depression scores in patients with major depressive disorder (Kazemi et al., 2019).

Many of the benefits derived from prebiotics are thought to be due to the increased production of SCFAs (Berding et al., 2021a), although the induction of SCFA production is not always sufficient to modulate host physiology (O'Connor et al., 2020). Other pathways implicated include tryptophan metabolism, the immune system, and the HPA axis (Burokas et al., 2017; Purton et al., 2021; Saulnier et al., 2013) with specific neuroimmune features of middle age reversed by a 10% oligofructose-enriched inulin (Boehme et al., 2020). As in the case of probiotics, it is also important to note that the effects observed are prebiotic- and context-specific. For example, resistant starch administration led to an increased abundance of the Actinobacteria phylum that was observed only in atypical antipsychotic users in adults with a diagnosis of bipolar disorder or schizophrenia (Flowers et al., 2019).

## Synbiotics

Synbiotics are defined by ISAPP as "a mixture comprising live microorganisms and substrate(s) selectively utilized by host microorganisms that confers a health benefit on the host" with a subcategory for synergistic synbiotics where the substrate is intended to be selectively utilized by the coadministered microorganisms. a mixture comprising live microorganisms and substrate(s) selectively utilized by host microorganisms that confer a health benefit on the host (Swanson et al., 2020). The prebiotic component of synergistic synbiotics can thus provide a source of fermentable fiber to improve the viability of the probiotic component as well as acting as a general prebiotic for resident host microorganisms (Long-Smith et al., 2020).

A combination of *L. plantarum* ATCC 8014 and inulin was associated with improvements in learning and memory in diabetic male Wistar rats (Morshedi et al., 2020). A symbiotic preparation containing *L. paracasei* HII01 and prebiotic xyloolidosaccharide as prebiotic restored cognitive function induced by chronic high-fat diet (HFD) consumption in male Wistar rats, associated with decreased microglial activation, improved hippocampal plasticity, and attenuated brain mitochondrial dysfunction (Chunchai et al., 2018). A diet containing a combination of PDX/GOS and LGG attenuated the effects of early-life stress exposure (maternal separation) on anxiety-like behavior and hippocampal-dependent learning (McVey Neufeld et al., 2019). Supplementation with *L. casei* 54-2-33 and inulin exerted anxiogenic effects in healthy juvenile male Sprague—Dawley rats (Barrera-Bugueno et al., 2017). In a transgenic humanized *Drosophila melanogaster* model of AD, a synbiotic containing *L. plantarum* NCIMB 8826 (Lp8826), *L. fermentum* NCIMB 5221 (Lf5221) and *B. longum* spp. *infantis* NCIMB 702255 (Bi702255), and a polyphenol plant extract increased survivability and motility and rescued amyloid-beta deposition and acetylcholinesterase activity (Westfall et al., 2019).

These promising preclinical results have been supported by a small number of studies in human subjects. A fermented milk drink containing *Streptococcus salivarius* subsp *thermophilus*, *Enterococcus faecium*, *L. rhamnosus* GG, *L. acidophilus*, *L. plantarum*, *L. paracasei*, *L. delbrueckii* subsp *bulgaricus*, and *B. breve* and *B. animalis* subsp *lactis*, and an FOS prebiotic was superior to placebo in superior in improving constipation in patients with Parkinson disease (Barichella et al., 2016). A small pilot study using *B. infantis* in combination with a bovine colostrum product (source of prebiotic oligosaccharides) reported that some participants reported a reduction in the frequency of gastrointestinal symptoms and a reduced occurrence aberrant behavior in children with ASD (Sanctuary et al., 2019).

## Postbiotics

ISAPP defines postbiotics as "a preparation of inanimate microorganisms and/or their components that confers a health benefit on the host" (Salminen et al., 2021a) although this is a definition subject to debate (Aguilar-Toala et al., 2021; Salminen et al., 2021b). Potential mechanisms of action of postbiotics include modulation of the resident microbiota, enhancement of epithelial barrier functions, modulation of immune or metabolic responses, modulation of systemic metabolic responses, and signaling via the nervous system with SCFAs an exemplar of microbial effector molecules (Salminen et al., 2021a). Preclinical studies based on oral supplementation of a mixture of SCFAs (acetate, propionate, and butyrate) in C57BL/6J male mice indicated an effect on anxiety- and depressive-like behaviors in unstressed control but not stressed animals (van de Wouw et al., 2018). The mice in this study undergoing psychosocial stress exhibited benefits of SCFA administration in terms of reduced stress–responsiveness for HPA-axis function and intestinal permeability. The delivery of acetate to the colon using acylated starches as a carrier resulted in a decrease in anxiety-like behavior compared with butyrylated or propionylated starches (Kimura-Todani et al., 2020). The translational relevance of these observations was demonstrated in a study in healthy men where colon–delivered SCFA attenuated the cortisol response to a psychosocial stress exposure (Dalile et al., 2020).

Paraprobiotics, inanimate microorganisms including nonviable probiotics (e.g., heat-killed probiotics) can also be included in the postbiotic concept where it is envisaged that biological activity derives from structural components of the microorganisms and/or microbial metabolites. Examples include ADR–159, a heat-killed fermentate generated by two Lactobacillus strains, which was associated with increased sociability and lower baseline corticosterone levels in mice (Warda et al., 2019). A heat-killed *L. paracasei* PS23 was associated with anxiolytic and antidepressant effects in the open field, FST and sucrose preference tests and reductions in dopamine levels in the hippocampus and prefrontal cortex in a chronic corticosterone-induced depression-like phenotype in mice (Wei et al., 2019). A 12-week intervention with a heat-killed, washed, paraprobiotic CP2305, was associated with reduced anxiety and improved sleep quality in Japanese medical students during a stressful examination period (Nishida et al., 2019) and during a cadaver dissection course (Nishida et al., 2017).

## Fermented foods and diet

ISAPP defines fermented foods as "foods made through desired microbial growth and enzymatic conversions of food components" (Marco et al., 2021). This includes kefir, a fermented dairy product arising from the symbiotic fermentation of milk by the lactic acid bacteria and yeasts contained within a complex called a kefir grain (Bourrie et al., 2016). In preclinical studies, kefir has been shown to modulate peripheral immunoregulation and reduce some ASD-related behavioral dysfunctions in the BTBR mouse strain (van de Wouw et al., 2021). Various kefir preparations also ameliorated stress-induced decreases in serotonergic signaling in the colon, reward-seeking behavior in the saccharin preference test, decreased repetitive behavior, and reduced stress-induced deficits in reward-seeking behavior in mice (van de Wouw et al., 2020).

In humans, a highly fermented food diet has recently been shown to exert an antiinflammatory effect in healthy adults (Wastyk et al., 2021). There is also an increasing focus on the gut microbiota as a mechanism of action by which diet may influence mental and brain health (Marx et al., 2021), framed within the growing field of nutritional psychiatry (Sarris et al., 2015). It is currently unclear if interventions such as the Mediterranean diet produce benefits by recruiting the gut microbiota but it is plausible that this or dietary interventions associated with improvements, for example, in major depression (Jacka et al., 2017) have a microbial mechanism of action via their prebiotic effects (Long-Smith et al., 2020).

## Fecal microbiota transplantation

FMT consists of transferring the fecal material from a donor into the gastrointestinal tract of a recipient (Settanni et al., 2021). In the preclinical literature, it is primarily used as a means of demonstrating a potential causal role for a disease-associated microbiota in symptom expression (Gheorghe et al., 2021). It has been demonstrated, for example, that the transfer of a depression-associated microbiota can recapitulate both the prominent behavioral phenotypes (anhedonia, comorbid anxiety) as well as some neurobiological hallmarks (altered tryptophan metabolism) (Kelly et al., 2016). From a clinical perspective and outside of the main indication for the treatment of recurrent *Clostridium difficile* infection, it has been evaluated in a research context with mixed results for a range of gastrointestinal and metabolic disorders but studies in psychiatric populations are currently sparse outside of hepatic encephalopathy (Ianiro et al., 2020). There are many regulatory hurdles as well as safety considerations around FMT administration and a superior understanding of the mechanism of action will be required before more widespread routine use can be envisaged (Merrick et al., 2020). Such mechanistic insights may pave the way for the development of designer microbial consortia (Kong et al., 2018). The preclinical research has however informed guidelines on donor screening and exclusion criteria for the use of FMT in clinical practice (Cammarota et al., 2017).

## Antibiotics

It is somewhat counterintuitive to include antibiotics in this discussion, but it is worth reflecting that the origins of modern antidepressants can be traced back to an antibacterial drug developed for treating tuberculosis (Butler et al., 2019). A number of psychotropic drugs, including antidepressants, are now known to also have antimicrobial effects (Cussotto et al., 2019a, 2019b) (See Chapter 10, this book). Meanwhile, the antibiotic rifaximin is indicated in the United States and Canada for the treatment of adults with IBS (a stress-related disorder of gut—brain axis interactions) with diarrhea, although the mechanism of action is open to debate (Chey et al., 2020). It remains to be seen what, if any, role is played by this antimicrobial action on mental health benefits or if this knowledge can be exploited toward new treatment options.

## Conclusions, perspectives, and future directions

Therapeutic targeting of host—microbe interactions in CNS disorders incorporates a range of interventions that affect microbiota—gut—brain axis signaling and represents an opportunity to identify of novel targets for drug development with superior safety profiles by containing the initial site of action within the gut. Most studies to date have focused on probiotics and this body of work represents the richest source of mechanistic insight as well as translational efforts. More recently other options, such as synbiotics, postbiotics and fermented foods have come under scrutiny while FMT may also hold promise for future applications. Further work is required to gain a more detailed understanding of the mechanisms involved in the action of psychobiotics as we move slowly toward the evidence base to support clinical translation. This may facilitate the matching of mechanisms of action to the underlying neurobiological signature of specific patient subgroups. Guidelines and strategies for the clinical deployment of psychobiotics, either as adjunctive therapies or direct options, may thus coalesce around individualized treatment strategies. There are, however, a number of open questions remaining, and carefully controlled clinical trials of scale will be required to bolster confidence in the use of psychobiotics for pathological conditions in a clinical setting.

# References

Abildgaard, A., et al., 2017. Probiotic treatment reduces depressive-like behaviour in rats independently of diet. Psychoneuroendocrinology 79, 40−48.

Agahi, A., et al., 2019. Corrigendum: does severity of Alzheimer's disease contribute to its responsiveness to modifying gut microbiota? A double blind clinical trial. Front Neurol 10, 31.

Aguilar-Toala, J.E., et al., 2021. Postbiotics - when simplification fails to clarify. Nat Rev Gastroenterol Hepatol 18 (11), 825−826.

Ait-Belgnaoui, A., et al., 2018. Bifidobacterium longum and Lactobacillus helveticus synergistically suppress stress-related visceral hypersensitivity through hypothalamic-pituitary-adrenal Axis modulation. J Neurogastroenterol Motil 24 (1), 138−146.

Akbari, E., et al., 2016. Effect of probiotic supplementation on cognitive function and metabolic status in Alzheimer's disease: a randomized, double-blind and controlled trial. Front Aging Neurosci 8, 256.

Akkasheh, G., et al., 2016. Clinical and metabolic response to probiotic administration in patients with major depressive disorder: a randomized, double-blind, placebo-controlled trial. Nutrition 32 (3), 315−320.

Allen, A.P., et al., 2017. Bifidobacterium infantis 35624 and other probiotics in the management of irritable bowel syndrome. Strain specificity, symptoms, and mechanisms. Curr Med Res Opin 33 (7), 1349−1351.

Allen, A.P., et al., 2016. Bifidobacterium longum 1714 as a translational psychobiotic: modulation of stress, electrophysiology and neurocognition in healthy volunteers. Transl Psychiatry 6 (11), e939.

Andersson, H., et al., 2016. Oral administration of Lactobacillus plantarum 299v reduces cortisol levels in human saliva during examination induced stress: a randomized, double-blind controlled trial. Internet J Microbiol 2016, 8469018.

Arseneault-Breard, J., et al., 2012. Combination of Lactobacillus helveticus R0052 and Bifidobacterium longum R0175 reduces post-myocardial infarction depression symptoms and restores intestinal permeability in a rat model. Br J Nutr 107 (12), 1793−1799.

Audira, G., et al., 2018. A versatile setup for measuring multiple behavior endpoints in zebrafish. Inventions 3 (4), 75.

Barichella, M., et al., 2016. Probiotics and prebiotic fiber for constipation associated with Parkinson disease: an RCT. Neurology 87 (12), 1274−1280.

Barrera-Bugueno, C., et al., 2017. Anxiogenic effects of a Lactobacillus, inulin and the synbiotic on healthy juvenile rats. Neuroscience 359, 18−29.

Benton, D., Williams, C., Brown, A., 2007. Impact of consuming a milk drink containing a probiotic on mood and cognition. Eur J Clin Nutr 61 (3), 355−361.

Berding, K., Carbia, C., Cryan, J.F., 2021a. Going with the grain: fiber, cognition, and the microbiota-gut-brain-axis. Exp Biol Med 246 (7), 796−811.

Berding, K., et al., 2021b. A specific dietary fibre supplementation improves cognitive performance-an exploratory randomised, placebo-controlled, crossover study. Psycho-pharmacology (Berl) 238 (1), 149−163.

Bharwani, A., et al., 2017. Oral treatment with Lactobacillus rhamnosus attenuates behavioural deficits and immune changes in chronic social stress. BMC Med 15 (1), 7.

Bharwani, A., et al., 2020. The vagus nerve is necessary for the rapid and widespread neuronal activation in the brain following oral administration of psychoactive bacteria. Neuropharmacology 170, 108067.

Boehme, M., et al., 2020. Mid-life microbiota crises: middle age is associated with pervasive neuroimmune alterations that are reversed by targeting the gut microbiome. Mol Psychiatr 25 (10), 2567−2583.

Borrelli, L., et al., 2016. Probiotic modulation of the microbiota-gut-brain axis and behaviour in zebrafish. Sci Rep 6, 30046.

Bourrie, B.C., Willing, B.P., Cotter, P.D., 2016. The microbiota and health promoting characteristics of the fermented beverage kefir. Front Microbiol 7, 647.

Bravo, J.A., et al., 2011. Ingestion of Lactobacillus strain regulates emotional behavior and central GABA receptor expression in a mouse via the vagus nerve. Proc Natl Acad Sci U S A 108 (38), 16050−16055.

Buffington, S.A., et al., 2016. Microbial reconstitution reverses maternal diet-induced social and synaptic deficits in offspring. Cell 165 (7), 1762−1775.

Buffington, S.A., et al., 2021. Dissecting the contribution of host genetics and the microbiome in complex behaviors. Cell 184 (7), 1740−1756.

Burokas, A., et al., 2017. Targeting the microbiota-gut-brain Axis: prebiotics have anxiolytic and antidepressant-like effects and reverse the impact of chronic stress in mice. Biol Psychiatr 82 (7), 472−487.

Butler, M.I., et al., 2019. From isoniazid to psychobiotics: the gut microbiome as a new antidepressant target. Br J Hosp Med 80 (3), 139−145.

Cammarota, G., et al., 2017. European consensus conference on faecal microbiota trans-plantation in clinical practice. Gut 66 (4), 569−580.

Ceccarelli, G., et al., 2017. Impact of high-dose multi-strain probiotic supplementation on neurocognitive performance and central nervous system immune activation of HIV-1 infected Individuals. Nutrients 9 (11).

Charbonneau, D., Gibb, R.D., Quigley, E.M., 2013. Fecal excretion of Bifidobacterium infantis 35624 and changes in fecal microbiota after eight weeks of oral supplementation with encapsulated probiotic. Gut Microb 4 (3), 201−211.

Chey, W.D., Shah, E.D., DuPont, H.L., 2020. Mechanism of action and therapeutic benefit of rifaximin in patients with irritable bowel syndrome: a narrative review. Therap Adv Gastroenterol 13, 1756284819897531.

Chong, H.X., et al., 2019. Lactobacillus plantarum DR7 alleviates stress and anxiety in adults: a randomised, double-blind, placebo-controlled study. Benef Microbes 10 (4), 355−373.

Chunchai, T., et al., 2018. Decreased microglial activation through gut-brain axis by pre-biotics, probiotics, or synbiotics effectively restored cognitive function in obese-insulin resistant rats. J Neuroinflammation 15 (1), 11.

Cowan, C.S., Callaghan, B.L., Richardson, R., 2016. The effects of a probiotic formulation (Lactobacillus rhamnosus and L. helveticus) on developmental trajectories of emotional learning in stressed infant rats. Transl Psychiatry 6 (5), e823.

Cryan, J.F., Sweeney, F.F., 2011. The age of anxiety: role of animal models of anxiolytic action in drug discovery. Br J Pharmacol 164 (4), 1129–1161.

Cryan, J.F., et al., 2019. The microbiota-gut-brain Axis. Physiol Rev 99 (4), 1877–2013.

Cuomo, M., et al., 2021. DNA methylation profiles of Tph1A and BDNF in gut and brain of L. Rhamnosus-treated zebrafish. Biomolecules 11 (2).

Cussotto, S., et al., 2019a. Psychotropics and the microbiome: a chamber of secrets. Psychopharmacology (Berl) 236 (5), 1411–1432.

Cussotto, S., et al., 2019b. Differential effects of psychotropic drugs on microbiome composition and gastrointestinal function. Psychopharmacology (Berl) 236 (5), 1671–1685.

Dalile, B., et al., 2020. Colon-delivered short-chain fatty acids attenuate the cortisol response to psychosocial stress in healthy men: a randomized, placebo-controlled trial. Neuropsychopharmacology 45 (13), 2257–2266.

Davis, D.J., et al., 2016. Lactobacillus plantarum attenuates anxiety-related behavior and protects against stress-induced dysbiosis in adult zebrafish. Sci Rep 6, 33726.

Dawson, S.L., et al., 2019. Targeting the infant gut microbiota through a perinatal educational dietary intervention: protocol for a randomized controlled trial. JMIR Res Protoc 8 (10), e14771.

Dawson, S.L., et al., 2021. Targeting the perinatal diet to modulate the gut microbiota increases dietary variety and prebiotic and probiotic food intakes: results from a randomised controlled trial. Publ Health Nutr 24 (5), 1129–1141.

Depommier, C., et al., 2019. Supplementation with Akkermansia muciniphila in overweight and obese human volunteers: a proof-of-concept exploratory study. Nat Med 25 (7), 1096–1103.

Derrien, M., van Hylckama Vlieg, J.E., 2015. Fate, activity, and impact of ingested bacteria within the human gut microbiota. Trends Microbiol 23 (6), 354–366.

Desbonnet, L., et al., 2008. The probiotic Bifidobacteria infantis: an assessment of potential antidepressant properties in the rat. J Psychiatr Res 43 (2), 164–174.

Desbonnet, L., et al., 2010. Effects of the probiotic Bifidobacterium infantis in the maternal separation model of depression. Neuroscience 170 (4), 1179–1188.

Dickerson, F., et al., 2018. Adjunctive probiotic microorganisms to prevent rehospitalization in patients with acute mania: a randomized controlled trial. Bipolar Disord 20 (7), 614–621.

Dinan, T.G., Stanton, C., Cryan, J.F., 2013. Psychobiotics: a novel class of psychotropic. Biol Psychiatr 74 (10), 720–726.

Flowers, S.A., et al., 2019. Effects of atypical antipsychotic treatment and resistant starch supplementation on gut microbiome composition in a cohort of patients with bipolar disorder or schizophrenia. Pharmacotherapy 39 (2), 161–170.

Gheorghe, C.E., et al., 2021. Investigating causality with fecal microbiota transplantation in rodents: applications, recommendations and pitfalls. Gut Microb 13 (1), 1941711.

Gibson, G.R., et al., 2017. Expert consensus document: the International Scientific Association for Probiotics and Prebiotics (ISAPP) consensus statement on the definition and scope of prebiotics. Nat Rev Gastroenterol Hepatol 14 (8), 491–502.

Grimaldi, R., et al., 2018. A prebiotic intervention study in children with autism spectrum disorders (ASDs). Microbiome 6 (1), 133.

Haas, G.S., et al., 2020. Probiotic treatment (Bifidobacterium longum subsp. longum 35624) affects stress responsivity in male rats after chronic corticosterone exposure. Behav Brain Res 393, 112718.

Hebert, J.C., et al., 2021. Mom's diet matters: maternal prebiotic intake in mice reduces anxiety and alters brain gene expression and the fecal microbiome in offspring. Brain Behav Immun 91, 230–244.

Hill, C., et al., 2014. Expert consensus document. The International Scientific Association for Probiotics and Prebiotics consensus statement on the scope and appropriate use of the term probiotic. Nat Rev Gastroenterol Hepatol 11 (8), 506−514.

Huang, R., Wang, K., Hu, J., 2016. Effect of probiotics on depression: a systematic review and meta-analysis of randomized controlled trials. Nutrients 8 (8).

Ianiro, G., et al., 2020. Fecal microbiota transplantation in gastrointestinal and extraintestinal disorders. Future Microbiol 15, 1173−1183.

Jacka, F.N., et al., 2017. A randomised controlled trial of dietary improvement for adults with major depression (the 'SMILES' trial). BMC Med 15 (1), 23.

Janik, R., et al., 2016. Magnetic resonance spectroscopy reveals oral Lactobacillus promotion of increases in brain GABA, N-acetyl aspartate and glutamate. Neuroimage 125, 988−995.

Kato, M., et al., 2018. Clostridium butyricum MIYAIRI 588 increases the lifespan and multiple-stress resistance of *Caenorhabditis elegans*. Nutrients 10 (12).

Kazemi, A., et al., 2019. Effect of probiotic and prebiotic vs placebo on psychological outcomes in patients with major depressive disorder: a randomized clinical trial. Clin Nutr 38 (2), 522−528.

Kelly, J.R., et al., 2017. Lost in translation? The potential psychobiotic Lactobacillus rhamnosus (JB-1) fails to modulate stress or cognitive performance in healthy male subjects. Brain Behav Immun 61, 50−59.

Kelly, J.R., et al., 2016. Transferring the blues: depression-associated gut microbiota induces neurobehavioural changes in the rat. J Psychiatr Res 82, 109−118.

Kimura-Todani, T., et al., 2020. Dietary delivery of acetate to the colon using acylated starches as a carrier exerts anxiolytic effects in mice. Physiol Behav 223, 113004.

Kong, W., et al., 2018. Designing microbial consortia with defined social interactions. Nat Chem Biol 14 (8), 821−829.

Kumar, A., et al., 2020. *Caenorhabditis elegans*: a model to understand host-microbe interactions. Cell Mol Life Sci 77 (7), 1229−1249.

Lewis, Z.T., et al., 2016. Validating bifidobacterial species and subspecies identity in commercial probiotic products. Pediatr Res 79 (3), 445−452.

Li, H., et al., 2019. Effects of regulating gut microbiota on the serotonin metabolism in the chronic unpredictable mild stress rat model. Neuro Gastroenterol Motil 31 (10), e13677.

Long-Smith, C., et al., 2020. Microbiota-Gut-Brain Axis: new therapeutic opportunities. Annu Rev Pharmacol Toxicol 60, 477−502.

Marco, M.L., et al., 2021. The international scientific association for probiotics and prebiotics (ISAPP) consensus statement on fermented foods. Nat Rev Gastroenterol Hepatol 18 (3), 196−208.

Marin, I.A., et al., 2017. Microbiota alteration is associated with the development of stress-induced despair behavior. Sci Rep 7, 43859.

Marx, W., et al., 2020. Prebiotics, probiotics, fermented foods and cognitive outcomes: a meta-analysis of randomized controlled trials. Neurosci Biobehav Rev 118, 472−484.

Marx, W., et al., 2021. Diet and depression: exploring the biological mechanisms of action. Mol Psychiatr 26 (1), 134−150.

McKernan, D.P., et al., 2010. The probiotic Bifidobacterium infantis 35624 displays visceral antinociceptive effects in the rat. Neuro Gastroenterol Motil 22 (9), 1029−1035 e268.

McVey Neufeld, K.A., et al., 2019. Neurobehavioural effects of Lactobacillus rhamnosus GG alone and in combination with prebiotics polydextrose and galactooligosaccharide in male rats exposed to early-life stress. Nutr Neurosci 22 (6), 425−434.

McVey Neufeld, K.A., et al., 2020. Lactobacillus rhamnosus GG soluble mediators ameliorate early life stress-induced visceral hypersensitivity and changes in spinal cord gene expression. Neuronal Signal 4 (4). NS20200007.

Merrick, B., et al., 2020. Regulation, risk and safety of faecal microbiota transplant. Infect Prev Pract 2 (3), 100069.

Messaoudi, M., et al., 2011a. Beneficial psychological effects of a probiotic formulation (Lactobacillus helveticus R0052 and Bifidobacterium longum R0175) in healthy human volunteers. Gut Microb 2 (4), 256—261.

Messaoudi, M., et al., 2011b. Assessment of psychotropic-like properties of a probiotic formulation (Lactobacillus helveticus R0052 and Bifidobacterium longum R0175) in rats and human subjects. Br J Nutr 105 (5), 755—764.

Mika, A., et al., 2017. Early life diets with prebiotics and bioactive milk fractions attenuate the impact of stress on learned helplessness behaviours and alter gene expression within neural circuits important for stress resistance. Eur J Neurosci 45 (3), 342—357.

Miquel, S., et al., 2013. Faecalibacterium prausnitzii and human intestinal health. Curr Opin Microbiol 16 (3), 255—261.

Miyaoka, T., et al., 2018. Clostridium butyricum MIYAIRI 588 as adjunctive therapy for treatment-resistant major depressive disorder: a prospective open-label trial. Clin Neuropharmacol 41 (5), 151—155.

Mohammadi, A.A., et al., 2016. The effects of probiotics on mental health and hypothalamic-pituitary-adrenal axis: a randomized, double-blind, placebo-controlled trial in petrochemical workers. Nutr Neurosci 19 (9), 387—395.

Mohammadi, G., et al., 2019. The effects of probiotic formulation pretreatment (Lactobacillus helveticus R0052 and Bifidobacterium longum R0175) on a lipopolysaccharide rat model. J Am Coll Nutr 38 (3), 209—217.

Moloney, G.M., et al., 2021. Improvements in sleep indices during exam stress due to consumption of a Bifidobacterium longum, Brain. Behavior, & Immunity - Health 10, 100174.

Morais, L.H., et al., 2020. Enduring behavioral effects induced by birth by caesarean section in the mouse. Curr Biol 30 (19), 3761—3774.

Morshedi, M., Saghafi-Asl, M., Hosseinifard, E.S., 2020. The potential therapeutic effects of the gut microbiome manipulation by synbiotic containing-Lactobacillus plantarum on neuropsychological performance of diabetic rats. J Transl Med 18 (1), 18.

Myles, E.M., et al., 2020. Supplementation with combined Lactobacillus helveticus R0052 and Bifidobacterium longum R0175 across development reveals sex differences in physiological and behavioural effects of western diet in long-evans rats. Microorganisms 8 (10).

Nagpal, J., Cryan, J.F., 2021. Microbiota-brain interactions: moving toward mechanisms in model organisms. Neuron.

Nakagawa, H., et al., 2016. Effects and mechanisms of prolongevity induced by Lactobacillus gasseri SBT2055 in *Caenorhabditis elegans*. Aging Cell 15 (2), 227—236.

Ng, Q.X., et al., 2018. A meta-analysis of the use of probiotics to alleviate depressive symptoms. J Affect Disord 228, 13—19.

Nishida, K., et al., 2019. Health benefits of Lactobacillus gasseri CP2305 tablets in young adults exposed to chronic stress: a randomized, double-blind, placebo-controlled study. Nutrients 11 (8).

Nishida, K., et al., 2017. Para-psychobiotic Lactobacillus gasseri CP2305 ameliorates stress-related symptoms and sleep quality. J Appl Microbiol 123 (6), 1561—1570.

O'Connor, K.M., et al., 2020. Prebiotic administration modulates gut microbiota and faecal short-chain fatty acid concentrations but does not prevent chronic intermittent hypoxia-induced apnoea and hypertension in adult rats. EBioMedicine 59, 102968.

O'Toole, P.W., Marchesi, J.R., Hill, C., 2017. Next-generation probiotics: the spectrum from probiotics to live biotherapeutics. Nat Microbiol 2, 17057.

Papalini, S., et al., 2019. Stress matters: randomized controlled trial on the effect of probiotics on neurocognition. Neurobiol Stress 10, 100141.

Park, M.R., et al., 2018. Probiotic Lactobacillus fermentum strain JDFM216 stimulates the longevity and immune response of *Caenorhabditis elegans* through a nuclear hormone receptor. Sci Rep 8 (1), 7441.

Pinto-Sanchez, M.I., et al., 2017. Probiotic Bifidobacterium longum NCC3001 reduces depression scores and alters brain activity: a pilot study in patients with irritable bowel syndrome. Gastroenterology 153 (2), 448–459 e8.

Pirbaglou, M., et al., 2016. Probiotic supplementation can positively affect anxiety and depressive symptoms: a systematic review of randomized controlled trials. Nutr Res 36 (9), 889–898.

Purton, T., et al., 2021. Prebiotic and probiotic supplementation and the tryptophan-kynurenine pathway: a systematic review and meta analysis. Neurosci Biobehav Rev 123, 1–13.

Romijn, A.R., et al., 2017. A double-blind, randomized, placebo-controlled trial of Lactobacillus helveticus and Bifidobacterium longum for the symptoms of depression. Aust N Z J Psychiatr 51 (8), 810–821.

Rudzki, L., et al., 2019. Probiotic Lactobacillus Plantarum 299v decreases kynurenine concentration and improves cognitive functions in patients with major depression: a double-blind, randomized, placebo controlled study. Psychoneuroendocrinology 100, 213–222.

Salminen, S., et al., 2021. The International Scientific Association of Probiotics and Prebiotics (ISAPP) consensus statement on the definition and scope of postbiotics. Nat Rev Gastroenterol Hepatol 18, 649–667.

Salminen, S., et al., 2021b. Reply to: postbiotics - when simplification fails to clarify. Nat Rev Gastroenterol Hepatol 18 (11), 827.

Sanctuary, M.R., et al., 2019. Pilot study of probiotic/colostrum supplementation on gut function in children with autism and gastrointestinal symptoms. PLoS One 14 (1), e0210064.

Sarkar, A., et al., 2016. Psychobiotics and the manipulation of bacteria-gut-brain signals. Trends Neurosci 39 (11), 763–781.

Sarris, J., et al., 2015. Nutritional medicine as mainstream in psychiatry. Lancet Psychiatr 2 (3), 271–274.

Saulnier, D.M., et al., 2013. The intestinal microbiome, probiotics and prebiotics in neurogastroenterology. Gut Microb 4 (1), 17–27.

Savignac, H.M., et al., 2014. Bifidobacteria exert strain-specific effects on stress-related behavior and physiology in BALB/c mice. Neuro Gastroenterol Motil 26 (11), 1615–1627.

Savignac, H.M., et al., 2015. Bifidobacteria modulate cognitive processes in an anxious mouse strain. Behav Brain Res 287, 59–72.

Schmidt, K., et al., 2015. Prebiotic intake reduces the waking cortisol response and alters emotional bias in healthy volunteers. Psychopharmacology (Berl) 232 (10), 1793–1801.

Settanni, C.R., et al., 2021. Gut microbiota alteration and modulation in psychiatric disorders: current evidence from fecal microbiota transplantation. Prog Neuro-Psychopharmacol Biol Psychiatry 109, 110258.

Sgritta, M., et al., 2019. Mechanisms underlying microbial-mediated changes in social behavior in mouse models of autism spectrum disorder. Neuron 101 (2), 246–259 e6.

Shapira, M., 2017. Host-microbiota interactions in *Caenorhabditis elegans* and their significance. Curr Opin Microbiol 38, 142–147.

Silk, D.B., et al., 2009. Clinical trial: the effects of a trans-galactooligosaccharide prebiotic on faecal microbiota and symptoms in irritable bowel syndrome. Aliment Pharmacol Ther 29 (5), 508–518.

Slattery, D.A., Cryan, J.F., 2012. Using the rat forced swim test to assess antidepressant-like activity in rodents. Nat Protoc 7 (6), 1009–1014.

Slykerman, R.F., et al., 2017. Effect of Lactobacillus rhamnosus HN001 in pregnancy on postpartum symptoms of depression and anxiety: a randomised double-blind placebo-controlled trial. EBioMedicine 24, 159–165.

Staudacher, H.M., et al., 2021. Gut microbiota associations with diet in irritable bowel syndrome and the effect of low FODMAP diet and probiotics. Clin Nutr 40 (4), 1861–1870.

Steenbergen, L., et al., 2015. A randomized controlled trial to test the effect of multispecies probiotics on cognitive reactivity to sad mood. Brain Behav Immun 48, 258–264.

Swanson, K.S., et al., 2020. The International Scientific Association for Probiotics and Prebiotics (ISAPP) consensus statement on the definition and scope of synbiotics. Nat Rev Gastroenterol Hepatol 17 (11), 687–701.

Takada, M., et al., 2016. Probiotic Lactobacillus casei strain Shirota relieves stress-associated symptoms by modulating the gut-brain interaction in human and animal models. Neuro Gastroenterol Motil 28 (7), 1027–1036.

Tian, P., et al., 2020. Towards a psychobiotic therapy for depression: Bifidobacterium breve CCFM1025 reverses chronic stress-induced depressive symptoms and gut microbial abnormalities in mice. Neurobiol Stress 12, 100216.

Tian, P., et al., 2021. Unraveling the microbial mechanisms underlying the psychobiotic potential of a Bifidobacterium breve strain. Mol Nutr Food Res 65 (8), e2000704.

Tillisch, K., et al., 2013. Consumption of fermented milk product with probiotic modulates brain activity. Gastroenterology 144 (7), 1394-401, 401 e1–4.

Tomasik, J., et al., 2015. Immunomodulatory effects of probiotic supplementation in schizophrenia patients: a randomized, placebo-controlled trial. Biomark Insights 10, 47–54.

Valcarce, D.G., et al., 2020. Probiotics reduce anxiety-related behavior in zebrafish. Heliyon 6 (5), e03973.

van de Wouw, M., et al., 2018. Short-chain fatty acids: microbial metabolites that alleviate stress-induced brain-gut axis alterations. J Physiol 596 (20), 4923–4944.

van de Wouw, M., et al., 2020. Distinct actions of the fermented beverage kefir on host behaviour, immunity and microbiome gut-brain modules in the mouse. Microbiome 8 (1), 67.

van de Wouw, M., et al., 2021. Kefir ameliorates specific microbiota-gut-brain axis impairments in a mouse model relevant to autism spectrum disorder. Brain Behav Immun 97, 119–134.

Wallace, C.J.K., Milev, R., 2017. The effects of probiotics on depressive symptoms in humans: a systematic review. Ann Gen Psychiatry 16, 14.

Wang, H., et al., 2019. Bifidobacterium longum 1714 strain modulates brain activity of healthy volunteers during social stress. Am J Gastroenterol 114 (7), 1152–1162.

Wang, T., et al., 2020. Probiotics modulate intestinal motility and inflammation in zebrafish models. Zebrafish 17 (6), 382–393.

Warda, A.K., et al., 2019. Heat-killed lactobacilli alter both microbiota composition and behaviour. Behav Brain Res 362, 213–223.

Wastyk, H.C., et al., 2021. Gut-microbiota-targeted diets modulate human immune status. Cell 184 (16), 4137–4153.

Wei, C.L., et al., 2019. Antidepressant-like activities of live and heat-killed Lactobacillus paracasei PS23 in chronic corticosterone-treated mice and possible mechanisms. Brain Res 1711, 202–213.

Westfall, S., Lomis, N., Prakash, S., 2019. A novel synbiotic delays Alzheimer's disease onset via combinatorial gut-brain-axis signaling in *Drosophila melanogaster*. PLoS One 14 (4), e0214985.

Zhou, L., Foster, J.A., 2015. Psychobiotics and the gut-brain axis: in the pursuit of happiness. Neuropsychiatric Dis Treat 11, 715−723.

# CHAPTER 12

# Microbiota-related biomarkers for precision medicine and drug discovery

## Introduction

In scientific discovery, it is often advantageous to look back to better consider where the future may go. While Hippocrates was credited with the statement "all diseases begin in the gut," the first documented observation of human-associated bacteria was provided in 1683 by Antonie van Leeuwenhoek to the Royal Society of London. The publication of the book, *A Flora and Fauna with Living Animals*, published 160 years later in 1853, is considered to some extent as the origin of microbiome research (Pariente, 2019). A recent multimedia collection, published online, provided a historical perspective of microbiome research and identified research milestones that have contributed to the advancement of this field (https://www.nature.com/collections/bhciihjhei). The use of germ-free (GF) animals to understand the importance of the microbiome to host systems was initiated in 1965 and contributed significantly to the expansion of the scope of research studies beyond microbiology. Since the development of sequence-based identification of microbiota in 1996, the development of novel multiomic and bioinformatic tools has continually developed (https://www.nature.com/collections/bhciihjhei). Within this context, microbiota—brain axis research in the past decade has moved to the forefront of basic and clinical neuroscience research. Microbes include viruses, fungi, parasites, protozoa, and bacteria—at this time, most research has studied gut bacteria, attention to other body sites and other microbes, for example, the virome, is needed (McGuinness et al., 2023). The majority of microbiota—brain research has been conducted in animal models; however, there is a need for more translational studies in clinical populations (Cryan and Mazmanian, 2022). Moreover, the potential of

*Microbiota Brain Axis*
ISBN 978-0-12-814800-6
https://doi.org/10.1016/B978-0-12-814800-6.00012-1

microbiome-related biomarkers to influence advances in precision health in the next decade is promising.

## Translational science

The observation that brain histamine levels were altered in GF rats compared to conventionally housed rats was reported in 1986 (Hegstrand and Hine, 1986); however, interest in microbiota—brain signaling in neuroscience was ignited 2 decades later by the observation that GF mice had an exaggerated response to stress (Sudo et al., 2004) and the observation that GF mice showed reduced anxiety-like behavior compared to conventionally housed mice (Clarke et al., 2013; Diaz Heijtz et al., 2011; Neufeld et al., 2011a, 2011b). Notably, the broader neuroscience community was slow to attend to these findings (Shen, 2015). Since then, animal studies have investigated microbiome-stress associations and identified key signaling pathways that play a role in microbiome-related stress responses including the vagus nerve, the immune system, and gut/host metabolites (Cryan et al., 2019). Evidence of the association between the microbiome and stress in healthy and clinical populations has been demonstrated in case/control studies, through probiotic intervention studies in healthy individuals and to some extent in clinical populations (see Chapter 8 and Chapter 11 (Foster and Clarke, 2023)). And yet, we still only have a limited understanding of the microbiota—brain mechanisms driving these changes. Investigations that focus on microbial function and the contribution of changes in microbial function to host systems are needed to translate mechanistic findings in basic research to clinical populations.

The evidence supporting a link between gut microbiota and anxiety-like behavior in animal models is robust. In clinical populations, reports have identified alterations in microbiome composition for individuals with general anxiety disorder (GAD) (Chen et al., 2019; Jiang et al., 2018) and social anxiety disorder (Butler et al., 2023). Several taxa were associated with clinical symptoms. Specifically, Prevotellaceae *UCG-001, Mollicutes RF39_norank, Succinivibrio, Mitsuokella, Prevotella 9, Subdoligranulum,* Ruminococcaceae_*NK4A214_group,* Ruminococcaceae_*UCG-014,* and *Eubacterium_coprostanoligenes* group were negatively associated with depressive and anxiety symptoms, whereas *Bacteroides* and *Escherichia-Shigella* were positively correlated with these symptoms (Chen et al., 2019). A limitation of these studies is that they did not consider the community nature of the gut microbiome and we know that it is not simply single bacteria taxa that change host systems. A recent study that adapted weight correlation network analysis (WCNA), a technique used for gene expression studies, for use with 16S rRNA sequencing data, considered both the composition and community networks of the microbial community. Community structure revealed three stable microbial communities in people with depression. An advantage of WCNA was that an individual microbial signature for each of these stable modules was exported as an "Eigentaxa" value and then used to investigate the association with clinical phenotype. Notably, one of these communities was significantly associated with clinical features of anxiety. Moreover, this clinically relevant microbial community was enriched with butyrate-producing bacteria, including several taxa belonging to the *Ruminococcaceae* family, such that individuals with less of this network of bacteria had higher levels of anxiety. This is important as butyrate, a short-chain fatty acid and the product of microbiota fermentation of dietary fiber, plays an important role in maintaining a healthy gut and brain barrier and is an important regulator of immune function (Chin Fatt et al., 2023).

As researchers strive to identify direct mechanisms that connect microbiota to the brain, microbial—metabolite pathways are emerging as a key area for translational studies (Foster, 2022). Microbial genes outnumber host genes by more than 100 times, such that the genetic and metabolic potential of the metagenome substantially add to the biochemical flexibility of the host (Gill et al., 2006; Sender et al., 2016). Microbial-derived molecules, including neurotransmitters, short-chain fatty acids, indoles, bile acids, choline metabolites, lactate, and vitamins, exert local effects in the gastrointestinal environment but also can enter the systemic circulation to act at remote sites, including the brain (Caspani et al., 2019). A recent translational study demonstrated that bacterial genes encoding enzymes involved in nicotinamide metabolism were reduced in the superoxide dismutase-1 mutant ($Sod1^{G93A}$) mouse model of amyotrophic lateral sclerosis (ALS), and in a small cohort of ALS patients observed reduced serum and cerebrospinal fluid concentrations of nicotinamide compared to healthy controls (Blacher et al., 2019). While additional studies are needed, this observational study is a good example of translational science and suggests a potential microbiota-related target for future investigations.

An extended body of preclinical and clinical evidence suggests a role for the microbiome in neurodevelopment and in particular, in autism (Vuong and Hsiao, 2017). Increase levels of detrimental microbial metabolites have been linked to behavioral differences and brain connectivity changes in mice (Hsiao et al., 2013; Needham et al., 2022). Case-control studies demonstrate alterations in microbiota composition as well as altered plasma and fecal metabolic phenotypes in individuals with autism compared to typically developing individuals (Liu et al., 2019a; Needham et al., 2021). To translate this work to the clinic, researchers showed that oral administration of a GI-restricted absorbent (AB-2004) that binds and sequesters aromatic metabolites as it passes through the GI tract was able to reduce the levels of detrimental metabolites in mice and ameliorate the behavioral deficits (Niwa et al., 1991; Stewart Campbell et al., 2022). In a phase I open-label clinical trial, the investigators demonstrated that oral administration of AB-2004 was safe and well tolerated in adolescents with autism and reduced both anxiety and irritability, particularly in individuals with high scores on these indices at baseline (Stewart Campbell et al., 2022). While preliminary in nature, these positive results argue for the need to continue to explore the potential for microbiota-targeted interventions to improve clinical outcomes in both psychiatry and neurology.

## Microbiome biomarkers for precision health

The need for biologically based models to improve treatment outcomes in clinical neuroscience is gaining momentum. It is well established that the composition, diversity, and function of gut microbes are influenced by genetics, lifestyle, and environmental factors (Hayes et al., 2020; Kurilshikov et al., 2021; Qin et al., 2022; Sanna et al., 2022; Valles-Colomer et al., 2019; Weersma et al., 2020). As such it is possible to conceptualize the microbiome of an individual as a biological representation of their host genetics (DNA, RNA, epigenetics, sex) in combination with environmental and lifestyle factors (diet, medications, geography, exercise, age, early life experiences). Just like no two fingerprints are ever alike, no two microbiomes are ever alike, aka—your microbiome is your own! Integrating the microbiome into biomarker discovery is advantageous as it provides an accessible and biologically based approach to generate more precise, clinically relevant biomarkers that consider the host and the environment in a comprehensive way. Leveraging microbiome-based biomarkers is especially important when considering microbiome-targeted treatments. In a broad sense, this approach will directly leverage individual biological differences and will advance our ability to tailor treatment strategies to the individual; identify at-risk individuals for whom early intervention could be made available; identify novel targets for drug development; and advance the development of microbiome-targeted therapies. The move toward more sophisticated sampling approaches that capture real-time alterations at distinct GI sites can only help expedite this development.

In the scope of microbiome-targeted therapies, there has been a lot of attention focused on the potential for prebiotics and probiotics to improve brain health (see Chapter 11 (Foster and Clarke, 2023). Postbiotics, as defined by the International Scientific Association of Probiotics and Prebiotics (ISAPP), is a "preparation of inanimate microorganisms and/or their components that confers a health benefit to the host" and is receiving more attention recently (Salminen et al., 2021). Notably, postbiotics do not include interventions using only microbial metabolites (Salminen et al., 2021). The potential use of postbiotics in clinical neuroscience has yet to be explored. Nutritional psychiatry and the potential for improved diet or nutritional interventions to impact brain health has started to attract more attention in the field of microbiota—brain research (Jacka et al., 2017; Marx et al., 2021). Similarly, the benefits of lifestyle changes including exercise and social connection are emerging as potential targets to improve an individual's microbiome and overall health. The suggestion that geography and ethnicity, both of which influence microbiota composition and function, may be important to consider in efforts to leverage the microbiome to improve clinical outcomes across disorders (Gaulke and Sharpton, 2018). In all cases, a challenge for the clinician is how do they determine who will benefit from a microbiome-targeted intervention. For example, in clinical studies, probiotic interventions have modest antidepressant effects in individuals with depression (Liu et al., 2019b; Sarkar et al., 2016). Across several studies, benefits observed to alleviate depressive symptoms are promising; however, the expectation that the results will demonstrate an antidepressant effect across the entire cohort is problematic. A shift in perspective is needed to advance precision medicine such that the selected biomarkers predict who will respond to a given tailored treatment approach at an individual, not population level.

Another promising microbiota-targeted intervention is fecal microbiota transplantation (FMT), which can be a treatment used to improve an individual's microbiome. FMT is a process through which a healthy patient donates a fecal sample or their "microbiome" which is transferred into the gastrointestinal tract of the recipient (Sherwin et al., 2018). Much research and benefits of FMT have been seen through the treatment for recurrent *Clostridium difficile*, where patients have exhibited changes to both their diversity and microbiota composition. In particular, patients have shown to have an increase in diversity as well as a shift in their microbial composition, resembling that of the donor (Rossen et al., 2015; Seekatz et al., 2014, 2018). Evidence has shown that those who undergo FMT through colonoscopy exhibit a 90% cure rate and oral administration resulting in a 76%—79% rate of cure (Gough et al., 2011; Vindigni and Surawicz, 2017). Although there has been success of FMT with recurrent *C. difficle*, FMT has also shown the potential to be successful in other diseases including irritable bowel syndrome, with 70% of patients exhibiting improvement in their symptoms (Pinn et al., 2014) as well as has been used to treat metabolic syndrome (Vrieze et al., 2012). Open questions relate to whether the effects observed in these more complex disorders will be transient or stable, and if booster inoculations or other strategies will be required to maintain the health benefits. To note, there are many initiatives to bypass the donor in FMT interventions with the development of standardized synthetic healthy microbiomes in the lab.

## Where to next?

Microbiota—brain research has penetrated all areas of clinical neuroscience and the potential to harness the microbiome to improve outcomes in neurology and psychiatry is attractive. As noted above, it is essential that the field moves from understanding and considering the impact of treatments and interventions on the individual based on their own biological makeup. Recent successful clinical outcomes in immunotherapy and metabolic disorders suggest that microbiome-related biomarkers can stratify clinical populations into more similar biological groups and have the potential to inform treatment decisions. Longitudinal studies that match fluctuations in microbial biomarkers to the waxing and waning of symptoms in complex heterogenous disorders will be essential in delivering a promising start. Moreover, recent successful clinical outcomes in other biomedical conditions using FMT suggest that microbiome-targeted therapies may provide new therapeutic options for neurology and psychiatry. While not the focus of this chapter, there is also a need to consider a more holistic and transdiagnostic approach in clinical neuroscience. Overlap in mechanisms contributing to the clinical presentation and progression of disease, in parallel, with the significant presence of comorbidities in mental health, metabolic disorders, and physical health conditions warrants a broader view of how we consider the role of the microbiome in health and disease and in the development of precision medicine approaches.

## References

Blacher, E., Bashiardes, S., Shapiro, H., Rothschild, D., Mor, U., Dori-Bachash, M., Kleimeyer, C., Moresi, C., Harnik, Y., Zur, M., Zabari, M., Brik, R.B., Kviatcovsky, D., Zmora, N., Cohen, Y., Bar, N., Levi, I., Amar, N., Mehlman, T., Brandis, A., Biton, I., Kuperman, Y., Tsoory, M., Alfahel, L., Harmelin, A., Schwartz, M., Israelson, A., Arike, L., Johansson, M.E.V., Hansson, G.C., Gotkine, M., Segal, E., Elinav, E., 2019. Potential roles of gut microbiome and metabolites in modulating ALS in mice. Nature 572, 474−480.

Butler, M.I., Bastiaanssen, T.F.S., Long-Smith, C., Morkl, S., Berding, K., Ritz, N.L., Strain, C., Patangia, D., Patel, S., Stanton, C., O'Mahony, S.M., Cryan, J.F., Clarke, G., Dinan, T.G., 2023. The gut microbiome in social anxiety disorder: evidence of altered composition and function. Transl Psychiatry 13, 95.

Caspani, G., Kennedy, S., Foster, J.A., Swann, J., 2019. Gut microbial metabolites in depression: understanding the biochemical mechanisms. Microb Cell 6, 454−481.

Chen, Y.H., Bai, J., Wu, D., Yu, S.F., Qiang, X.L., Bai, H., Wang, H.N., Peng, Z.W., 2019. Association between fecal microbiota and generalized anxiety disorder: severity and early treatment response. J Affect Disord 259, 56−66.

Chin Fatt, C.R., Asbury, S., Jha, M.K., Minhajuddin, A., Sethuram, S., Mayes, T., Kennedy, S.H., Foster, J.A., Trivedi, M.H., 2023. Leveraging the microbiome to understand clinical heterogeneity in depression: findings from the T-RAD study. Transl Psychiatry 13, 139.

Clarke, G., Grenham, S., Scully, P., Fitzgerald, P., Moloney, R.D., Shanahan, F., Dinan, T.G., Cryan, J.F., 2013. The microbiome-gut-brain axis during early life regulates the hippocampal serotonergic system in a sex-dependent manner. Mol Psychiatr 18, 666−673.

Cryan, J.F., Mazmanian, S.K., 2022. Microbiota-brain axis: context and causality. Science 376, 938−939.

Cryan, J.F., O'Riordan, K.J., Cowan, C.S.M., Sandhu, K.V., Bastiaanssen, T.F.S., Boehme, M., Codagnone, M.G., Cussotto, S., Fulling, C., Golubeva, A.V., Guzzetta, K.E., Jaggar, M., Long-Smith, C.M., Lyte, J.M., Martin, J.A., Molinero-Perez, A., Moloney, G., Morelli, E., Morillas, E., O'Connor, R., Cruz-Pereira, J.S., Peterson, V.L., Rea, K., Ritz, N.L., Sherwin, E., Spichak, S., Teichman, E.M., van de Wouw, M., Ventura-Silva, A.P., Wallace-Fitzsimons, S.E., Hyland, N., Clarke, G., Dinan, T.G., 2019. The microbiota-gut-brain axis. Physiol Rev 99, 1877−2013.

Diaz Heijtz, R., Wang, S., Anuar, F., Qian, Y., Bjorkholm, B., Samuelsson, A., Hibberd, M.L., Forssberg, H., Pettersson, S., 2011. Normal gut microbiota modulates brain development and behavior. Proc Natl Acad Sci U S A 108, 3047−3052.

Foster, J., Clarke, G., 2023. Microbiota Brain Axis. Elsevier.

Foster, J.A., 2022. Modulating brain function with microbiota. Science 376, 936−937.

Gaulke, C.A., Sharpton, T.J., 2018. The influence of ethnicity and geography on human gut microbiome composition. Nat Med 24, 1495−1496.

Gill, S.R., Pop, M., Deboy, R.T., Eckburg, P.B., Turnbaugh, P.J., Samuel, B.S., Gordon, J.I., Relman, D.A., Fraser-Liggett, C.M., Nelson, K.E., 2006. Metagenomic analysis of the human distal gut microbiome. Science 312, 1355−1359.

Gough, E., Shaikh, H., Manges, A.R., 2011. Systematic review of intestinal microbiota transplantation (fecal bacteriotherapy) for recurrent *Clostridium difficile* infection. Clin Infect Dis 53, 994–1002.

Hayes, C.L., Peters, B.J., Foster, J.A., 2020. Microbes and mental health: can the microbiome help explain clinical heterogeneity in psychiatry? Front Neuroendocrinol 100849.

Hegstrand, L.R., Hine, R.J., 1986. Variations of brain histamine levels in germ-free and nephrectomized rats. Neurochem Res 11, 185–191.

Hsiao, E.Y., McBride, S.W., Hsien, S., Sharon, G., Hyde, E.R., McCue, T., Codelli, J.A., Chow, J., Reisman, S.E., Petrosino, J.F., Patterson, P.H., Mazmanian, S.K., 2013. Microbiota modulate behavioral and physiological abnormalities associated with neurodevelopmental disorders. Cell 155, 1451–1463.

Jacka, F.N., O'Neil, A., Opie, R., Itsiopoulos, C., Cotton, S., Mohebbi, M., Castle, D., Dash, S., Mihalopoulos, C., Chatterton, M.L., Brazionis, L., Dean, O.M., Hodge, A.M., Berk, M., 2017. A randomised controlled trial of dietary improvement for adults with major depression (the 'SMILES' trial). BMC Med 15, 23.

Jiang, H.Y., Zhang, X., Yu, Z.H., Zhang, Z., Deng, M., Zhao, J.H., Ruan, B., 2018. Altered gut microbiota profile in patients with generalized anxiety disorder. J Psychiatr Res 104, 130–136.

Kurilshikov, A., Medina-Gomez, C., Bacigalupe, R., Radjabzadeh, D., Wang, J., Demirkan, A., Le Roy, C.I., Raygoza Garay, J.A., Finnicum, C.T., Liu, X., Zhernakova, D.V., Bonder, M.J., Hansen, T.H., Frost, F., Ruhlemann, M.C., Turpin, W., Moon, J.Y., Kim, H.N., Lull, K., Barkan, E., Shah, S.A., Fornage, M., Szopinska-Tokov, J., Wallen, Z.D., Borisevich, D., Agreus, L., Andreasson, A., Bang, C., Bedrani, L., Bell, J.T., Bisgaard, H., Boehnke, M., Boomsma, D.I., Burk, R.D., Claringbould, A., Croitoru, K., Davies, G.E., van Duijn, C.M., Duijts, L., Falony, G., Fu, J., van der Graaf, A., Hansen, T., Homuth, G., Hughes, D.A., Ijzerman, R.G., Jackson, M.A., Jaddoe, V.W.V., Joossens, M., Jorgensen, T., Keszthelyi, D., Knight, R., Laakso, M., Laudes, M., Launer, L.J., Lieb, W., Lusis, A.J., Masclee, A.A.M., Moll, H.A., Mujagic, Z., Qibin, Q., Rothschild, D., Shin, H., Sorensen, S.J., Steves, C.J., Thorsen, J., Timpson, N.J., Tito, R.Y., Vieira-Silva, S., Volker, U., Volzke, H., Vosa, U., Wade, K.H., Walter, S., Watanabe, K., Weiss, S., Weiss, F.U., Weissbrod, O., Westra, H.J., Willemsen, G., Payami, H., Jonkers, D., Arias Vasquez, A., de Geus, E.J.C., Meyer, K.A., Stokholm, J., Segal, E., Org, E., Wijmenga, C., Kim, H.L., Kaplan, R.C., Spector, T.D., Uiterlinden, A.G., Rivadeneira, F., Franke, A., Lerch, M.M., Franke, L., Sanna, S., D'Amato, M., Pedersen, O., Paterson, A.D., Kraaij, R., Raes, J., Zhernakova, A., 2021. Large-scale association analyses identify host factors influencing human gut microbiome composition. Nat Genet 53, 156–165.

Liu, F., Li, J., Wu, F., Zheng, H., Peng, Q., Zhou, H., 2019a. Altered composition and function of intestinal microbiota in autism spectrum disorders: a systematic review. Transl Psychiatry 9, 43.

Liu, R.T., Walsh, R.F.L., Sheehan, A.E., 2019b. Prebiotics and probiotics for depression and anxiety: a systematic review and meta-analysis of controlled clinical trials. Neurosci Biobehav Rev 102, 13–23.

Marx, W., Lane, M., Hockey, M., Aslam, H., Berk, M., Walder, K., Borsini, A., Firth, J., Pariante, C.M., Berding, K., Cryan, J.F., Clarke, G., Craig, J.M., Su, K.P., Mischoulon, D., Gomez-Pinilla, F., Foster, J.A., Cani, P.D., Thuret, S., Staudacher, H.M., Sanchez-Villegas, A., Arshad, H., Akbaraly, T., O'Neil, A., Segasby, T., Jacka, F.N., 2021. Diet and depression: exploring the biological mechanisms of action. Mol Psychiatr 26, 134–150.

McGuinness, A.J., Loughman, A., Foster, J.A., Jacka, F., 2023. Mood disorders: the gut bacteriome and beyond. Biol Psychiatr. S0006-3223(23)01532-9.

Needham, B.D., Adame, M.D., Serena, G., Rose, D.R., Preston, G.M., Conrad, M.C., Campbell, A.S., Donabedian, D.H., Fasano, A., Ashwood, P., Mazmanian, S.K., 2021. Plasma and fecal metabolite profiles in autism spectrum disorder. Biol Psychiatr 89, 451–462.

Needham, B.D., Funabashi, M., Adame, M.D., Wang, Z., Boktor, J.C., Haney, J., Wu, W.L., Rabut, C., Ladinsky, M.S., Hwang, S.J., Guo, Y., Zhu, Q., Griffiths, J.A., Knight, R., Bjorkman, P.J., Shapiro, M.G., Geschwind, D.H., Holschneider, D.P., Fischbach, M.A., Mazmanian, S.K., 2022. A gut-derived metabolite alters brain activity and anxiety behaviour in mice. Nature 602, 647–653.

Neufeld, K.A., Kang, N., Bienenstock, J., Foster, J.A., 2011a. Effects of intestinal microbiota on anxiety-like behavior. Commun Integr Biol 4, 492–494.

Neufeld, K.M., Kang, N., Bienenstock, J., Foster, J.A., 2011b. Reduced anxiety-like behavior and central neurochemical change in germ-free mice. Neuro Gastroenterol Motil 23, 255–264, e119.

Niwa, T., Emoto, Y., Maeda, K., Uehara, Y., Yamada, N., Shibata, M., 1991. Oral sorbent suppresses accumulation of albumin-bound indoxyl sulphate in serum of haemodialysis patients. Nephrol Dial Transplant 6, 105–109.

Pariente, N., 2019. A field is born. In: Nature Milestones in Human Microbiota Research. https://www.nature.com/immersive/d42859-019-00041-z/index.html:Springer.Nature.

Pinn, D.M., Aroniadis, O.C., Brandt, L.J., 2014. Is fecal microbiota transplantation the answer for irritable bowel syndrome? A single-center experience. Am J Gastroenterol 109, 1831–1832.

Qin, Y., Havulinna, A.S., Liu, Y., Jousilahti, P., Ritchie, S.C., Tokolyi, A., Sanders, J.G., Valsta, L., Brozynska, M., Zhu, Q., Tripathi, A., Vazquez-Baeza, Y., Loomba, R., Cheng, S., Jain, M., Niiranen, T., Lahti, L., Knight, R., Salomaa, V., Inouye, M., Meric, G., 2022. Combined effects of host genetics and diet on human gut microbiota and incident disease in a single population cohort. Nat Genet 54, 134–142.

Rossen, N.G., Fuentes, S., van der Spek, M.J., Tijssen, J.G., Hartman, J.H., Duflou, A., Lowenberg, M., van den Brink, G.R., Mathus-Vliegen, E.M., de Vos, W.M., Zoetendal, E.G., D'Haens, G.R., Ponsioen, C.Y., 2015. Findings from a randomized controlled trial of fecal transplantation for patients with ulcerative colitis. Gastroenterology 149, 110–118.e114.

Salminen, S., Collado, M.C., Endo, A., Hill, C., Lebeer, S., Quigley, E.M.M., Sanders, M.E., Shamir, R., Swann, J.R., Szajewska, H., Vinderola, G., 2021. The International Scientific Association of Probiotics and Prebiotics (ISAPP) consensus statement on the definition and scope of postbiotics. Nat Rev Gastroenterol Hepatol 18, 649–667.

Sanna, S., Kurilshikov, A., van der Graaf, A., Fu, J., Zhernakova, A., 2022. Challenges and future directions for studying effects of host genetics on the gut microbiome. Nat Genet 54, 100–106.

Sarkar, A., Lehto, S.M., Harty, S., Dinan, T.G., Cryan, J.F., Burnet, P.W.J., 2016. Psychobiotics and the manipulation of bacteria-gut-brain signals. Trends Neurosci 39, 763–781.

Seekatz, A.M., Aas, J., Gessert, C.E., Rubin, T.A., Saman, D.M., Bakken, J.S., Young, V.B., 2014. Recovery of the gut microbiome following fecal microbiota transplantation. mBio 5 e00893-00814.

Seekatz, A.M., Theriot, C.M., Rao, K., Chang, Y.M., Freeman, A.E., Kao, J.Y., Young, V.B., 2018. Restoration of short chain fatty acid and bile acid metabolism following fecal microbiota transplantation in patients with recurrent *Clostridium difficile* infection. Anaerobe 53, 64–73.

Sender, R., Fuchs, S., Milo, R., 2016. Revised estimates for the number of human and bacteria cells in the body. PLoS Biol 14, e1002533.

Shen, H.H., 2015. News feature: microbes on the mind. Proc Natl Acad Sci U S A 112, 9143—9145.

Sherwin, E., Dinan, T.G., Cryan, J.F., 2018. Recent developments in understanding the role of the gut microbiota in brain health and disease. Ann N Y Acad Sci 1420, 5—25.

Stewart Campbell, A., Needham, B.D., Meyer, C.R., Tan, J., Conrad, M., Preston, G.M., Bolognani, F., Rao, S.G., Heussler, H., Griffith, R., Guastella, A.J., Janes, A.C., Frederick, B., Donabedian, D.H., Mazmanian, S.K., 2022. Safety and target engagement of an oral small-molecule sequestrant in adolescents with autism spectrum disorder: an open-label phase 1b/2a trial. Nat Med 28, 528—534.

Sudo, N., Chida, Y., Aiba, Y., Sonoda, J., Oyama, N., Yu, X.N., Kubo, C., Koga, Y., 2004. Postnatal microbial colonization programs the hypothalamic-pituitary-adrenal system for stress response in mice. J Physiol 558, 263—275.

Valles-Colomer, M., Falony, G., Darzi, Y., Tigchelaar, E.F., Wang, J., Tito, R.Y., Schiweck, C., Kurilshikov, A., Joossens, M., Wijmenga, C., Claes, S., Van Oudenhove, L., Zhernakova, A., Vieira-Silva, S., Raes, J., 2019. The neuroactive potential of the human gut microbiota in quality of life and depression. Nat Microbiol 4, 623—632.

Vindigni, S.M., Surawicz, C.M., 2017. Fecal microbiota transplantation. Gastroenterol Clin N Am 46, 171—185.

Vrieze, A., Van Nood, E., Holleman, F., Salojarvi, J., Kootte, R.S., Bartelsman, J.F., Dallinga-Thie, G.M., Ackermans, M.T., Serlie, M.J., Oozeer, R., Derrien, M., Druesne, A., Van Hylckama Vlieg, J.E., Bloks, V.W., Groen, A.K., Heilig, H.G., Zoetendal, E.G., Stroes, E.S., de Vos, W.M., Hoekstra, J.B., Nieuwdorp, M., 2012. Transfer of intestinal microbiota from lean donors increases insulin sensitivity in individuals with metabolic syndrome. Gastroenterology 143, 913—916.e917.

Vuong, H.E., Hsiao, E.Y., 2017. Emerging roles for the gut microbiome in autism spectrum disorder. Biol Psychiatr 81, 411—423.

Weersma, R.K., Zhernakova, A., Fu, J., 2020. Interaction between drugs and the gut microbiome. Gut 69 (8), 1510—1519.

# Abbreviations

| | |
|---|---|
| 3'UTR | 3'-untranslated region |
| ACTH | Adrenocorticotrophic hormone |
| AD | Alzheimer's disease |
| ADHD | Attention deficit hyperactivity disorder |
| AH | After-hyperpolarizing |
| AHR | Aryl hydrocarbon receptor |
| ALS | Amyotrophic lateral sclerosis |
| APP | Aβ-precursor protein |
| ARNTL | Aryl hydrocarbon receptor nuclear translocator-like |
| ASD | Autism spectrum disorder |
| ASDs | Autism spectrum disorders |
| AVPR1b | Arginine vasopressin receptor 1b |
| BD | Bipolar disorder |
| BDI | Beck Depression Inventory |
| BMI | Body mass index |
| CC | Collaborative Cross |
| CCK | Cholecystokinin |
| CNS | Central nervous system |
| CORT | Corticosterone |
| CRH | Corticotropin-releasing hormone |
| CS | Cesarean section |
| CUMS | Chronic unpredictable mild stress |
| DASS | Depression, anxiety, and stress scale |
| DCX | Doublecortin |
| DTI | Diffusion tensor imaging |
| EAE | Experimental autoimmune encephalomyelitis |
| EBV | Epstein—Barr virus |
| EEG | Electroencephalography |
| EFSA | European Food Safety Authority |
| ENS | Enteric nervous system |
| EPM | Elevated plus maze |
| fMRI | Functional magnetic resonance imaging |
| FMT | Fecal microbiota transplantation |
| FOSs | Fructo-oligosaccharides |
| FST | Forced swim test |
| GABA | Gamma-aminobutyric acid |
| GAD | General anxiety disorder |
| GBM | Gut—brain module |
| GF | Germ-free |
| GHSR | Growth hormone secretagogue receptor |
| GI | Gastrointestinal |
| GLP-1 | Glucagonlike peptide-1 |
| GOSs | Galacto-oligosaccharides |

| | |
|---|---|
| **GWAS** | Genome-wide association studies |
| **HAB** | High anxiety-like behavior |
| **HAMD** | Hamilton Depression Score |
| **HC** | Healthy control |
| **HD** | Huntington's disease |
| **HFD** | High-fat diet |
| **HMP** | Human Microbiome Project |
| **HN** | Hypothalamic—neurohypophyseal |
| **HPA** | Hypothalamic—pituitary—adrenal |
| **HPG** | Hypothalamic—pituitary—gonadal |
| **HPT** | Hypothalamic—pituitary—thyroid |
| **HR** | High behavioral response |
| **IBD** | Inflammatory bowel disease |
| **IBQ-R** | Infant Behavior Questionnaire-Revised |
| **IBS** | Irritable bowel syndrome |
| **IECs** | Intestinal epithelial cells |
| **IFAP** | Intestinal fatty acid binding protein |
| **IFN-$\gamma$** | Interferon-$\gamma$ |
| **IL-1$\beta$** | Interleukin-1$\beta$ |
| **ISAPP** | International Scientific Association for Probiotics and Prebiotics |
| **LAC** | Lactoferrin |
| **LD** | Light—dark |
| **LPS** | Lipopolysaccharide |
| **LR** | Low behavioral response |
| **LTP** | Long-term potentiation |
| **MCAO** | Middle cerebral artery occlusion |
| **MDD** | Major depressive disorder |
| **MHC** | Major histocompatibility complex |
| **MHFD** | Maternal high-fat diet |
| **MIA** | Maternal immune activation |
| **miRNAs** | MicroRNAs |
| **MMSE** | Mini—Mental State Exam |
| **MPTP** | Methyl-4-phenyl-1,2,3,6-tetrahydropyridine |
| **MRI** | Magnetic resonance imaging |
| **MS** | Maternal separation |
| **MS** | Multiple sclerosis |
| **NAB** | Normal anxiety-like behavior |
| **NTS** | Nucleus tractus solitarius |
| **OF** | Open field |
| **OTUs** | Operational taxonomic units |
| **PANSS** | Positive and Negative Syndrome Scale |
| **PD** | Parkinson's disease |
| **PDX** | Polydextrose |
| **PFC** | Prefrontal cortex |
| **PRRs** | Pattern recognition receptors |
| **PRS** | Polygenic risk scores |
| **PULs** | Polysaccharide utilization loci |

| | |
|---|---|
| **PYY** | Peptide YY |
| **QTL** | Quantitative trait loci |
| **RCT** | Randomized controlled trial |
| **SCFA** | Short-chain fatty acid |
| **SNP** | Single nucleotide polymorphism |
| **SPF** | Specific pathogen-free |
| **SPT** | Sucrose preference test |
| **sRNA** | Small regulatory RNA |
| **SSRIs** | Selective serotonin reuptake inhibitors |
| **SZ** | Schizophrenia |
| **T2D** | Type 2 diabetes |
| **TCR** | T-cell receptor |
| **TLRs** | Tolllike receptors |
| **TMAO** | Trimethylamine-N-oxide |
| **TYM** | Test your memory |
| **WCNA** | Weight correlation network analysis |

# Index

Printed in the United States
by Baker & Taylor Publisher Services